Computational Methods in Condensed Matter: Electronic Structure

Computational Methods in Condensed Matter: Electronic Structure

A. A. Katsnelson
V. S. Stepanyuk
A. I. Szász
O. V. Farberovich

Translated by
Kevin Hendzel

Library of Congress Cataloging-in-Publication Data

Ėlektronnai͡a teorii͡a kondensirovannykh sred. English.
Computational Methods in Condensed Matter: Electronic Structure /
A. A. Katsnelson... [et al.].
p. cm.
Translation of: Ėlektronnai͡a teorii͡a kondensirovannykh sred.
Includes bibliographical references.
ISBN 978-0-88318-865-1
1. Condensed matter. I. Kat͡snel'son, Al'bert Anatol'evich. II. Title.
QC173.4.C65E4313 1992
530.4'1–dc20 91-39752
CIP

Contents

Chapter 4. Application of electronic theory to calculation of physical properties

Biographical Notes

Dr. Albert A. Katsnelson received both his candidate and doctorate degrees from Moscow State University where he is currently a professor of physics. His research interests include pseudopotential theory, ordering problems in solids, and x-ray scattering from nearly perfect crystals. He has published more than 250 papers, is a co-author of two monographs, *Short-Range Order in Solid Solutions* and *Pseudopotential Theory of Crystal Structures*, and is the author of the textbook *Introduction to Solid State Physics*. In 1980 he was awarded the Academy of Sciences' Fedorov Prize.

Dr. Valerij S. Stepanyuk received his MSc and PhD degrees from Moscow State University. Recently he is a senior scientist at the Department of Low Temperature Physics of Moscow State University. He is author of more than 60 scientific works, and has written a university textbook *Interparticle Interactions and Properties of Metals*. His scientific interest is centered on the electronic theory of crystals and nonordered systems, physics of clusters, as well as their computational models.

Dr. Andras Szász is currently associate professor and head of the laboratory of Surface and Interface Physics in Eotvos University, Budapest. He holds his MSc, PhD, and CSc degrees in solid state and surface physics. His recent research interest is concentrated on the interdependence of the stability and electronic structure of solids and interface systems. He published more than 170 scientific papers and is co-author of three books.

Dr. Oleg V. Farberovich received his PhD degree from Voronezh State University. His DSc was performed at Moscow Institute of Steel and Alloys. He is author of more than 100 scientific papers. He is the professor of the Physics Department of Voronezh State University. His scientific attention is focused on the electronic structure of crystals, optical and magnetic properties of solids, and physics of small atomic clusters.

Foreword

In recent years we have witnessed the extensive development of theoretical and computational methods of calculating the electronic structure and properties of condensed matter. There are evidently two factors that have facilitated this development of the theory of the electron states of condensed matter. The first is the increasingly broad utilization of density-functional theory, which makes it possible from a unified conceptual basis to describe electron states in various kinds of matter, and the second is the progress in the field of computational physics which has made possible a quantitative analysis of the physical properties of complex systems.

During the same period the range of fundamental issues in condensed matter physics has expanded significantly, encompassing liquids, polymers, vitreous and amorphous matter, alloys, and solid solutions and crystal surfaces. Such a variety of objects and the interesting phenomena occurring within such objects (from different spectral properties through high-temperature superconductivity) has posed the challenge of a theoretical description of such phenomena from a unified viewpoint based on electronic theory. Unfortunately, to date there has been no general theory of the electronic properties of condensed matter consisting of atomic clusters configured at random points in space. Hence the primary object of electronic quantum theory has recently included the properties of ideal or near-ideal crystalline systems. The problem of investigating their electronic structure is significantly more complicated for disordered systems. However, using a variety of computational techniques makes it possible in this case to solve several important problems of condensed matter theory. Such methods include cluster modeling, calculation of the electronic structure of clusters, as well as techniques based on Green's function theory.

Given the lack of a unified conceptual approach, substantial difficulties will arise to a varying degree with respect to both ideal crystals and disordered systems in considering the various theoretical techniques for calculating the electronic structure of condensed matter. The descriptions of available electronic structure calculation techniques (both classical techniques and those that have just recently appeared) is spread over a wide variety of articles, surveys, and books, which substantially complicates their analysis. Hence, our purpose was to combine in a single book, and, from a single viewpoint, the most important electronic structure calculation techniques and to describe their implementation and field of application.

The present work consists of four chapters. Chapter 1 is devoted to a description of electronic states in condensed matter based on the modern, well-developed density-functional theory. The formalism of this theory makes it possible to incorporate all correlation effects while conserving the conceptual simplicity of the one-electron approach. The primary achievement of this theory is its universality. This theory can be used with equal success in atomic, molecular, and condensed matter theory.

Chapter 2 is devoted to a description of the principal theoretical computational methods of calculating the electronic structure of condensed matter. We have especially focused on the most recent achievements in electronic structure calculations which include linear techniques in band theory which utilize the general form of the crystal potential, methods of calculating the electronic structure of crystal surfaces, and a universal approach based on molecular dynamics and density-functional theory. Techniques based on Green's function theory, which are widely used for crystals as well as cluster methods, are given rather complete treatment. Of course, in order to remain within the limits of the space allotted for our use in the present book, we were forced to give only cursory treatment to a number of techniques here.

Chapter 3 is devoted to an analysis of the primary problems of the pseudopotential method which is widely used in conjunction with a diffraction model for a theoretical analysis of the structure and properties of ordered and disordered metallic alloys. The attraction of this method lies in its simplicity and therefore the current calculation techniques as well as the capable and highly developed methods can be used in day-to-day practice by experimentalists having access to modern computer technology.

Chapter 4 examines the different applications of the electronic theory.

In conclusion, we wish to express our gratitude to those scientific colleagues and groups whose cooperation over the last 15 years has led to the formation of our views on solid-state electronic theory and development of this theory. Above all this includes the teams of which we have been a part, as well as The Institute of the Physics of Metals of the Urals Division of the USSR Academy of Sciences; Tomsk University; the Institute for the Physics of Metallic Surfaces of the Siberian Division of the USSR Academy of Sciences; the Institute of Metal Physics of the Academy of Sciences of the Ukranian SSR; the Lebedev Physics Institute of the USSR Academy of Sciences; the Moscow Institute of Steel and Alloys; and Leningrad University.

We especially wish to express our gratitude to V. E. Panin, V. L. Shirovkovskiy, V. I. Rezer, R. F. Yegorov, V. N. Antonov, V. E. Yegorushkin, E. G. Maksimov, Yu. Kh. Vekilov, E. P. Doiashevskaya, Yu. A. Uspenskiy, I. I. Mazin, S. I. Kurganskiy, E. V. Kozlov, Yu. A. Khon, D. L. Fuks, Z. A. Gurskiy, L. I. Yastrebov, S. V. Vlasov, K. A. Kikoin, V. M. Silonov, G. P. Nizhnikova, and A. O. Mekhrabov. Some of the data reported here were used by agreement with the authors from dissertations by V. A. Gorbunov and I. A. Anishchenko to whom we are also deeply grateful. We express our thanks to E. I. Bodnev and O. A. Pachikov for extensive assistance in preparing the monograph.

Chapter 1

Electronic states in condensed matter

1.1. The many-body problem in condensed matter theory

In considering the motion of electrons in condensed matter we are dealing with the problem of describing the motion of an enormous number of electrons and nuclei ($\sim 10^{23}$) obeying the laws of quantum mechanics. The observed properties of solids can be explained based on such a description whose implementation requires appropriate incorporation in the relativistic Hamiltonian of the electrostatic and electromagnetic interactions between particles. An exact solution of this problem is impossible and hence it is necessary to rely on a wide range of simplifying approximations.

1.1.1. Adiabatic approximation [1]

Since the nuclear mass M far exceeds electron mass m, we can naturally limit the analysis to a model for electrons traveling in a fixed field of nuclei. In this approximation the electron wave function is determined by the instantaneous position of the nuclei, while the wave function of ions is determined by the averaged electron field.

We consider the stationary Schrödinger equation in the crystal

$$H\Psi_{\text{cr}}=E\Psi_{\text{cr}}, \tag{1.1}$$

where

$$H=-\sum_{\alpha}\frac{\hbar^2}{2M_\alpha}\nabla_\alpha^2-\sum_i\frac{\hbar^2}{2m}\nabla_i^2+\frac{1}{2}\sum_{i\neq j}\frac{e^2}{|\mathbf{r}_i-\mathbf{r}_j|}+\frac{1}{2}\sum_{\alpha\neq\beta}\frac{z_\alpha z_\beta e^2}{|\mathbf{R}_\alpha-\mathbf{R}_\beta|}$$
$$-\sum_{i,\alpha}\frac{z_\alpha e^2}{|\mathbf{r}_i-\mathbf{R}_\alpha|}. \tag{1.2}$$

Here M_α is nuclear mass, m is electron mass, $\mathbf{r}_i$ is the radius vector of the ith electron, $\mathbf{R}_\alpha$ is the radius vector of the αth nucleus, and z_α is the atomic number. Since we have separated nuclear and electron motion, we can write

$$\Psi_{\text{cr}}(\mathbf{r},\mathbf{R})=\Psi(\mathbf{r},\mathbf{R})\Phi(\mathbf{R}). \tag{1.3}$$

Substituting Eq. (1.3) into Eq. (1.1) we obtain two equations:

$$H_e\Psi(\mathbf{r},\mathbf{R})=\varepsilon\Psi(\mathbf{r},\mathbf{R}), \tag{1.4}$$

$$H_{\rm nu}\,\Phi(\mathbf{R})=\varepsilon_{\rm nu}\Phi(\mathbf{R}), \tag{1.5}$$

where

$$H_e=-\sum_i \frac{\hbar^2}{2m}\nabla_i^2+\frac{1}{2}\sum_{i\neq j}\frac{e^2}{|\mathbf{r}_i-\mathbf{r}_j|}-\sum_{i,\alpha}\frac{z_\alpha e^2}{|\mathbf{r}_i-\mathbf{R}_\alpha|}, \tag{1.6}$$

$$H_{\rm nu}=-\sum_\alpha \frac{\hbar^2}{2M_\alpha}\nabla_\alpha^2+\frac{1}{2}\sum_{\alpha\neq\beta}\frac{z_\alpha z_\beta e^2}{|\mathbf{R}_\alpha-\mathbf{R}_\beta|}. \tag{1.7}$$

In Eq. (1.5) we neglect the term

$$\sum_\alpha \frac{\hbar^2}{2M_\alpha}\left[\Phi\int d\mathbf{r}\,\Psi^*\nabla_\alpha^2\Psi-2\left(\nabla_\alpha\Phi\int d\mathbf{r}\,\Psi^*\nabla_\alpha\Psi\right)\right]. \tag{1.8}$$

For this estimate we premultiply by Φ^* and integrate with respect to the nuclear coordinates. We obtain

$$\Delta\varepsilon=\sum_\alpha\frac{\hbar^2}{2M_\alpha}\int d\mathbf{r}\,\Psi^*\nabla_\alpha^2\Psi+\sum_\alpha\frac{\hbar^2}{M_\alpha}\left(\int d\mathbf{R}\,\Phi^*\nabla_\alpha\Phi\int d\mathbf{r}\,\Psi^*\nabla_\alpha\Psi\right).$$

For the first term we have

$$\sum_\alpha\frac{\hbar^2}{2M_\alpha}\int d\mathbf{r}\,\Psi^*\nabla_\alpha^2\Psi=-\sum_{i,\alpha}\frac{m}{M_\alpha}\int d\mathbf{r}\,\Psi^*\left(-\frac{\hbar^2}{2m}\right)\Psi=-\sum_{i,\alpha}\frac{m}{M_\alpha}\langle T_i\rangle,$$

where $\langle T_i\rangle$ is the average kinetic energy of a single electron. Thus

$$\sum_\alpha\frac{\hbar^2}{2M_\alpha}\int d\mathbf{r}\,\Psi^*\nabla_\alpha^2\Psi=-ZN\frac{m}{M}\langle T_i\rangle.$$

Since $m/M\sim 10^{-5}$ then we neglect this term compared to ε.

The second term in Eq. (1.8) is estimated analogously:

$$\frac{\hbar^2}{M_\alpha}\left(\int d\mathbf{R}\,\Phi^*\nabla_\alpha\Phi\int d\mathbf{r}\,\Psi^*\nabla_\alpha\Psi\right)=-\frac{1}{M_\alpha}(\langle P_\alpha\rangle\langle P_i\rangle).$$

In thermodynamic equilibrium for the case of classical statistics we have

$$\left\langle\frac{P_i^2}{2m}\right\rangle=\left\langle\frac{P_\alpha^2}{2M_\alpha}\right\rangle$$

and

$$\langle P_e\rangle^2=\frac{8}{3\pi}\langle P_e^2\rangle.$$

Then

$$\langle P_e\rangle\sim\sqrt{\frac{m}{M}}\langle P_\alpha\rangle$$

and the second term in Eq. (1.8) is of the order of $\sqrt{m/M}$ of the total crystal energy.

Therefore, discarding both corrections in Eq. (1.5) we make an energy error of less than $\sqrt{m/M}$. The discarded terms characterize the internal nonadiabaticity of the system which is expressed as an effect of nuclear motion on their interaction with the electrons. Essentially the second correction describes electron–phonon interactions. Therefore, electron–phonon interactions are neglected in the electronic structure calculations of crystals from the very outset. Thermal motion can be accounted for only as a perturbation that sets up a specific electron state distribution.

Therefore, we consider a solution of a simplified problem of electron motion in condensed matter with fixed nuclei, that is, we solve Eq. (1.4) with the Hamiltonian (1.6).

1.1.2. The self-consistent-field method

Explicitly isolating the electronic subsystem in the crystal, a microscopic theory is required that would explain the observed properties. Here we rely solely on electronic properties. What information on electronic properties must underlie the theory? First, electron motion in an ideal crystal occurs in an ionic periodic potential (the "solid-state effect" or more precisely the crystallinity effect). Second, it is necessary to know how the electrons interact (electron–electron interaction). When the influence of the crystallinity effect is accounted for this will yield Bloch functions and a new quantum number wave vector **k** and therefore a calculation of the electronic states in reciprocal space (the Brillouin zone). This issue is largely related to the symmetry aspects of the problem and will be examined below.

We now consider a many-electron system with electron–electron interaction described by the nonrelativistic Hamiltonian*

$$H=\sum_i (-\nabla_i^2)+2\sum_{i\neq j}\frac{1}{|\mathbf{r}_i-\mathbf{r}_j|}. \tag{1.9}$$

The corresponding multielectron problem lies far beyond the scope of current computational capabilities. Moreover, it is assumed in Eq. (1.9) that electron–electron interaction reduces to the sum of "bare" pair interactions, each of which is described by the potential $v_{ij}^0=1/|\mathbf{r}_i-\mathbf{r}_j|$. At the same time there is no reason to doubt that the interaction between two electrons remains unchanged in the presence of the remaining electrons. Therefore, the interaction between two electrons (the *i*th and *j*th electrons) will also depend on the surrounding electrons, that is, the problem essentially becomes a multiparticle problem whose complete solution is hardly possible at the present time. Hence the only escape from this situation is to introduce a certain effective interaction between electrons v_{ij}^{eff}. This interaction which is a two-particle interaction in form is essentially a multiparticle interaction since it roughly accounts for the effect of the surrounding electrons on the interaction between the *i*th and *j*th electrons. Introduction of the effective interaction results in the substitution of the Hamiltonian (1.9) with the following:

---

*We use the system of atomic units: $m=1/2$, $\hbar=e=1$ with energy expressed in Rydbergs: 1 Ry = 13.6 eV.

$$H = \sum_i (-\nabla_i^2) + \sum_{i \neq j} v_{ij}^{eff}. \tag{1.10}$$

Use of the effective interaction assumes a significant reduction in the number of degrees of freedom. We consider as an example[2] the state

$$\Psi = \begin{pmatrix} \varphi_1 \\ \varphi_2 \end{pmatrix},$$

which is a solution of the Schrödinger equation

$$H\Psi = \varepsilon\Psi$$

and

$$\begin{pmatrix} H_{11} & H_{12} \\ H_{21} & H_{22} \end{pmatrix} \begin{pmatrix} \varphi_1 \\ \varphi_2 \end{pmatrix} = \varepsilon \begin{pmatrix} \varphi_1 \\ \varphi_2 \end{pmatrix},$$

from which we obtain a system of two equations to find φ_1 and φ_2:

$$H_{11}\varphi_1 + H_{12}\varphi_2 = \varepsilon\varphi_1,$$

$$H_{21}\varphi_1 + H_{22}\varphi_2 = \varepsilon\varphi_2.$$

Solving this equation we find

$$\varphi_2 = -(H_{22} - \varepsilon)^{-1} H_{21}\varphi_1$$

and

$$[H_{11} - H_{12}(H_{22} - \varepsilon)^{-1} H_{21}]\varphi_1 = \varepsilon\varphi_1. \tag{1.11}$$

This equation takes the form of the Schrödinger equation for the first state although the role of the Hamiltonian is played by the operator

$$H_{\text{eff}} = H_{11} - H_{12}(H_{22} - \varepsilon)^{-1} H_{21}. \tag{1.12}$$

We therefore have the following Schrödinger equation with effective interaction:

$$H_{\text{eff}}\varphi_1 = \varepsilon\varphi_1, \tag{1.13}$$

where

$$\begin{aligned} H_{\text{eff}} &= -\nabla_1^2 + v_{\text{eff}}, \\ v_{\text{eff}} &= V_{11} - H_{12}(H_{22} - \varepsilon)^{-1} H_{21}. \end{aligned} \tag{1.14}$$

This means that the degree of freedom in the system (the transition from the φ_1 state to the φ_2 state) produces an auxiliary term in Eq. (1.12). In order to avoid an explicit analysis of this degree of freedom we must substitute the potential v_{11} by v_{eff}. Here, if we use some approximation for v_{eff} we obtain a problem that is much simpler than the initial problem. Moreover, even the approximate value for the effective potential more accurately describes the electron system compared to the initial potential v_{11} which does not contain the information on the discarded degree of freedom.

The primary element of the theory describing the electronic subsystem in the crystal is the formulation of the effective interaction which can be constructed by two fundamentally different techniques:

(a) formulation of v_{eff} based solely on the "bare" electron interaction (the microscopic approach) and (b) a phenomenological representation of v_{eff} which avoids an explicit relation to the interaction of the free electrons (the macroscopic approach).

After the effective potential is formulated in some manner it is necessary to solve the Schrödinger equation with the Hamiltonian (1.10) and the multiparticle wave function $\Psi(\mathbf{r}_1,\mathbf{r}_2,...,\mathbf{r}_N)$. The solution to this problem requires one additional significant approximation. Specifically, an attempt is made to construct in an appropriate manner a multiparticle function from certain single-particle functions $\varphi_\mu(\mathbf{r})$. Here we must search out the best approximation among all single-particle approximations, that is, we must use the variational principle. Thus, the exact wave function Ψ is replaced by the approximate function $\tilde{\Psi}(\mathbf{r}_1,\mathbf{r}_2,...,\mathbf{r}_N)$, which is formulated from certain as yet arbitrary single-particle functions $\{\varphi_\mu(\mathbf{r})\}$. Of course, such a solution of the Schrödinger equation Ψ does not belong to the class of such functions $\tilde{\Psi}$. However, it is also possible to find such $\{\varphi_\mu\}$ that the approximate solution $\tilde{\Psi}$ formulated will be the best solution in the sense of the variational principle:

$$E=\min\left(\int d\xi\, \tilde{\Psi}^* H_{eff}\tilde{\Psi} \Big/ \int d\xi\, \tilde{\Psi}^*\tilde{\Psi}\right), \tag{1.15}$$

where E is the system energy.

Such a formulation of the problem yields a system of equations for the single-particle functions φ_μ. Each of these equations can be interpreted as a Schrödinger equation for the μth particle moving in a certain average (self-consistent) field produced by the remaining particles.

We consider a system of noninteracting electrons in a certain external field $v_{ext}(\mathbf{r})$. The multiparticle wave function for the noninteracting electrons can be given as the simple product[3]

$$\tilde{\Psi}(\mathbf{r}_1,\mathbf{r}_2,...,\mathbf{r}_N)=\varphi_1(\mathbf{r}_1)\varphi_2(\mathbf{r}_2)...\varphi_N(\mathbf{r}_N). \tag{1.16}$$

Then substituting Eq. (1.16) into Eq. (1.15), allowing the derived expression to vary in φ_μ and setting δE_k equal to zero we obtain with an orthonormalizable function $\{\varphi_\mu\}$ the following system of integrodifferential equations:

$$\left[-\nabla^2+v_{ext}(\mathbf{r})+\sum_\lambda{}'\int d\mathbf{r}\,\varphi_\lambda^*(\mathbf{r}')v_{eff}(\mathbf{r},\mathbf{r}')\varphi_\lambda(\mathbf{r}')\right]\varphi_\mu(\mathbf{r})=\varepsilon\varphi_\mu(\mathbf{r}). \tag{1.17}$$

Equation (1.17) is analogous to the Schrödinger equation for noninteracting electrons in an external field $v_{ext}(\mathbf{r})$ with the exception of the term

$$v^H(\mathbf{r})=\int d\mathbf{r}'\,\rho(\mathbf{r}')v_{eff}(\mathbf{r},\mathbf{r}'), \tag{1.18}$$

where $\rho(\mathbf{r}) = \Sigma_\lambda|\varphi_\lambda(\mathbf{r})|^2$ is the electron density.

$v^H(\mathbf{r})$ is called the Hartree potential and this represents the average field acting on the given electron from the remaining electrons. The "prime" on the sum in Eq. (1.17) means that the summation excludes the term with $\lambda = \mu$ (i.e., the electron has no effect on itself and the self-interaction effect is excluded). Since the field with potential $v^H(\mathbf{r})$ is determined by the states of the electrons traveling within the field, in this sense it is a self-consistent field.

This approximation, which is based on the use of a multiplicative function, is called the Hartree approximation.[3] The Hartree method underlies all self-consistent theories that yield an expression for electron energy as a functional of the electron density. Moreover, the Hartree self-consistent-field method is sufficiently accurate and convenient for numerical calculations. It is obvious that correlations between electron positions are ignored in the Hartree method.

If we now use an approximation in the form of a Slater determinant for the wave function we can obtain the well-known Hartree–Fock approximation[3–5]

$$[-\nabla^2 + v_{\text{ext}}(\mathbf{r}) + v^H(\mathbf{r})]\varphi_\mu(\mathbf{r}) - 2\sum_{\nu\neq\mu}\delta(m_{\sigma\mu},m_{\sigma\nu}) \times\varphi_\nu(\mathbf{r})\int d\mathbf{r}'\frac{\varphi_\nu^*(\mathbf{r}')\varphi_\mu(\mathbf{r}')}{|\mathbf{r}-\mathbf{r}'|} = \varepsilon_\mu\varphi_\mu(\mathbf{r}). \tag{1.19}$$

The Hartree–Fock equation (1.19) differs from the Hartree equation (1.17) in the additive term

$$v_{\text{ex}}(\mathbf{r}) \equiv -2\sum_{\nu\neq\mu}\delta(m_{\sigma\mu},m_{\sigma\nu})\varphi_\nu(\mathbf{r})\int d\mathbf{r}'\frac{\varphi_\nu^*(\mathbf{r}')\varphi_\mu(\mathbf{r}')}{|\mathbf{r}-\mathbf{r}'|}, \tag{1.20}$$

which is called the exchange term. The Kronecker symbol $\delta(m_{\sigma\mu},m_{\sigma\nu})$ designates that summation is carried out over the single-particle states and over the same spin. The potential entering into Eq. (1.20):

$$U(\mathbf{r},\mathbf{r}') = \sum_\nu \frac{\delta(m_{\sigma\mu},m_{\sigma\nu})\varphi_\nu^*(\mathbf{r}')\varphi_\nu(\mathbf{r})}{|\mathbf{r}-\mathbf{r}'|} \tag{1.21}$$

is called the nonlocal potential. Subject to Eq. (1.21), Eq. (1.19) takes the form

$$[-\nabla^2 + v_{\text{ext}}(\mathbf{r}) + v^H(\mathbf{r})]\varphi_\mu(\mathbf{r}) - \int d\mathbf{r}\, U(\mathbf{r},\mathbf{r}')\varphi_\mu(\mathbf{r}') = \varepsilon_\mu\varphi_\mu(\mathbf{r}). \tag{1.22}$$

We introduce

$$\rho(\mathbf{r},\mathbf{r}') = \sum_\nu n_\nu\varphi_\nu^*(\mathbf{r})\varphi_\nu(\mathbf{r}'), \tag{1.23}$$

which is called the density matrix, which has to be satisfied. In the following sum rule:

$$\int d\mathbf{r}'\,\rho(\mathbf{r},\mathbf{r}') = -1. \tag{1.24}$$

(The appropriate set of n_ν has been chosen that way.) In the density matrix language

$$U(\mathbf{r},\mathbf{r}')=\frac{\rho(\mathbf{r},\mathbf{r}')}{|\mathbf{r}-\mathbf{r}'|}. \tag{1.25}$$

The average exchange potential

$$\bar{v}_{ex}(\mathbf{r})=\int d\mathbf{r}\; U(\mathbf{r},\mathbf{r}') \tag{1.26}$$

can be interpreted as the potential at point **r** which results from the absence of a single electron [due to Eq. (1.24)]. Therefore, exchange interaction is due to electron interaction with the charge distribution of electrons of the same spin. This distribution is produced by a charge that is one less than that of the total number of electrons. In other words, it is as though the electron "drags along" the hole which is ordinarily an exchange (Fermi) hole.[5]

Since integral (1.24) is independent of **r** we can assume that the hole shape is also independent of electron position at coordinate **r**, that is,

$$\rho(\mathbf{r},\mathbf{r}')\cong-\rho(\mathbf{r})G\left(\frac{\mathbf{r}-\mathbf{r}'}{a}\right), \tag{1.27}$$

where the function G describes the hole shape and a describes its characteristic dimension. It follows from the equality $[\rho(\mathbf{r},\mathbf{r}')=-\rho(\mathbf{r})]$ that $G(0)=1$. Using Eq. (1.24) we obtain

$$8\pi a^3\rho(\mathbf{r})\int_0^\infty d\mathbf{R}\; G(\mathbf{R})\mathbf{R}^2=1, \tag{1.28}$$

where $\mathbf{R}=|\mathbf{r}-\mathbf{r}'|$ and $G(\mathbf{R})=(1/4\pi)\int d\Omega\,[G(\mathbf{r}-\mathbf{r}')/a]$ is the spherical average value of the function G. The average exchange potential (1.26), subject to Eq. (1.27), will appear as

$$\bar{v}_{ex}(\mathbf{r})=-8\pi a^2\rho(\mathbf{r})\int_0^\infty d\mathbf{R}\; G(\mathbf{R})\mathbf{R}. \tag{1.29}$$

Finding a from Eq. (1.28) and substituting it into Eq. (1.29) we obtain

$$\bar{v}_{ex}(\mathbf{r})=-\alpha[8\pi\rho(\mathbf{r})]^{1/3}, \tag{1.30}$$

where

$$\alpha=\left[\int_0^\infty d\mathbf{R}\; G(\mathbf{R})\mathbf{R}\right]\Big/\left[\int_0^\infty d\mathbf{R}\; G(\mathbf{R})\mathbf{R}^2\right]^{2/3}. \tag{1.31}$$

Therefore, substitution of the function $G[(\mathbf{r}-\mathbf{r}')/a]$ by its spherical average yields an exchange potential proportional to the cubed root of the electron density, while the proportionality factor α will depend on the shape of the exchange hole. The potential (1.30) is the same as the exchange potential for a free electron gas[5]:

$$\bar{v}_{ex}(\mathbf{r})=-4\left[\frac{3\rho(\mathbf{r})}{\bar{n}}\right]^{1/3}. \tag{1.32}$$

In other words, spherical averaging of the exchange hole will yield the same results as though we used the exchange potential for the electron gas (1.32) with local density $\rho(\mathbf{r})$. Hence such an approximation can still be called a local density approximation (LDA). Thus, in the LDA the hole will always be spherically symmetrical.

The exchange energy can be given as

$$E_{ex}=\frac{1}{2}\int d\mathbf{r}\,\rho(\mathbf{r})\int d\mathbf{r}'\,\frac{\rho(\mathbf{r},\mathbf{r}')}{|\mathbf{r}-\mathbf{r}'|}$$
$$=\frac{1}{2}\int d\mathbf{r}\,\rho(\mathbf{r})\int d\mathbf{r}'\,U(\mathbf{r},\mathbf{r}')$$
$$=\frac{1}{2}\int d\mathbf{r}\,\rho(\mathbf{r})\mathrm{v}_{ex}(\mathbf{r}). \qquad (1.33)$$

It is clear that neglecting the nonspherical components in $\rho(\mathbf{r},\mathbf{r}')$ is meaningless since only the spherical average over the exchange hole makes a contribution to E_{ex}. Hence although the exact hole is, generally speaking, severely aspherical, no approximations are required for describing the aspherical part of the hole. A very important fact is that the sum rule (1.24) holds in the LDA. These factors mean that the LDA yields excellent results for actual systems with large variations in electron density.

The Hartree–Fock method has found broad application in atomic theory.[6] Among its different modifications, one is referred to as the Hartree–Fock–Rutan method, which is used in quantum chemistry to calculate the electronic structure of molecules.[7] At the same time it has only limited suitability for the majority of applications to condensed matter. For condensed matter theory the area of special interest concerns low density valence electrons for which correlations of electrons with antiparallel spins, neglected in the Hartree–Fock method, yield effects of the same order as exchange. Thus, for example, using the Hartree–Fock method for band structure calculations of ideal crystals will cause the density of states on the Fermi level in metals to vanish and yield anomalously large band gaps in semiconductors and insulators.[8]

The correlation effects in Hartree–Fock theory can be introduced through the configuration interaction corrections.[9] The multiconfigurational Hartree–Fock method (HFMM) has found broad application recently in atomic theory.[6] However, the complexity of this method and its extreme sensitivity to the basis selection have hindered the use of the HFMM in condensed matter theory.

1.1.3. X_α method

In 1951 Slater proposed[10] a simplification of the exchange term (the Slater approximation) in order to provide a simpler form of Eq. (1.22) for crystal calculations; Slater used this approximation to reduce Eq. (1.22) to a Schrödinger-type equation.

If the last integral on the left-hand side of Eq. (1:22) is given as

$$\mathrm{v}_{ex}^{\mu}(\mathbf{r})\varphi_\mu(\mathbf{r})=-\int d\mathbf{r}'\,U(\mathbf{r},\mathbf{r}')\varphi_\mu(\mathbf{r}'), \qquad (1.34)$$

then

$$v_{ex}^{\mu}(\mathbf{r}) = -\int d\mathbf{r}'\, U(\mathbf{r},\mathbf{r}')\, \frac{\varphi_\mu(\mathbf{r}')}{\varphi_\mu(\mathbf{r})} \tag{1.35}$$

will be the effective exchange potential for the μth electron. Now averaging Eq. (1.35) over all occupied states in the free electron gas approximation

$$v_{ex}(\mathbf{r}) \equiv \langle v_{ex}^{\mu}(\mathbf{r})\rangle_{av} = \int_0^{k_F} d\mathbf{k}_\mu\, v_{ex}^{\mu}(\mathbf{r}) \Big/ \int_0^{k_F} d\mathbf{k}_\mu, \tag{1.36}$$

we obtain

$$v_{ex}(\mathbf{r}) = -\frac{4}{\rho(\mathbf{r})} \int d\mathbf{r}'\, \frac{\rho^2(\mathbf{r},\mathbf{r}')}{|\mathbf{r}-\mathbf{r}'|}. \tag{1.37}$$

Calculating $\rho(\mathbf{r},\mathbf{r}')$ in this free electron gas approximation we finally obtain

$$v_{ex}(\mathbf{r}) = -3\left[\frac{3}{\pi}\rho(\mathbf{r})\right]^{1/3}. \tag{1.38}$$

It is clear from a comparison of Eqs. (1.32) and (1.38) that

$$v_{ex}(\mathbf{r}) = \tfrac{3}{4}\bar{v}_{ex}(\mathbf{r}). \tag{1.39}$$

Equation (1.38) is known as the Slater approximation. It therefore follows from Eq. (1.39) that the Slater approximation can be treated as the LDA in Hartree–Fock theory. Hence, in this approximation we have a self-consistent-field method utilizing the Schrödinger equation

$$[-\nabla^2 + v_{ext}(\mathbf{r}) + v^H(\mathbf{r}) + v_{ex}(\mathbf{r})]\varphi_\mu(\mathbf{r}) = \varepsilon_\mu \varphi_\mu(\mathbf{r}). \tag{1.40}$$

This so-called X_α method began to be widely used in 1965; this method solves Eq. (1.40) with an exchange potential of the following type:

$$v_{X_\alpha}(r) = -3\alpha\left[\frac{3}{\pi}\rho(\mathbf{r})\right]^{1/3}, \tag{1.41}$$

where $2/3 < \alpha < 1$ is a certain fitting parameter. The principal drawback of the X_α method is its dependence on the parameter α. The total energy E_{X_α} is not a variational function with respect to α. To a very good approximation, E_{X_α} is a linear, strongly decreasing function of α.[5] However, if we use the orbitals $\varphi_\mu^{X_\alpha}$ which are self-consistent solutions of the one-electron equations of the X_α method from the expression

$$E[\varphi_\mu^{X_\alpha}] = \langle \Phi(\varphi_\mu^{X_\alpha}) | \hat{H} | \Phi(\varphi_\mu^{X_\alpha}) \rangle \tag{1.42}$$

this expression will be a variational functional with respect to α, that is,

$$\left.\frac{\partial E[\varphi_\mu^{X_\alpha}]}{\partial \alpha}\right|_{\alpha=\alpha_{min}} = 0. \tag{1.43}$$

This approach was used in Ref. 11. It was determined that the optimum value of α is a decreasing function of the ordinal number of the element, varying from 0.77 for light atoms through 0.69 for heavy atoms.

One of the most obvious methods of determining α follows from the coincidence of the total energies E_{X_α} and $E_{X_{\rm HF}}$. The resulting table of values of α for the different atoms is provided in Refs. 11 and 12.

The procedures described for determining α were used only for neutral atoms. In the condensed matter case, as well as the molecular case, the primary difficulty lies in finding an exact solution for systems without spherical symmetry. Hence the atomic values of the parameter α are used in the X_α method for virtually all calculations of the electronic structure of molecules and condensed matter which at least yields proper behavior of the solution in the limit of isolated atoms.

We now determine the physical meaning of the quantities E_μ entering into Eqs. (1.19) and (1.40). The total energy is

$$E_{\rm HF}=\langle H_{\rm HF}\rangle \quad \text{where } H_{\rm HF}=\sum_\mu f_\mu+\sum_{\mu,\nu} g_{\mu\nu}$$

(f_μ is the one-electron operator and $g_{\mu\nu}$ is the two-electron operator representing Coulomb repulsion of the μ and ν electrons) is equal to

$$E_{\rm HF}=\sum_\mu n_\mu\langle\mu|\,f_\mu|\mu\rangle-\sum_{\mu<\nu} n_\mu n_\nu[\langle\mu\nu|g_{\mu\nu}|\mu\nu\rangle-\langle\mu\nu|g_{\mu\nu}|\nu\mu\rangle]. \tag{1.44}$$

From here we have

$$E_{\rm HF}(n_\mu=1)-E_{\rm HF}(n_\mu=0)=\langle\mu|\,f_\mu|\mu\rangle$$
$$+\sum_\nu n_\nu[\mu\nu|g_{\mu\nu}|\mu\nu\rangle-\langle\mu\nu|g_{\mu\nu}|\nu\mu\rangle]. \tag{1.45}$$

If Eq. (1.19) is now multiplied by $\varphi_\mu^*(\mathbf{r})$ and volume integration is carried out, we obtain

$$\varepsilon_\mu^{\rm HF}=\langle\mu|\,f_\mu|\mu\rangle+\sum_\nu n_\nu[\langle\mu\nu|g_{\mu\nu}|\mu\nu\rangle-\langle\mu\nu|g_{\mu\nu}|\nu\mu\rangle]. \tag{1.46}$$

Comparing Eqs. (1.45) and (1.46) we find

$$\varepsilon_\mu^{\rm HF}=E_{\rm HF}(n_\mu=1)-E_{\rm HF}(n_\mu=0). \tag{1.47}$$

Equation (1.47) formulates Koopman's theorem: $-$ $\varepsilon_\mu^{\rm HF}$ which is equal to the energy difference of the ion ($n_\mu=0$) and atom ($n_\mu=1$) where the ion and atom are assumed to have orbitals $\varphi_\mu(\mathbf{r})$ that are a solution of Eq. (1.19) for the atom.

The eigenvalues in the X_α method are fundamentally different from the eigenvalues in the Hartree–Fock method. They do not satisfy Koopman's theorem:

$$E_{X_\alpha}(n_\mu=0)-E_{X_\alpha}(n_\mu=1)=-\varepsilon_\mu^{X_\alpha}-\tfrac{1}{2}\langle\mu|\mathbf{v}_{X_\alpha}|\mu\rangle. \tag{1.48}$$

There is a relationship between the eigenvalues and the total energy functional in the X_α method. To differentiate E_{X_α} with respect to the μth occupation number n_μ, we obtain

$$\frac{\partial E_{X_\alpha}}{\partial n_\mu} = \varepsilon_\mu. \tag{1.49}$$

Therefore, the eigenvalue in the X_α method corresponds to the derivative of the total energy functional with respect to the occupation number and not the difference between the two total energy values with occupation numbers n_μ differing by unity.

Now solving Eq. (1.40) for all possible n_μ we can define the set n_μ yielding the maximum total energy value, that is,

$$\delta\left[E_{X_\alpha} + \lambda \sum_\mu n_\mu\right] = 0, \tag{1.50}$$

where λ is the Lagrange multiplier. Variation with respect to n_μ yields

$$\frac{\partial E_{X_\alpha}}{\partial n_\mu} = \lambda, \tag{1.51}$$

so the minimum energy would correspond to the state in which all one-electron energies ε_μ are identical. Introducing the auxiliary constraint $0 \leqslant n_\mu \leqslant 1$ we find that the state with the lowest energy corresponds to Fermi statistics, that is, $\varepsilon_\mu = \lambda$, $n_\mu = 1$, or 0. For levels with $\varepsilon_\mu < \lambda$, $n_\mu = 1$, while for levels with $\varepsilon_\mu > \lambda$, $n_\mu = 0$. The Lagrange multiplier λ therefore plays the role of the Fermi level: $\lambda = \varepsilon_F$. We note that the fractional value of the occupation numbers follows from Eq. (1.49).

Solving the equations of the X_α method we obtain the one-electron states of the system in the ground state. In order to obtain the excited states we can use Eq. (1.49). Here we must drop the conservation of particle number conditions in the system. We consider a Taylor series for the total energy near the ground state[3]:

$$E_{X_\alpha} = E^0_{X_\alpha} + \sum_\mu \left(\frac{dE_{X_\alpha}}{dn_\mu}\right)_0 \Delta n_\mu + \frac{1}{2}\sum_{\nu,\mu}\left(\frac{d^2E_{X_\alpha}}{dn_\nu dn_\mu}\right)_0 \Delta n_\nu \Delta n_\mu + \cdots. \tag{1.52}$$

The total energy derivative with respect to the occupation numbers can be given as

$$\frac{dE_{X_\alpha}}{dn_\mu} = \left(\frac{\partial E_{X_\alpha}}{\partial n_\mu}\right)\varphi_\mu + \sum_\nu \left(\frac{\partial E_{X_\alpha}}{\partial \varphi_\nu}\right)\frac{\partial \varphi_\nu}{\partial n_\mu}. \tag{1.53}$$

If we now use the condition $\partial E_{X_\alpha}/\partial\varphi_\nu = 0$ which arises from minimization of the total energy in φ_ν, we obtain the following expression for the excitation energy δE_{X_α}:

$$\delta E_{X_\alpha} = E_{X_\alpha} - E^0_{X_\alpha} = \sum_\mu \left(\frac{dE_{X_\alpha}}{\partial n_\mu}\right)\Delta n_\mu + \frac{1}{2}\sum_{\nu,\mu}\left(\frac{\partial^2 E_{X_\alpha}}{\partial n_\nu \partial n_\mu}\right)\Delta n_\nu \Delta n_\mu + \cdots. \tag{1.54}$$

Therefore, the excitation energy becomes a function of the occupation numbers $\delta E_{X_\alpha}(n)$ and the first term in Eq. (1.54) satisfies condition (1.49), that is, it can be defined as the eigenvalue of Eq. (1.40):

$$\delta E_{X_\alpha}(n) = \sum_\mu \varepsilon_\mu \Delta n_\mu + \frac{1}{2} U_{\mu\nu} \Delta n_\nu \Delta n_\mu + \cdots, \tag{1.55}$$

where

$$U_{\mu\nu} \equiv \frac{\partial^2 E_{X_\alpha}}{\partial n_\mu \partial n_\nu} = \frac{\partial \varepsilon_\mu}{\partial n_\nu} = \frac{\partial \varepsilon_\nu}{\partial n_\mu}.$$

We now formally represent the occupation number of the μth state of the electron as

$$n_\mu = n_\mu^{(0)} + \lambda_\mu m_\mu, \tag{1.56}$$

where $n_\mu^{(0)}$ is the occupation number of the state $|\mu\rangle$ in the ground state of the system; m_μ is the occupation number of the "quasiparticle" state that is genetically related to the ground state $|\mu\rangle$, that is

$$m_\mu = \begin{cases} +1, & \text{for addition of an electron to } |\mu\rangle; \\ 0, & \text{for the ground state;} \\ -1, & \text{for subtraction of an electron from } |\mu\rangle. \end{cases}$$

The parameter λ_μ may vary continuously over the range $0 \leqslant \lambda_\mu \leqslant 1$, that is, introducing the *parameter* λ_μ means that essentially the occupation numbers in the X_α method adopt fractional values. Since $n_\mu^{(0)}$ are determined from a ground state calculation, the parameter λ can be assumed to be independent of the state, that is, under any single-particle excitation,[13]

$$\delta E_{X_\alpha} = E_{X_\alpha}(\lambda=1) - E_{X_\alpha}(\lambda=0) = \int_0^1 d\lambda \frac{dE_{X_\alpha}}{d\lambda}. \tag{1.57}$$

Now using Eq. (1.56) for expansion (1.54) near the ground state $\lambda = \lambda_0$ we obtain*

$$E_{X_\alpha} \cong E_{X_\alpha}(\lambda_0) + \sum_\mu \left(\frac{\partial E_{X_\alpha}}{\partial \lambda} \right)_0 (\lambda - \lambda_0) m_\mu + \frac{1}{2} \sum_{\mu,\nu} \left(\frac{\partial^2 E_{X_\alpha}}{\partial \lambda^2} \right)_0 (\lambda - \lambda_0)^2 m_\mu m_\nu. \tag{1.58}$$

Differentiating Eq. (1.58) with respect to λ we find

$$\frac{dE_{X_\alpha}}{d\lambda} = \sum_\mu \left(\frac{\partial E_{X_\alpha}}{\partial \lambda} \right)_0 m_\mu + \sum_{\mu,\nu} \left(\frac{\partial^2 E_{X_\alpha}}{\partial \lambda^2} \right)_0 (\lambda - \lambda_0) m_\mu m_\nu. \tag{1.59}$$

Then substituting Eq. (1.59) into Eq. (1.57) we obtain for the excitation energy

---

*We limit the analysis to second order terms in expansion (1.52).

$$\delta E_{X_\alpha} \cong \sum_\mu \varepsilon_\mu(\lambda_0) m_\mu + \left(\frac{1}{2} - \lambda_0\right) \sum_{\mu,\nu} U_{\mu,\nu} m_\mu m_\nu, \tag{1.60}$$

where

$$\varepsilon_\mu(\lambda_0) = \left(\frac{\partial E_{X_\alpha}}{\partial \lambda}\right)_{\lambda=\lambda_0} = \left(\frac{\partial E_{X_\alpha}}{\partial n_\mu}\right)_{n_\mu = n_\mu^{(0)}} \tag{1.61}$$

and

$$U_{\mu\nu} = \left(\frac{\partial^2 E_{X_\alpha}}{\partial \lambda^2}\right)_{\lambda=\lambda_0} = \left(\frac{\partial^2 E_{X_\alpha}}{\partial n_\mu \partial n_\nu}\right)_{\substack{n_\mu = n_\mu^{(0)}.\\ n_\nu = n_\nu^{(0)}}} \tag{1.62}$$

Comparing Eqs. (1.55) and (1.60) we see the obvious analogy between the one-electron energies of the ground state ε_μ and the quasiparticle energies $\varepsilon_\mu(\lambda_0)$ which coincide in the ground state. By definition for the quasiparticle excitations

$$\delta E = \sum_\mu \varepsilon_\mu(\lambda_0) m_\mu. \tag{1.63}$$

It is clear from Eq. (1.60) that Eq. (1.63) holds if we set $\lambda_0 = 1/2$, that is, if we use the Slater transition state.[5] Analyzing Eq. (1.60) we come to a very important conclusion. For a sufficiently weak electron–electron interaction $U_{\mu\nu}$ the second term can be ignored. Then the one-electron spectrum calculated in the X_α approximation coincides with the excitation spectrum. Specifically this explains the success of the present theory when applied to simple metals and standard semiconductors. At the same time the one-electron ground state spectrum is quite different from the excitation spectrum for systems with strong electron–electron interaction which includes, for example, d and f metals and their compounds. However, the X_α one-electron spectrum coincides with the excitation spectrum if we set $\lambda_0 = 1/2$ in Eq. (1.60). This satisfies Eq. (1.63).

In deriving formula (1.60) we were limited to second order terms in expansion (1.52), that is, we retained only pair interactions. Such a breaking off of the series is quite justified[5] since third order terms are sufficiently small ($\sim$0.01 a.u.).

We examine in greater detail the strongly interacting electron limit. The hole and electron quasiparticle bands are separated by an energy gap U for the Hubbard model[14] with the intraatomic Coulomb correlation parameter U. For the hole band $m_\mu = 1$ we obtain from Eq. (1.60),

$$\delta E^h_{X_\alpha} = -\varepsilon_\mu(\lambda_0 = 0) + \tfrac{1}{2} U_{\mu\mu}, \tag{1.64}$$

while from Eq. (1.63) we have

$$\delta E^h = -\varepsilon_\mu(\lambda_0 = -1/2). \tag{1.65}$$

For the electron band with $m_\mu = +1$ we have by analogy

$$\delta E^e_{X_\alpha} = \varepsilon_\mu(\lambda_0 = 0) + \tfrac{1}{2} U_{\mu\mu} \tag{1.66}$$

and

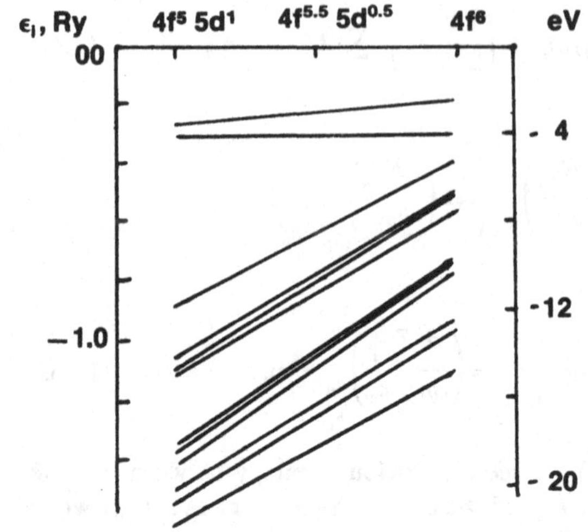

FIG. 1.1. One-electron energies of samarium plotted as a function of occupation numbers.

$$\delta E^e = \varepsilon_\mu(\lambda_0 = 1/2). \tag{1.67}$$

Adding Eq. (1.65) to Eq. (1.67) we find

$$\delta E^2 + \delta E^h = \varepsilon_\mu(\lambda_0 = 1/2) - \varepsilon_\mu(\lambda_0 = -1/2) = U_{\mu\mu}. \tag{1.68}$$

Thus using the transition state method we can rather easily estimate the Hubbard correlation energy $U_{\mu\mu}$.

Figure 1.1 shows ε_μ plotted as a function of n_μ for several levels of the rare earth atom samarium. As we see from this figure the strongest dependence is observed for the $4f$ states. We easily obtain $U_{\mu\mu} = 6.7$ eV from the graph. This theoretical value of U_{4f-4f} for Sm is in good agreement with the experimental value of ~7 eV.[16]

Using Eq. (1.60) we can also obtain the ionization energy I_μ:

$$I_\mu \cong \varepsilon_\mu(\lambda_0 = 1/2). \tag{1.69}$$

Equation (1.69) is an analog of Koopman's theorem although unlike the Hartree–Fock method, it is physically better substantiated since it partially accounts for relaxation effects at the same time that the Hartree–Fock method utilizes a "frozen" core representation. Therefore, the concept of the transition state was also applied to the Hartree–Fock method,[17] which made it possible to substantially improve the interpretation of the energy eigenvalues compared to Koopman's theorem.

Clausius' virial theorem and the Hellmann–Feynman theorem play important roles in condensed matter theory. It can be shown[18] that the total energy functional in the X_α model satisfies these two theorems exactly.

The X_α method was widely used for calculating the electronic structure of atoms, molecules, and condensed matter.[5] However, the applicability of the X_α model to such systems has only been proven empirically. Hence the rigorous Hohenberg–Kohn–Sham approach[19] is of indisputable interest. We examine this method below.

1.2. Uniform electron gas

1.2.1. Correlations in uniform electron gas theory

Uniform electron gas theory has been examined in sufficient detail in the literature (see, e.g., Ref. 20) and hence we shall examine only its fundamental principles required for further analysis of uniform electron gas theory.

We use the term uniform electron gas to refer to any system of interacting electrons traveling in the absence of external fields. It is always assumed that such a system has a uniform distribution of a spatially "smeared" positive charge whose density is equal in magnitude yet opposite in sign to the average electron charge density so that the overall system remains electrically neutral. Such a model was first proposed by Sommerfeld to describe the properties of simple metals. The Sommerfeld model makes it possible to rather easily obtain a number of important characteristics of electron–electron interactions which we will analyze.

We will consider a system of N electrons in a cube of side L; moreover, we assume that a positive charge of density Ne/L^3 is uniformly and continuously distributed in this box. The average volume per electron is equal to

$$\frac{4\pi r_s^3}{3}=\frac{L^3}{N}=\frac{1}{\rho}, \tag{1.70}$$

where ρ is the average electron density. We find from Eq. (1.70),

$$r_s=\left(\frac{3}{4\pi\rho}\right)^{1/3}. \tag{1.71}$$

The parameter r_s characterizes the average distance between electrons. The Hamiltonian of the system takes the form

$$H=-\sum_{i=1}^{N}\nabla^2+2\sum_{i<j}\frac{1}{|\mathbf{r}_i-\mathbf{r}_j|}. \tag{1.72}$$

In order to use perturbation theory we carry out the following substitution of variables in Eq. (1.72):

$$\mathbf{x}=\left(\frac{3}{4\pi}\right)^{1/3}\frac{\mathbf{r}}{r_s}. \tag{1.73}$$

We then obtain

$$H'=-\sum_{i=1}^{N}\nabla_{\mathbf{x}_i}^2+\lambda\sum_{i<j}\frac{2}{|\mathbf{x}_i-\mathbf{x}_j|}, \tag{1.74}$$

where

$$H=\frac{1}{\lambda^2}H',\quad \lambda=\left(\frac{4\pi}{3}\right)^{1/3}r_s. \tag{1.75}$$

The parameter λ in Eq. (1.74) functions as the coupling constant:

$$H=H_0+\lambda V. \tag{1.76}$$

If the coupling constant is small then λV is a perturbation. The unperturbed Schrödinger equation takes the form

$$H_0\Phi_0=\varepsilon_0\Phi_0, \tag{1.77}$$

where Φ_0 is the ordinary Slater determinant from plane waves:

$$\varphi_{\mathbf{k}}(\mathbf{x})=\frac{1}{\sqrt{\Omega}}\exp(i\mathbf{kx})\chi_\sigma. \tag{1.78}$$

Here $\Omega = L^3$ and χ_σ is the spin function.

We have the ground state energy of the system per single particle in the zeroth (lowest) order of perturbation theory; this energy is equal to the average kinetic energy per electron:

$$E_0=\frac{1}{N}\langle\Phi_0|H_0|\Phi_0\rangle=\frac{2.21}{r_s^2}\ \text{Ry}. \tag{1.79}$$

We obtain in the first order of perturbation theory the exchange energy per single particle[21]:

$$\Delta E^{(1)}=\frac{1}{N}\langle\Phi_0|V|\Phi_0\rangle=-\frac{0.916}{r_s}\ \text{Ry}. \tag{1.80}$$

The sum $E_0 + E^{(1)}$ yields the ground state energy per single electron in the Hartree–Fock approximation

$$E_{\text{HF}}=\frac{2.21}{r_s^2}-\frac{0.0916}{r_s}\ \text{Ry}. \tag{1.81}$$

For a more detailed analysis of electron behavior we introduce the pair correlation function[22]

$$g(\mathbf{r})=\frac{1}{N(N-1)}\left\langle\Phi_0\left|\sum_{i<j}\delta(\mathbf{r}+\mathbf{r}_i-\mathbf{r}_j)\right|\Phi_0\right\rangle, \tag{1.82}$$

which gives the probability of finding a particle within a characteristic volume $v_0 = 1/(N - 1)$ over a distance $\mathbf{r}$ from a certain point $\mathbf{r}_0$ when another particle is already at this point. It is obvious that with no correlation between the positions of the two different particles the function $g(\mathbf{r})$ is equal to unity. The function $g(\mathbf{r})$ is related to the structure factor $S(\mathbf{k})$:

$$g(\mathbf{r})=\frac{1}{N-1}\sum_k[S(\mathbf{k})-1]e^{i\mathbf{kr}}, \tag{1.83}$$

which is the Fourier component of the density–density correlator

$$S(\mathbf{r})=\frac{1}{N}\langle\Phi_0|\rho(\mathbf{r}+\mathbf{r}')\rho(\mathbf{r}')|\Phi_0\rangle \tag{1.84}$$

and

$$S(\mathbf{k})=\int d\mathbf{r}\,S(\mathbf{r})e^{i\mathbf{kr}}=\frac{1}{N}\langle\Phi_0|\rho_{\mathbf{k}}^+\rho_{\mathbf{k}}|\Phi_0\rangle. \tag{1.85}$$

The frequency-dependent Van Hove correlation function $S(\mathbf{k},\omega)$ is a generalization of the structure factor:

$$S(\mathbf{k}) = \int_0^\infty d\omega\, S(\mathbf{k}\omega). \tag{1.86}$$

In the Hartree–Fock approximation,[22]

$$S_{\mathrm{HF}}(\mathbf{k}) = \begin{cases} N\delta_{\mathbf{k},0} + \dfrac{3}{4}\dfrac{k}{k_F} - \dfrac{1}{16}\dfrac{k^3}{k_F^3}, & k < 2k_F \\ 1, & k > 2k_F \end{cases} \tag{1.87}$$

and

$$g_{\mathrm{HF}}(\mathbf{r}) = 1 - \frac{9}{2}\left[\frac{\sin k_F r - k_F r \cos k_F r}{k_F^3 r^3}\right]^2. \tag{1.88}$$

In the Hartree–Fock approximation $g_{\mathrm{HF}}^{\uparrow\downarrow}(\mathbf{r}) = 1$ and $g_{\mathrm{HF}}^{\uparrow\uparrow}(0) = 0$, that is, there is no correlation between electrons with different spins and two particles with identical spins cannot occupy the same point according to the Pauli exclusion principle. It can be demonstrated that an exchange hole is near the electron, that is, a domain in which the probability of finding another electron with the same spin is small [see Eq. (1.26)]. It is clear from Eq. (1.88) that the exchange hole radius is equal to $k_F^{-1} = (4/g\pi)^{1/3} r_s$.

Therefore the Coulomb interaction, which is also responsible for electron repulsion, is not reflected in the first perturbation order. Analyzing the second and higher orders of perturbation theory we account for the Coulomb correlations. Two processes—direct and exchange processes—make a contribution to $\Delta E^{(2)}$:

$$\Delta E^{(2)} = \Delta E^{(2)}_{\mathrm{dir}} + \Delta E^{(2)}_{\mathrm{ex}}. \tag{1.89}$$

The exchange process is easily calculated and is equal to 0.046 Ry. The contribution of the direct processes $\Delta E^{(2)}_{\mathrm{dir}}$ diverges logarithmically[21]:

$$\lim_{\mathbf{k}\to 0} \Delta E^{(2)}_{\mathrm{dir}} \sim -\ln \mathbf{k}. \tag{1.90}$$

This logarithmic feature is directly related to the long-range character of the Coulomb forces. Hence, in order to calculate the correlation energy with a high degree of accuracy it is necessary to more accurately reflect the features associated with the long-range character of the Coulomb forces.

One possible approach to solving this problem was proposed by Wigner,[23] while another one was proposed by Bohm and Pines.[24]

Wigner wrote the wave function of the system as a product of two determinants constructed from the single-particle functions with "up" and "down" spins: $\psi = |\varphi_\uparrow| \cdot |\varphi_\downarrow|$. The correlations attributable to the Coulomb interaction between electrons with antiparallel spins was explicitly accounted for by means of the assumption that the single-particle wave functions of electrons with up spins depend on the coordinates of all electrons with down spins. Then the Wigner wave function formulated in this manner was used to calculate the average energy E. Numerical calculation revealed that for $r_s = 1$ the correlation electron energy is

$$E_{\text{corr}} = -\frac{0.88}{7.8}\ \text{Ry} = -0.11\ \text{Ry}. \tag{1.91}$$

Wigner considered a low density electron gas ($r_s \gtrsim 20$) where Coulomb interaction has the predominant effect on electron behavior. In this case we can assume that the electrons form a stable lattice against a uniform positive charge background since the potential energy will drive the electrons to the maximum possible distances while the kinetic energy is insufficient to overcome electron localization. The ground state energy in the low density limit can be given as an expansion in powers of $r_s^{1/2}$ (Ref. 25):

$$\frac{E}{N} = -\frac{1.792}{r_s} + \frac{2.66}{r_s^{3/2}} + \frac{a}{r_s^2} + \frac{b}{r^{5/2}} + \cdots. \tag{1.92}$$

Here the first term represents the potential energy of electrons localized at specific points, while the second term represents the null vibration energy of the electrons near equilibrium; the third and last terms of the expansion are due to the anharmonicity of lattice vibrations.

Therefore, in the low density limit and accurate to terms of the order of $r_s^{-3/2}$ we have

$$\frac{E_{\text{corr}}}{N} = \frac{E}{N} - \frac{E_{\text{HF}}}{N} = -\frac{1.792}{r_s} + \frac{0.916}{r_s} = -\frac{0.88}{r_s}\ \text{Ry}. \tag{1.93}$$

Wigner formulated an interpolation formula for the correlation energy relating the low and high density regions[23] (excluded are phase transitions):

$$\frac{E_{\text{corr}}}{N} = -\frac{0.88}{r_s + 7.8}\ \text{Ry}. \tag{1.94}$$

The difficulties associated with applying perturbation theory to a uniform electron gas led Bohm and Pines[24] to develop a plasma collective vibration theory. The primary physical effect in this case is polarization of the medium, that is, screening of effective interaction between a particle pair due to the motion of other particles in this medium. If we treat an electron gas as a quantum plasma in the Fermi–Thomas approximation we can easily obtain the following equation[25] for the total potential in the plasma v(r):

$$\text{v}(r) = \frac{q}{r}\exp[-k_{\text{FT}} r], \tag{1.95}$$

where k_{FT} is the Fermi–Thomas inverse screening radius

$$k_{\text{FT}} = \left(\frac{6\pi N e^2}{\varepsilon_F}\right)^{1/2}. \tag{1.96}$$

Therefore, the polarization generated by charge q causes screening of the field of this charge at distances of the order of $\lambda_{\text{FT}} = k_{\text{FT}}^{-1}$. Since $\lambda_{\text{FT}} \sim 1$ Å in good metals, then the effect of Coulomb interaction is very rapidly screened in practice. If an excess positive charge arises in some domain of the electron gas then the electrons in screening the gas will begin to move in the direction of this domain. Here, generally speaking, the electrons will not immediately assume an equilibrium in this domain but rather will assume an equilibrium after certain oscillations, that is, oscillations in the space charge density arise. The equation of motion for this process takes the form[25]

$$\ddot{\rho}_{\mathbf{k}} = -\sum_i (\mathbf{k}\cdot\mathbf{y}_i + k^2)^2 e^{-i\mathbf{k}\mathbf{r}_i} - \sum_q \frac{16\pi}{q^2} (\mathbf{k}\cdot\mathbf{q}) \rho_{\mathbf{k}-\mathbf{q}} \rho_{\mathbf{q}} \tag{1.97}$$

[$\rho_{\mathbf{k}}$ is the Fourier component of electron density $\rho(\mathbf{r})$]. The first term on the right hand side of Eq. (1.97) is related to the kinetic energy of the particles while the second term is related to interaction between the particles. Factoring out the component with $q = k$ in the second term, we recast Eq. (1.97) as

$$\ddot{\rho}_{\mathbf{k}} + \omega_p \rho_{\mathbf{k}} = -\sum_i (\mathbf{k}\cdot\mathbf{v}_i + k^2)^2 e^{-i\mathbf{k}\mathbf{r}_i} - \sum_{q\neq k} \frac{16\pi}{q^2} (\mathbf{q}\cdot\mathbf{k}) \rho_{\mathbf{k}-\mathbf{q}} \rho_{\mathbf{q}}, \tag{1.98}$$

where

$$\omega_p = (16\pi N)^{1/2} \tag{1.99}$$

is the so-called plasma frequency of the electron gas. We consider the second term on the right-hand side of Eq. (1.98). For $\mathbf{q}\neq 0$, the quantity $\rho_{\mathbf{q}} = \Sigma \exp(-i\mathbf{q}\cdot\mathbf{r}_i)$ represents the sum of exponents with randomly varying phases with an average equal to zero. Hence the second term in Eq. (1.98) is only a small correction to the equation of motion for ρ_k and can be dropped. This approximation is called the random phase approximation (RPA). Equation (1.98) (without the second term) demonstrates that the quantities ρ_k oscillate roughly with an angular frequency ω_p for very small k (the long-wavelength approximation). With large k these oscillations will decay due to thermal electron motion. It is clear from Eq. (1.98) that plasma oscillations will occur only when $k \ll k_c$, where $k_c = \omega_p/\mathrm{v}_F$ (v_F is the electron velocity on the Fermi surface).

Dropping the second term on the right-hand side of Eq. (1.98) we ignore the relation between density fluctuations with different wavelengths. In other words, in calculating any characteristic of the system dependent on **k** we retain within the framework of the RPA only those Coulomb interaction components that correspond to the same momentum **k**.

The Hamiltonian of the electron system in the RPA neglecting interaction between electrons and plasmons takes the following form[25]:

$$H = \sum_i \mathbf{P}_i^2 + \sum_{\mathbf{k}<\mathbf{k}_c} \left\{ \frac{1}{2} (\mathbf{P}_{\mathbf{k}}^+ \mathbf{P}_{\mathbf{k}} + \omega_p \mathbf{Q}_{\mathbf{k}} \mathbf{Q}_{\mathbf{k}}) - \frac{4\pi N}{\Omega k^2} \right\} + \sum_{\mathbf{k}>\mathbf{k}_c} \frac{4\pi}{\Omega k^2} (\rho_{\mathbf{k}}^+ \rho_{\mathbf{k}} - N). \tag{1.100}$$

Here $\mathbf{P}_{\mathbf{k}}$ and $\mathbf{Q}_{\mathbf{k}}$ are the canonically conjugate momentum and coordinate operators for the oscillator. The third sum in Eq. (1.100) describes screened interaction between electrons:

$$H_{s,r} = \sum_{k>k_c} \frac{4\pi}{\Omega k^2} (\rho_{\mathbf{k}}^+ \rho_{\mathbf{k}} - N) = \sum_{i<j}^{N} \frac{1}{r_{ij}} F(k_c r_{ij}), \tag{1.101}$$

where

$$F(k_c r)=1-\frac{2}{\pi}\int_0^{k_c r} dx\,\frac{\sin x}{x}. \tag{1.102}$$

Thus the potential effective interaction energy between electrons at a distance of r is equal to $2/rF(k_c r)$ which decreases to zero for $k_c r > 2$, that is, the effective interaction radius is of the order $2k_c^{-1}$.

The second term in Eq. (1.100) represents the "long-range" part of Coulomb interaction which is entirely contained in the null plasma oscillation energy. Therefore, the RPA retains a system of electrons interacting through short-range forces of radius $2k_c^{-1}$.

In plasma theory the correlation energy is equal to[25]

$$E_{\text{corr}}=0.0622\ln r_s+B+O(r_s). \tag{1.103}$$

Terms of the order of r_s are ignored in Eq. (1.103).

Gell-Mann and Brueckner[26] made the next major step in analyzing the gas of interacting electrons. They demonstrated that perturbation theory techniques could be applied to an electron gas only if one does not stop at the first diverging term but rather continues the summation and sums the strongly diverging terms from each term in the series (the so-called ring diagrams). The ground state energy calculated in this manner is equal to

$$\frac{E}{N}=\frac{2.21}{r_s^2}-\frac{0.916}{r_s}+0.062\ln r_s-0.096+ar_s+br_s\ln r_s+cr_s^2+\cdots. \tag{1.104}$$

Equation (1.104) holds for $r_s \ll 1$, that is, in the high density limit. If we compare the correlation energy from Eq. (1.104):

$$\frac{E_{\text{corr}}}{N}=0.062\ln r_s-0.096+O(r_s) \tag{1.105}$$

to Eq. (1.103) obtained from the RPA it is clear that they coincide. Therefore, the approximation involving summation of the ring diagrams in the Gell-Mann and Brueckner theory is equivalent to the random phase approximation in the Bohm and Pines theory.

Equations (1.92) and (1.104) represent the two extreme cases: $r_s \gg 1$ (low densities) and $r_s \ll 1$ (high densities). There is a significant intermediate coupling range between these two limiting cases. This range includes, for example, electron concentrations in real metals ($2 < r_s < 5.5$). Evidently it is more correct to speak of an electron liquid in this intermediate range.

One of the first equations calculating electron correlation in the intermediate range r_s was, obviously, the Wigner formula (1.94). The calculation of E_{corr} for intermediate values of r_s remains one of the most difficult problems of theoretical physics today. We will briefly examine the primary studies that have made the strongest contribution to this problem.

The primary applicability condition of the RPA is $k \ll k_c$, that is, the RPA works with low momenta. It was demonstrated[27] that under this condition the RPA can even be applied to real electron densities in metals ($1.8 \lesssim r_s \lesssim 5.5$). The following approximate equation was obtained for these densities[27]:

$$\varepsilon_c(r_s) \equiv \frac{E_{\text{corr}}}{N} \approx 0.031 \ln r_s - 0.115 \text{ Ry}. \tag{1.106}$$

Here we introduced the quantity $\varepsilon_c(r_s)$: the so-called correlation energy density.

Note that with large values of k the contribution of exchange scattering to the diagram tends to cause a twofold decrease in the contribution from forward scattering.[28] Hence replacing the contribution of the primary polarization diagram $\chi_0(\mathbf{k},\omega)$ with the equation

$$\chi_H(\mathbf{k},\omega) = \frac{\chi_0(\mathbf{k},\omega)}{1 - (8\pi/k^2)(1-G)\chi_0(\mathbf{k},\omega)}, \tag{1.107}$$

where

$$G(k) = \frac{k^2}{2(k^2 + k_F^2)} \tag{1.108}$$

was proposed. The function $\chi_0(\mathbf{k},\omega)$ is called the polarizability of the system of noninteracting electrons (the Lindhard function) which is related to the dielectric function in the Hartree–Fock approximation $\varepsilon_{\text{HF}}(\mathbf{k},\omega)$ by the following equation:

$$\frac{1}{\varepsilon_{\text{HF}}(\mathbf{k},\omega)} = 1 + \frac{8\pi}{k^2}\chi_0(\mathbf{k},\omega) \tag{1.109}$$

In the RPA

$$\chi_{\text{RPA}}(\mathbf{k},\omega) = \frac{\chi_0(\mathbf{k},\omega)}{1 - (8\pi/k^2)\chi_0(\mathbf{k},\omega)}. \tag{1.110}$$

In the long-wavelength limit for $k \ll k_F$, Eq. (1.107) becomes Eq. (1.110). On the other hand, for $k \gg k_F$, we have $G \approx 1/2$, and expanding the denominator in Eq. (1.107) we find

$$\chi_H(\mathbf{k},\omega) \approx \chi_0(\mathbf{k},\omega) + \frac{8\pi}{k^2}\frac{\chi_0(\mathbf{k},\omega)}{2} + \cdots. \tag{1.111}$$

The first term in this expansion yields the exchange energy in the Hartree–Fock approximation; the second term which yields the direct interaction in the second order of perturbation theory for the ground state energy with the contribution from large k decreased by a factor of 2.

The variational technique is quite useful for calculating the ground state properties of many-particle systems. According to the variational principle for any function $\Psi_T(\mathbf{R})$ the variational energy

$$E_T = \int d\mathbf{R}\, \Psi_T^*(\mathbf{R}) H(\mathbf{R}) \Psi_T(\mathbf{R}) \Big/ \int d\mathbf{R} \left|\Psi_T(\mathbf{R})\right|^2 \tag{1.112}$$

will have a minimum if Ψ_T is the wave function of the ground state of the Schrödinger equation. In Eq. (1.112) $\mathbf{R}$ represents the coordinates of N particles while $H(\mathbf{R})$ is the Hamiltonian. $\Psi_T(\mathbf{R})$ will be either symmetrical or antisymmetrical with respect to coordinate permutation depending on the statistics. The variational method involves formulating a set of functions $\Psi_T(\mathbf{R},p)$ followed by optimization of the parameter p such that E_T is minimized for $p=p^*$. The variational energy represents the exact upper limit for the ground state energy, while $\Psi_T(\mathbf{R},p^*)$ will be a good approximation for the ground state wave function.

In the case of a Fermi liquid the wave function is selected as

$$\Psi_T(\mathbf{R})=D(\mathbf{R})\exp\left[-\sum_{i<j} U(|\mathbf{r}_i-\mathbf{r}_j|)\right], \tag{1.113}$$

where $D(\mathbf{R})$ is the Slater determinant from plane waves, $U(r)$ is the so-called "pseudopotential" which is a function of the distance between particles only, that is, it accounts for correlation effects.

The variational method was first applied by Gaskell[29] to calculate the ground state energy of a homogeneous electron gas. The results for $r_s \leqslant 2.66$ were very close to the data calculated by Eqs. (1.111), (1.106), and (1.94). For $r_s > 2.66$ the correlation function $g(\mathbf{r})$ [see Eq. (1.83)] was negative with small $\mathbf{r}$ which indicates the fact that certain approximations used in Gaskell's calculations[29] were incorrect.

One of the primary problems complicating variational calculations was the evaluation of multidimensional integrals in Eq. (1.112). It was first noted[30] that such integrals can be evaluated by the Metropolis method[31] [the $M(RT)^2$ algorithm]. The energy E_T in Eq. (1.112) is the average value of the operator $H(\mathbf{R})$, that is,

$$E_T\equiv\langle H\rangle=\int d\mathbf{R}\,\Psi_T^*(\mathbf{R})H\Psi_T(\mathbf{R})\Big/\int d\mathbf{R}\left|\Psi_T(\mathbf{R})\right|^2. \tag{1.112a}$$

Let $\{\mathbf{R}_c\}$ be the set of points representing a sample of the probability distribution

$$P(\mathbf{R})=|\Psi_T(\mathbf{R})|^2\Big/\int d\mathbf{R}\left|\Psi_T(\mathbf{R})\right|^2. \tag{1.114}$$

Then the central limit probability theorem yields

$$\lim_{M\to\infty}\frac{1}{M}\sum_{i=1}^{M}\Psi_T^{-1}(\mathbf{R}_i)H(\mathbf{R}_i)\Psi_T(\mathbf{R}_i)=\langle H\rangle. \tag{1.115}$$

The algorithm $M(RT)^2$ consists of a random walk in configuration space; ordinarily this is realized by moving all particles one after another to renew their positions uniformly distributed within a cube of side L. Each displacement is either adopted or rejected depending on the value of the new trial function compared to the previous value. We assume that $\mathbf{R}$ is the previous position while $\mathbf{R}'$ is the new position. Then if $|\Psi_T(\mathbf{R}')|^2 \geqslant |\Psi_T(\mathbf{R})|^2$, the new point is used. Otherwise the new point is used with a probability $q=|\Psi_T(\mathbf{R}')|^2/|\Psi_T(\mathbf{R})|^2$. This approach is one variant of the Monte Carlo method.[32]

A number of variational calculations of the ground state energy by the $M(RT)^2$ method were carried out for a uniform electron gas in two and three dimensions.[33,34] Two types of pseudopotentials were used in this case:

(a) the Gaskell pseudopotential in the RPA[29]:

$$2U_{\text{RPA}}(\mathbf{k}) = -\frac{1}{S_0(\mathbf{k})} + \left(\frac{1}{S_0^2(\mathbf{k})} + \frac{2\text{v}(\mathbf{k})}{k^2}\right)^{1/2}, \tag{1.116}$$

where $\text{v}(k)$ is the Fourier component of the interparticle potential and $S_0(\mathbf{k})$ is the structure function of the ideal Fermi gas;

(b) the Yukawa pseudopotential:

$$U_Y(r) = A[1 - \exp(-r/F)]/r. \tag{1.117}$$

The calculations were carried out over a very large electron density range ($1 \leqslant r_s \leqslant 500$). The following approximation was used for the two limits:

(a) $r_s \ll 1$:

$$E_0 = \frac{h_0}{r_s^2} + \frac{h_1}{r_s} + O(\ln r_s). \tag{1.118}$$

(b) $r_s \gg 1$:

$$E_0 = \frac{f_0}{r_s} + \frac{f_1}{r_s^{3/2}} + \frac{f_2}{r_s^2} + O(r_s^{-5/2}). \tag{1.119}$$

In the intermediate range the Padé approximation was used:

$$E(r_s) = \frac{\sum_{i=0}^{4} \alpha_i x^i}{r_s^2(1 + \sum_{i=1}^{2} \beta_i x^i)}, \tag{1.120}$$

where $\alpha_0 = h_0; \alpha_1 - h_0\beta_1; \alpha_2 = h_1 + \beta_2 h_0; \alpha_3 = h_1\beta_1; x = r_s^{1/2}$. The three parameters β_1, β_2, and α_4 were determined by the least-squares method from a fit to the energies calculated by the variational method using the $M(RT)^2$ algorithm with large r_s.

It was determined from these calculations that the variational method of calculating the ground state energy of a uniform electron gas by the $M(RT)^2$ algorithm is more accurate and reliable over the range $r_s > 5$. In the range $r_s \leqslant 5$ this method is somewhat less accurate than the method based on the so-called cluster-bound approximation[35] (in this range of r_s the correlation energies obtained by the variational method are on the average 0.002 Ry poorer than those calculated in Ref. 34).

The essence of the cluster-bound or the so-called $\exp(S)$ method[35] is that the direct contribution to the correlation energy ΔE_{dir} [see Eq. (1.89)] is calculated in the RPA while the exchange contribution ΔE_{ex} is calculated as the screened contribution, that is, the diagram of the ordinary exchange process is replaced by the "screened" diagrams on page 26.

The wave function of the $\exp(S)$ method is given as

$$\Psi = e^T \Phi_0,$$

where

$$T = \sum_{j=1}^{N} T_j; \quad T_j = \frac{1}{j!} \sum_{\substack{\alpha_1 \ldots \alpha_j \\ r_1 \ldots r_j}} t_{\alpha_1 \ldots \alpha_j}^{r_1 \ldots r_j} \prod_{i=1}^{j} C_{r_i}^{+} C_{\alpha_i}. \tag{1.121}$$

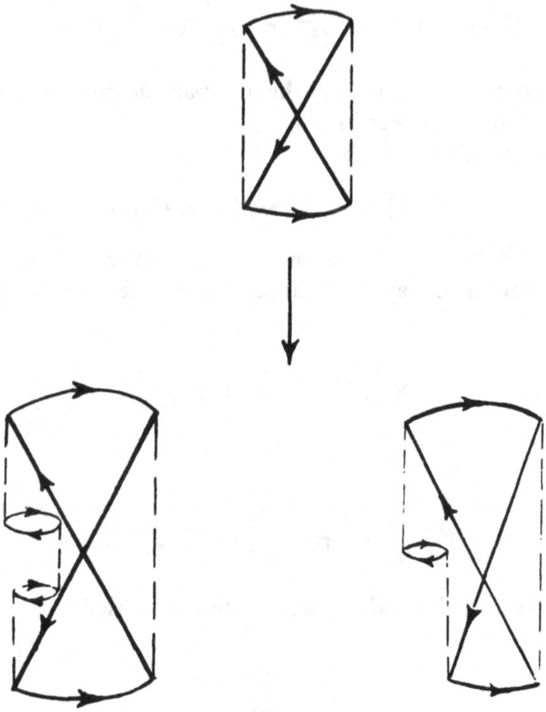

$C^+_{r_i}$ is the particle creation operator in the r_i state and C_{α_i} is the particle destruction operator in the α_i state. The following represents the fundamental approximation of the method:

$$T \approx T_2,$$

where

$$T_2 = \sum_{k_i, k_j, q} t_q(k_i k_j) C^+_{k_i+q} C^+_{k_j-q} C_{k_j} C_{k_i}. \qquad (1.122)$$

Here we have discussed the correlation energy calculation based on the susceptibility properties of the screened electron gas (the Hubbard approximation). This method was further developed.[36–38] We will examine in gre‑.er detail this method of calculating the correlation energy.

1.2.2. Screening in electron gas. Linear response

As noted above, electron–electron interaction yields polarization, that is, a fluctuating induced charge $\delta\rho_{\text{ind}}(\mathbf{r})$ arises in the system. In the general case it is possible to consider the external charge density (the perturbation density) $\rho_{\text{ext}}(k,\omega)$ which is dependent on frequency ω; this yields a dependence on ω of the macroscopic field induction D and field strength E. The total charge density is given as the sum

$$\rho_{\text{tot}}(\mathbf{k},\omega) = \rho_{\text{ext}}(\mathbf{k},\omega) + \delta\rho_{\text{ind}}(\mathbf{k},\omega). \qquad (1.123)$$

The screening properties of the electronic system are determined by the permittivity $\varepsilon(k,\omega)$*:

$$D(\mathbf{k},\omega)=\varepsilon(\mathbf{k},\omega)E(\mathbf{k},\omega), \tag{1.124}$$

$$\rho_{\text{ext}}(\mathbf{k},\omega)=\varepsilon(\mathbf{k},\omega)\rho_{\text{tot}}(\mathbf{k},\omega). \tag{1.125}$$

In the linear response case (a linear approximation of screening theory which is sufficient for weak perturbation):

$$\delta\rho_{\text{ind}}(\mathbf{k},\omega)=8\,\frac{\pi}{k^2}\,\rho_{\text{ext}}(\mathbf{k},\omega)\chi(\mathbf{k},\omega). \tag{1.126}$$

Using Eqs. (1.123), (1.125), and (1.126) we find

$$\frac{1}{\varepsilon(\mathbf{k},\omega)}=1+\frac{8\pi}{k^2}\,\chi(\mathbf{k},\omega). \tag{1.127}$$

Equation (1.127) is a generalization of Eq. (1.109). If we introduce the screened response function $\Pi(\mathbf{k},\omega)$ which establishes the relation between the induced and total charges

$$\delta\rho_{\text{ind}}(\mathbf{k},\omega)=-\frac{8\pi}{k^2}\,\Pi(\mathbf{k},\omega)\rho_{\text{tot}}(\mathbf{k},\omega). \tag{1.128}$$

We obtain

$$\varepsilon(\mathbf{k},\omega)=1+\frac{8\pi}{k^2}\,\Pi(\mathbf{k},\omega). \tag{1.129}$$

The function $\Pi(\mathbf{k},\omega)$ functions as the polarization operator $[-i\Pi(\mathbf{k},\omega)]$ which is the sum of all irreducible polarization diagrams; $\Pi(\mathbf{k},\omega) = \mathbf{k}^2\alpha(\mathbf{k},\omega)$, where $\alpha(\mathbf{k},\omega)$ is the polarizability.

The calculation of $\varepsilon(\mathbf{k},\omega)$ represents a central problem in uniform electron gas theory. It leads to Eq. (1.109) in the Hartree–Fock approximation. The neglect of Coulomb repulsion in this approximation causes strong potential overlapping which is the reason for the divergence of the total potential with large r.

A more realistic approach is the RPA which determines the response of the electron system to the total screened potential rather than an external field, as in the Hartree–Fock approximation. The induced charge density is

$$\delta\rho_{\text{ind}}(\mathbf{k},\omega)=\frac{8\pi}{k^2}\,\chi_0(\mathbf{k},\omega)\rho_{\text{tot}}(\mathbf{k},\omega). \tag{1.130}$$

In the RPA the susceptibility can be evaluated[22]:

$$\varepsilon_{\text{RPA}}(k,0)=1+\frac{4k_F}{\pi k^2}\left(\frac{1}{2}+\frac{4k_F^2-k^2}{8k_Fk}\ln\left|\frac{k+2k_F}{k-2k_F}\right|\right). \tag{1.131}$$

In the long-wavelength limit ($k \ll k_F$) we have

*We follow Ref. 39 in this discussion.

$$\varepsilon_{\mathrm{RPA}}(k,0) \to 1 + \frac{4k_F}{\pi k^2}, \tag{1.132}$$

that is, $\varepsilon_{\mathrm{RPA}}$ tends toward the susceptibility of the Fermi–Thomas approximation.[22]

The polarizability of the system in the RPA is calculated by Eq. (1.110). The total potential is equal to

$$\mathrm{v}_{\mathrm{tot}}(\mathbf{k},\omega) = \mathrm{v}_{\mathrm{ext}}(\mathbf{k},\omega) + \delta \mathrm{v}_{\mathrm{ind}}(\mathbf{k},\omega) \tag{1.133}$$

or

$$\mathrm{v}_{\mathrm{tot}}(\mathbf{k},\omega) = \mathrm{v}_{\mathrm{ext}}(\mathbf{k},\omega) + \frac{8\pi}{k^2}\, \delta\rho_{\mathrm{ind}}(\mathbf{k},\omega). \tag{1.134}$$

Accounting for the relation between $\delta\rho_{\mathrm{ind}}$ and ρ_{tot} we obtain

$$\mathrm{v}_{\mathrm{tot}}(\mathbf{k},\omega) = \mathrm{v}_{\mathrm{ext}}(\mathbf{k},\omega) + \frac{8\pi}{k^2}\, \chi_0(\mathbf{k},\omega)\mathrm{v}_{\mathrm{tot}}(\mathbf{k},\omega). \tag{1.135}$$

From here we have

$$\mathrm{v}_{\mathrm{tot}}(\mathbf{k},\omega) = \frac{\mathrm{v}_{\mathrm{ext}}(\mathbf{k},\omega)}{\varepsilon_{\mathrm{RPA}}(\mathbf{k},\omega)}. \tag{1.136}$$

Equation (1.136) permits a satisfactory description of the electronic system with small k only ($k \ll k_F$), that is, in the long-wavelength limit. With large k, as we have stated previously, it is necessary to go over to Eq. (1.107) or to the following expression for the potential:

$$\mathrm{v}_{\mathrm{tot}}(\mathbf{k},\omega) = \mathrm{v}_{\mathrm{ext}}(\mathbf{k},\omega) + \frac{8\pi}{k^2}\,[1 - G(\mathbf{k},\omega)]\rho_{\mathrm{ind}}(\mathbf{k},\omega), \tag{1.137}$$

where $G(k,\omega)$ is the local field correction. The local field correction was first introduced by Hubbard[28] [see Eq. (1.108)].

References 36, 37, and 40 derived formulas for the function G expressed through correlation function (1.82). In the Singwi–Tosi–Laud–Sjolander (STLS) approximation[36] we have

$$G(\mathbf{k}) = 1 - \int \frac{d\mathbf{k}'}{(2\pi)^3} \frac{\mathbf{k}\cdot\mathbf{k}'}{(k')^2}\, g(\mathbf{k} - \mathbf{k}'). \tag{1.138}$$

The first modification of the STLS approximation was presented in Ref. 37 [the Singwi–Sjolander–Tosi–Laud (SSTL) approximation]. The Coulomb interaction is screened in the SSTL approximation compared to the STLS approximation, that is,

$$G(\mathbf{k}) = 1 - \int \frac{dk'}{(2\pi)^3} \frac{\mathbf{k}\cdot\mathbf{k}'}{(\mathbf{k}')^2 \varepsilon(\mathbf{k}',0)}\, g(\mathbf{k}\cdot\mathbf{k}'). \tag{1.139}$$

A subsequent study[40] [the Vashishta–Singwi (VS) approximation] using the following expression for the pair correlation energy for small frequencies:

$$g(\mathbf{r},\mathbf{r}',t) = g(|\mathbf{r} - \mathbf{r}'|) + \delta\rho_{\mathrm{ind}} \frac{\partial}{\partial \rho} g(|\mathbf{r} - \mathbf{r}'|) \tag{1.140}$$

obtained

$$G(\mathbf{k})=\left(1-a\rho\frac{\partial}{\partial\rho}\right)\left[1-\int\frac{d\mathbf{k}'}{(2\pi)^3}\frac{\mathbf{k}\cdot\mathbf{k}'}{(k')^2}g(\mathbf{k}-\mathbf{k}',\rho)\right]. \quad (1.141)$$

Here a is the parameter selected based on satisfaction of the compressibility theorem.

The pair correlation function is related to the structure factor by Eq. (1.83). The structure factor in turn can be expressed through the permittivity[22]:

$$S(\mathbf{k})=-\frac{k^2}{8\pi^2\rho}\int_0^\infty d\omega\ \mathrm{Im}\frac{1}{\varepsilon(\mathbf{k},\omega)}. \quad (1.142)$$

Then from Eq. (1.137) we find

$$\varepsilon(\mathbf{k},\omega)=1-\frac{(8\pi/k^2)\chi_0(\mathbf{k},\omega)}{1+G(k)(8\pi/k^2)\chi_0(\mathbf{k},\omega)}. \quad (1.143)$$

Therefore, we have a self-consistent system of equations for finding $G(k)$. The self-consistent solution was obtained for three approximations: STLS, SSTL, and VS. With $a=2/3$ for $k<2k_F$ the function $G(k)$ can be approximated in the SSTL and VS approximations by the following equation:

$$G(k)=A(r_s)[1-\exp(-B(r_s)k^2/k_F^2)]. \quad (1.144)$$

Knowing the function $G(k)$ we can easily calculate the structure factor from Eqs. (1.142) and (1.143) and then calculate the correlation function

$$\varepsilon_c=\frac{1}{r_s^2}\int_0^{r_s}dr_s\left[-\frac{4}{\pi\alpha}\gamma(r_s)+0.916\right]\mathrm{Ry}, \quad (1.145)$$

where

$$\gamma(r_s)=-\frac{1}{2k_F}\int_0^\infty d\mathbf{k}[S(\mathbf{k})-1],$$
$$\alpha=(4/9\pi)^{1/3}=0.521. \quad (1.146)$$

In the VS approximation the correlation energy can be calculated by the following formula:

$$\varepsilon_c=(0.0335\ln r_s-0.112)\mathrm{Ry}. \quad (1.147)$$

Table 1.1 provides the correlation energies obtained in different approximations.

1.2.3. Generalization to the spin-polarized case

The uniform electron gas theory to the spin-polarized case was generalized by Ref. 41. In analogy to this we introduce the following designations: the indices P and F will refer to the paramagnetic and ferromagnetic states of the electron gas, respectively; $\rho_\uparrow$ is the up spin electron density while $\rho_\downarrow$ is the down spin electron density. Reference 41 obtained the parametrized equations for the correlation energy of spin-polarized electron gas which take the following form:

TABLE 1.1. 1—Correlation energies in different approximations; 2—RPA; 3—Hubbard (Ref. 38); 4—Freeman (Ref. 35); and 5—variational method (Ref. 34).

	$r_s=1$	$r_s=2$	$r_s=3$	$r_s=4$	$r_s=5$	$r_s=6$
RPA	157.6	123.6	105.5	93.6	25	78.2
Hubbard (Ref. 38)	···	92	75	64	···	···
STLS (Ref. 36)	125	97	80	70	63	57
SSTL (Ref. 37)	124	92	75	64	56	50
VS (Ref. 40)	130	98	81	70	62	56
Freeman (Ref. 35)	118	88.4	73.3	63.6	56.7	51.5
Variational method (Ref. 34)	122	87.4	72.2	62.4	55.0	49.8

$$\varepsilon_c^P = -C^P F\left(\frac{r_s}{r^P}\right), \tag{1.148a}$$

$$\varepsilon_c^F = -C^F F\left(\frac{r_s}{r^F}\right), \tag{1.148b}$$

where

$$F(x) = (1+x^3)\ln\left(1+\frac{1}{x}\right) + \frac{x}{2} - x^2 - \frac{1}{3}. \tag{1.149}$$

The coefficients $C^{P/F}$ and $r^{P/F}$ were found from a fit to the RPA results: $C^P = 0.0504$; $C^F = 0.0254$; $r^P = 30$; and $r^F = 75$.

Probably the most interesting study from a practical viewpoint is the work of Vosko *et al.*,[42] in which all published data have been analyzed on the techniques and results from correlation energy calculations and formulated a parametrized equation for $\varepsilon_C(r_s,\varphi)$ that is suitable for virtually any electron density range of spin-polarized uniform electron gas. The authors of this study noted that the results of von Barth and Hedin[41] were not exact and hence they carried out a new, more exact parametrization of the correlation energy using the RPA results. Moreover, they accounted[42] for the fact that the function

$$f(\varphi) = \frac{[(1+\varphi)^{4/3} + (1-\varphi)^{4/3} - 2]}{2(2^{1/3}-1)}, \tag{1.150}$$

entering into the exchange energy

$$\varepsilon_x(r_s\varphi) = \varepsilon_x^P(r_s) + [\varepsilon_x^F(r_s) - \varepsilon_x^P(r_s)]f(\varphi), \tag{1.151}$$

incorrectly yields the dependence on φ in the most important polarization range: $0<\varphi<0.5$. Reference 42 carried out one of the best interpolation approximations for the correlation energy. In this scheme new RPA results obtained accurate to better than 1% for all r_s and φ were combined with the variational calculation results[34] produced by the $M(RT)^2$ method using Padé's two-point approximation equation

TABLE 1.2. Interpolation parameters for energy-correlation parameters.

	x_0	b	c	A
ε_c^P	− 0.10498	3.72744	12.9352	0.0621814
ε_c^F	− 0.32500	7.06042	18.0578	0.0310907
α_c	− 0.0047584	1.13107	13.0045	− 1/3

$$r_s \frac{d\varepsilon_c}{dr_s} = A \frac{1 + b_1 x}{1 + b_1 x + b_2 x^2 + b_3 x^3}, \tag{1.152}$$

where $x = \sqrt{r_s}$. Unlike Eq. (1.120) this equation also includes the coefficient A from the equation

$$\varepsilon_c(r_s) = A \ln r_s + c + r_s(A \ln r_s + c) + \cdots \quad (r_s \ll 1).$$

Therefore, the four parameters A, b_1, b_2, and b_3 allow an accurate "joining" of the two limiting ranges of ρ: $r_s \ll 1$ and $r_s \gg 1$. For this procedure the following approximation formula has been obtained[42] for the correlation energy:

$$\varepsilon_c(r_s) = A \left\{ \ln \frac{x^2}{X(x)} + \frac{2b}{Q} \operatorname{arctg} \frac{Q}{2x + b} - \frac{b x_0}{X(x_0)} \times \left[\ln \frac{(x - x_0)^2}{X(x)} + \frac{2(b + 2x_0)}{Q} \operatorname{arctg} \frac{Q}{2x + b} \right] \right\}, \tag{1.153}$$

where $X = x^2 + bx + c$; $Q = \sqrt{4c - b^2}$; A, x_0, b, and c are the fitting parameters given in Table 1.2.

An analysis[42] of the spin dependence of ε_c revealed that the spin-dependent part $\Delta\varepsilon_c(r_s, \xi)$ even in the RPA approximation is best approximated by an equation of the type

$$\Delta\varepsilon_c(f_s, \varphi) = \alpha_c(r_s) \frac{f(\varphi)}{f''(0)} [1 + \beta_c(r_s)\varphi^4], \tag{1.154}$$

where $\alpha_c(r_s)$ is the so-called spin displacement function related to the spin susceptibility by the relation

$$\alpha_c(r_s) = \frac{3\alpha\pi}{2r_s^2} \left\{ \frac{\chi_0(r_s)}{\chi_h(r_s)} - 1 \right\} \tag{1.155}$$

(χ_h and χ_0 are the susceptibilities of the interacting uniform gas and the noninteracting gas, respectively) while the function

$$\beta_c(r_s) = \frac{f''(0)\Delta\varepsilon_c(r_s, 1)}{\alpha_c(r_s)} - 1. \tag{1.156}$$

The function $\alpha_c(r_s)$ is interpolated by Eq. (1.153) with the parameters shown in Table 1.2.

The relativistic generalization of the uniform electron gas theory was developed by Akhiezer and Peletminskiy.[43] The incorporation of electron interaction through the quantized electromagnetic field yields very important relativistic corrections to the nonrelativistic (NR) exchange-correlated energy density $\varepsilon_{\rm exc}^{\rm NR}(r_s)$. Specifically, in the relativistic case, it is necessary to consider exchange between electrons through the transversely polarized photons which has no analogy in the nonrelativistic case. Using the results from relativistic uniform gas theory,[43] MacDonald and Vosko obtained the following approximation equation for the exchange energy in the relativistic case:

$$\varepsilon_x^R(r_s)=\left\{1-\frac{3}{2}\left[\beta(1+\beta^2)^{1/2}-\frac{\ln[\beta+(1+\beta^2)]^{1/2}}{\beta^2}\right]\right\}\varepsilon_x^{\rm NR}(r_s), \tag{1.157}$$

where $\beta = 1/car_s$ (c is the speed of light).

It is very difficult to account for relativistic effects in the correlation energy.[45] Hence in practice the exchange energy is ordinarily calculated by Eq. (1.157) while nonrelativistic approximations are used as the correlation energy, that is, exchange is the dominant effect at densities where relativistic effects become significant (approximately 50 times greater than the correlation contribution to the total energy).

1.3. Nonuniform electron gas

The conceptual basis for describing a nonuniform electron gas is the density-functional theory (DFT). The DFT is the basis for calculating the electronic structure of various types of condensed matter. Hence substantial attention will be devoted to this theory here.

1.3.1. Density-functional theory

The DFT in its early version (called the Thomas–Fermi method[46,47]) was developed at nearly the same time as the quantum mechanics. The modern version of this theory was proposed by Hohenberg, Kohn, and Sham in the mid-1960s.[48,49] The theory is based on two theorems which define particle density as a fundamental variable for describing any multielectron system. Following the calculation of Hohenberg and Kohn,[48] we consider a system of electrons in a large box driven by an external local potential v(r) and mutual Coulomb repulsion. The Hamiltonian of such a system takes the form

$$\hat{H}=\hat{T}+\hat{V}+\hat{U}, \tag{1.158}$$

where

$$\hat{T}=\int d\mathbf{r}\,\nabla\Psi^*(\mathbf{r})\nabla\Psi(\mathbf{r}),$$

$$\hat{V}=\int d\mathbf{r}\,v(\mathbf{r})\Psi^*(\mathbf{r})\Psi(\mathbf{r}), \tag{1.159}$$

$$\hat{U}=\frac{1}{2}\int d\mathbf{r}\,d\mathbf{r}'\,\frac{\Psi^*(\mathbf{r})\Psi^*(\mathbf{r}')\Psi(\mathbf{r}')\Psi(\mathbf{r})}{|\mathbf{r}-\mathbf{r}'|}. \tag{1.160}$$

The first theorem states that the total energy E of the ground state of any multielectron system is a functional of the single-particle density $\rho(\mathbf{r})$. In this context different multielectron systems will differ solely in the local external potential $\mathrm{v}(\mathbf{r})$ acting on the electrons. Then separating the explicit interaction with $\mathrm{v}(\mathbf{r})$ from the total energy the theorem also states that the remainder is a universal functional of the density $\rho(\mathbf{r})$, that is, it will depend on the external potential:

$$E[\rho]=\int d\mathbf{r}\, \mathrm{v}(\mathbf{r})\rho(\mathbf{r})+F[\rho]. \tag{1.161}$$

The functional F in Eq. (1.161) is dependent only on ρ and not on v.

The second theorem states that for any system, that is, for any $\mathrm{v}(\mathbf{r})$, the functional $E[\rho]$ has a minimum equal to the ground state energy when

$$N[\rho]\equiv\int d\mathbf{r}\, \rho(\mathbf{r})=N, \tag{1.162}$$

where N is the number of particles in the system.

These theorems, which are rather abstract in their own nature, have played a significant role in the development of the DFT. Assuming nondegeneracy of the ground state and so-called v representability of the particle density, this role was proved.[48] We note that the particle density is called v-representable if it represents the density of the ground state of a multielectron system in a certain external potential $\mathrm{v}(\mathbf{r})$.

The approach proposed by Levy[50] is clearer from the physical viewpoint. We will analyze this approach in greater detail. We select the density which is N-representable, that is, the density is the average of the density operator in a certain N-particle wave function. In the Hohenberg and Kohn approach, the functional $F[\rho]$ is limited only by those ρ in the v representation. Hence it is demonstrated[50] that there exists a universal variational functional $Q[\rho]$ that does not require v representability for ρ; N representation is sufficient. The existence of $Q[\rho]$ explains the DFT for all ρ in the N representation where $F[\rho]=Q[P]$ if the density is given in the v representation.

We therefore consider a density $\rho(\mathbf{r})$ that is N-representable. Let $M(\rho)$ be the set of wave functions that define this density. We then define the functional[50]

$$Q[\rho]=\inf_{|\Psi_\rho\rangle\in M(\rho)}\langle\Psi_\rho|\hat{Q}|\Psi_\rho\rangle, \tag{1.163}$$

where $\hat{Q}$ is an operator corresponding to a certain physically observable energy such as kinetic energy and the Coulomb interaction energy or their sum. Therefore, for each selected density we search all wave functions determining this density in order to find the lowest average of the selected operator. We denote the ground state energy and density of a system of N electrons in an external potential $\mathrm{v}(\mathbf{r})$ as E_0 and $\rho_0(\mathbf{r})$. Then the ground state will be degenerate. In this case $\rho_0(\mathbf{r})$ is one possible ground state density. If we use the operator $\hat{Q}=\hat{T}+\hat{U}$ as $\hat{Q}$ we can easily prove the following two theorems[50]:

Theorem 1:

$$\int d\mathbf{r}\, \mathrm{v}(\mathbf{r})\rho(\mathbf{r})+Q[\rho]\geqslant E_0. \tag{1.164}$$

Theorem 2:

$$\int d\mathbf{r}\, v(\mathbf{r})\rho_0(\mathbf{r}) + Q[\rho_0] = E_0. \tag{1.165}$$

An important property derives from the N representability: the wave functions of the ground state Ψ_0 can be found even if the external fields are unknown. If the ground state is degenerate, all wave functions of the ground state can be found.

Contrary to the Hohenberg and Kohn study,[48] the Levy approach[50] can easily be generalized to spin-polarized systems where generally there is no one-to-one correlation between ρ_0 and v. It is also possible to construct other generalizations of the DFT. Metropolis *et al.*[31] propose a DFT for each symmetry type S as a method of obtaining the energy of the lower excited states whose symmetry differs from that of the ground state. It is possible to obtain a generalization of the DFT to finite temperatures.[51] In this case the averages are substituted with ensemble averages

$$\tilde{Q} = \inf \mathrm{Sp}(\rho\hat{Q}). \tag{1.166}$$

The search in Eq. (1.166) is not carried out over the wave functions but rather over the density matrices $\tilde{\rho}$.

The importance of the theorems discussed above is that they make it possible for us to formulate an equivalent single-particle formulation of a complex many-particle problem. According to this approach we must search the minimum of the total energy functional $E[\rho]$ in a set of all N-representable densities $\rho(\mathbf{r})$. Using μ as the Lagrange multiplier together with condition (1.162) we obtain the Euler equation

$$\frac{\delta E[\rho]}{\delta\rho(\mathbf{r})} = \mu. \tag{1.167}$$

However, the explicit form of the functional $E[\rho]$ is unknown and we rewrite the total energy as

$$E[\rho] = T_s[\rho] + \frac{1}{2}\int d\mathbf{r}\, d\mathbf{r}' \frac{\rho(\mathbf{r})\rho(\mathbf{r}')}{|\mathbf{r}-\mathbf{r}'|} + \int d\mathbf{r}\, v(\mathbf{r})\rho(\mathbf{r}) + E_{\mathrm{exc}}[\rho], \tag{1.168}$$

where $T_s[\rho]$ is the kinetic energy functional of a noninteracting electron gas which is determined by substitution of the operator Q in Eq. (1.163) by the kinetic energy operator $\hat{T} = -\Sigma_i\nabla_i^2$. Equation (1.168) is obtained from Eq. (1.161) if the classical Coulomb energy is subtracted from $F[\rho]$ in Eq. (1.161), that is,

$$E_{\mathrm{exc}}[\rho] = F[\rho] - T_s[\rho] - \frac{1}{2}\int d\mathbf{r}\, d\mathbf{r}' \frac{\rho(\mathbf{r})\rho(\mathbf{r}')}{|\mathbf{r}-\mathbf{r}'|} (F = \langle\hat{T}\rangle + \langle\hat{U}\rangle). \tag{1.169}$$

Equation (1.169) is in fact the definition of the exchange-correlation functional $E_{\mathrm{exc}}[\rho]$. The Euler equation takes the form

$$\frac{\delta T_s[\rho]}{\delta\rho(\mathbf{r})} + v_{\mathrm{eff}}(\mathbf{r}) = \mu, \tag{1.170}$$

where

$$v_{eff}(\mathbf{r}) = v_i^*(\mathbf{r}) - 2 \int d\mathbf{r} \frac{\rho(\mathbf{r})}{|\mathbf{r} - \mathbf{r}'|} + v_{exc}(\mathbf{r}) \tag{1.171}$$

and

$$v_{exc}(\mathbf{r}) \equiv \delta E_{exc}(\rho)/\delta\rho(\mathbf{r}). \tag{1.172}$$

Therefore, the Euler equation (1.170) is an equation for noninteracting particles in the effective potential $v_{eff}(\mathbf{r})$. Hence by setting up $v_{eff}(\mathbf{r})$ we can calculate $\rho(\mathbf{r})$ and $T_s[\rho]$ solving the single-particle Schrödinger equation with the potential $v_{eff}(\mathbf{r})$:

$$[-\nabla^2 + v_{eff}(\mathbf{r})]\Psi_i(\mathbf{r}) = \varepsilon_i \Psi_i(\mathbf{r}), \tag{1.173}$$

and

$$\rho(\mathbf{r}) = \sum_{i=1}^{N} |\Psi_i(\mathbf{r})|^2. \tag{1.174}$$

Equations (1.173), (1.171), and (1.174) form the so-called self-consistent Kohn–Sham equations.[49]

Now the primary problem is that we do not know $E_{exc}[\rho]$ or consequently, $v_{exc}(\mathbf{r})$. If we set $v_{exc} = 0$ in Eq. (1.171) we obtain Hartree equations (1.17) which already yield results of sufficient accuracy compared to, for example, the Fermi–Thomas method. One of the primary reasons is we are utilizing the exact kinetic energy in Eq. (1.173) while the Fermi–Thomas theory utilizes an approximate local form for the kinetic energy.

In the case of a spin-polarized system we obtain a Pauli-type theory in which the effective single-particle orbitals become two-component spinors with components $\varphi_{i\sigma}(\mathbf{r})$ obeying Schrödinger pair equations[52]:

$$\sum_{\sigma'} \left\{ -\nabla^2 \delta_{\sigma\sigma'} + 2\delta_{\sigma\sigma'} \int d\mathbf{r}' \frac{\rho(\mathbf{r}')}{|\mathbf{r} - \mathbf{r}'|} + v_{\sigma\sigma'}(\mathbf{r}) + v_{\sigma\sigma'}^{exc}(\mathbf{r}) \right\} \varphi_{i\sigma} = \varepsilon_i \varphi_{i\sigma}(\mathbf{r}), \tag{1.175}$$

where

$$v_{\sigma\sigma'}^{exc}(\mathbf{r}) = \delta E_{exc}[\rho]/\delta\rho_{\sigma\sigma'}(\mathbf{r}) \tag{1.176}$$

and

$$\rho_{\sigma\sigma'}(\mathbf{r}) = \sum_{i=1}^{N} \varphi_{i\sigma}(\mathbf{r}) \varphi_{i\sigma}^*(\mathbf{r}). \tag{1.177}$$

The total energy is given as

$$E[\rho] = \sum_{i=1}^{N} \varepsilon_i - \frac{1}{2} \int d\mathbf{r}\, d\mathbf{r}' \frac{\rho(\mathbf{r})\rho(\mathbf{r}')}{|\mathbf{r} - \mathbf{r}'|} - \sum_{\sigma\sigma'} \int d\mathbf{r}\, v_{\sigma\sigma'}^{exc}(\mathbf{r}) \rho_{\sigma\sigma'}(\mathbf{r}) + E_{exc}[\rho]. \tag{1.178}$$

In examining the electronic structure of condensed matter consisting of atoms of heavy elements it is necessary to account for relativistic effects. It has been demonstrated[51] that spin–orbit interaction is easily incorporated in the DFT described above by simply adding this interaction to the kinetic energy functional. One drawback of this routine is that it does not yield a technique for finding the coefficient in front of the spin-orbital term. A rigorous relativistic generalization of the DFT was carried out.[53] In this case the fundamental variable is not the density but rather the four-current density **J**. By analogy to the nonrelativistic case the generalization yields[53] an effective single-particle Dirac-type equation with effective scalar and vector potentials determined through the current density:

$$\{-i\alpha\nabla + m(1-\beta) + \mathrm{v}_{\mathrm{eff}}[\mathbf{r},\mathbf{J}_\mu]\}\Psi_i(\mathbf{r}) = \varepsilon_i\Psi_i(\mathbf{r}), \tag{1.179}$$

where

$$\mathrm{v}_{\mathrm{eff}}[\mathbf{r},\mathbf{J}_\mu] = -\left[\mathrm{v}(\mathbf{r}) + 2\int d\mathbf{r}'\,\frac{\rho(\mathbf{r}')}{|\mathbf{r}-\mathbf{r}'|} + \frac{\delta E_{\mathrm{exc}}[\rho,\mathbf{J}_\mu]}{\delta\rho}\right.$$
$$\left. -\alpha\left[A(\mathbf{r}) + \int d\mathbf{r}\,\frac{\mathbf{J}_\mu(\mathbf{r}')}{|\mathbf{r}-\mathbf{r}'|} + \frac{\delta E_{\mathrm{exc}}[\rho,\mathbf{J}_\mu]}{\delta\mathbf{J}_\mu}\right]\right]. \tag{1.180}$$

Here α and β are the Dirac matrices; $\mathbf{J}_\mu$ is the current density; **A** is the external electromagnetic potential.

As noted above the primary problem of the DFT is the definition of the functional $E_{\mathrm{exc}}[\rho]$ (or $F[\rho]$). We consider a variety of approximations. In analogy to Hohenberg and Kohn[48] we introduce a functional:

$$G[\rho] \equiv F[\rho] - \frac{1}{2}\int d\mathbf{r}\,d\mathbf{r}'\,\frac{\rho(\mathbf{r})\rho(\mathbf{r}')}{|\mathbf{r}-\mathbf{r}'|} = E_{\mathrm{exc}}[\rho] + T_s[\rho] \tag{1.181}$$

and consider a gas of near-constant density, that is

$$\rho(\mathbf{r}) = \rho_0 + \tilde{\rho}(\mathbf{r}) \tag{1.182}$$

where

$$\tilde{\rho}(\mathbf{r})/\rho_0 \ll 1, \text{ and } \int d\mathbf{r}\,\tilde{\rho}(\mathbf{r}) = 0.$$

In this case we can carry out the formal expansion

$$G[\rho] = G[\rho_0] + \int d\mathbf{r}\,d\mathbf{r}'\,K(\mathbf{r}-\mathbf{r}')\tilde{\rho}(\mathbf{r})\tilde{\rho}(\mathbf{r}')$$
$$+ \int d\mathbf{r}\,d\mathbf{r}''\,d\mathbf{r}'\,L(\mathbf{r},\mathbf{r}',\mathbf{r}'')\tilde{\rho}(\mathbf{r})\tilde{\rho}(\mathbf{r}')\tilde{\rho}(\mathbf{r}'') + \cdots. \tag{1.183}$$

The kernel in the quadratic term can be given as the Fourier expansion

$$K(\mathbf{r}-\mathbf{r}') = \frac{1}{\Omega}\sum_{\mathbf{q}} K(\mathbf{q})e^{-i\mathbf{q}(\mathbf{r}-\mathbf{r}')}. \tag{1.184}$$

We neglect the higher order terms. The functional G is conveniently given through the density function g_r

$$g_r[\rho] = g_0[\rho_0] + \int d\mathbf{r}'\,K(\mathbf{r}')\tilde{\rho}\left(\mathbf{r} + \frac{1}{2}\mathbf{r}'\right)\tilde{\rho}\left(\mathbf{r} - \frac{1}{2}\mathbf{r}'\right) + \cdots, \tag{1.185}$$

where $g_0[\rho_0]$ is the density function of the uniform electron gas of density ρ_0, which is, equal to $g_0 = t_s + \varepsilon_{exc}$ [t_s is the kinetic energy density from Eq. (1.79)]. Using different approximations for ε_{exc} obtained in uniform electron gas theory we can easily calculate the density $g_0[\rho_0]$. The kernel k can be expressed through the static permittivity $\varepsilon(q)$ dependent on q for an electron gas of density ρ_0 (Ref. 48):

$$K(q) = \frac{2\pi}{q^2} \frac{1}{\varepsilon(q) - 1}. \tag{1.186}$$

Using the SSTL results[37,40] which calculated $\varepsilon(q)$ we can also calculate $k(q)$.

The kernel $k(q)$ is also easily represented as[54]

$$K(q) = K(0) + \frac{\pi}{4k_F^4} Z(q) q^2, \tag{1.186a}$$

where $k(0)$ corresponds to the so-called local density approximation and in the high density limit can be calculated exactly[54]:

$$K(0) = -\frac{\pi}{k_F^2}\left[1 + \frac{0.521 r_s}{\pi}(1 - \ln 2)\right]. \tag{1.187}$$

$Z(q)$ is a certain size quantity, which calculates $Z(q)$ in the high density limit, by Langreth and Vosko.[54]

In the case of an electron gas with a slowly varying density, $G[\rho]$ has the gradient expansion

$$G[\rho] = \int d\mathbf{r}\{g_0(\rho) + g_2^{(2)}(\rho)|\nabla\rho|^2 + g_4^{(2)}(\rho)(\nabla^2\rho)^2 + g_4^{(3)}(\rho)\nabla^2\rho|\nabla\rho|^2 + g_4^{(4)}(\rho)|\nabla\rho|^4 + \cdots + O(\nabla_i^6)\}. \tag{1.188}$$

The quantities $g_0(\rho)$, $g_i(\rho)$ which are functions (not functionals) of ρ can be found if we identify these with the coefficients in the expansion of the inverse static polarizability $\varepsilon^{-1}(q)$ in small q.[48]

1.3.2. The local density approximation

The most commonly used technique for calculating the electronic properties of condensed matter within the framework of the DFT is the local density approximation (LDA). In this approximation the nonuniform system is treated as a certain set of small boxes containing a uniform interacting electron gas. Such a gas is completely characterized by two (up and down) densities, respectively, which are the eigenvalues of the local density matrix in each box. In other words, in each box we generally have the density matrix $\rho_{\sigma\sigma'}(\mathbf{r})$ which we initially diagonalize in order to obtain the local spin direction which we treat as the direction of polarization of the locally uniform gas. The total exchange-correlation energy is the sum of the contributions of all boxes:

$$E_{exc}[\rho_{\sigma\sigma'}] = \int d\mathbf{r}\, \varepsilon_{exc}(\rho_\uparrow, \rho_\downarrow)\rho(\mathbf{r}), \tag{1.189}$$

where $\rho_\uparrow$ and $\rho_\downarrow$ are the eigenvalues of the single-particle density matrix $\rho_{\sigma\sigma'}$; $\rho(\mathbf{r}) = \rho_\uparrow(\mathbf{r}) + \rho_\downarrow(\mathbf{r})$ is the electron density; and $\varepsilon_{exc}(\rho_\uparrow \rho_\downarrow)$ is the exchange-correlation energy per particle of the uniform spin-polarized gas with spin densities $\rho_\uparrow$ and $\rho_\downarrow$. In the case of ferromagnets and antiferromagnets the density matrix can be reduced to diagonal form, that is, it is possible in turn to simplify equation set (1.175); these equations now become two independent equations, each with its own exchange-correlated potential $\mathrm{v}_\sigma^{exc}(\mathbf{r}) = \delta E_{exc}/\delta\rho_\sigma(\mathbf{r})$. In the LDA the exchange-correlation potential is expressed as

$$\mathrm{v}_{exc}(\mathbf{r}) = \frac{d}{d\rho}\{\varepsilon_{exc}(\rho)\cdot\rho(\mathbf{r})\} \equiv \mathrm{v}_{exc}(\rho). \tag{1.190}$$

Then the total energy in the LDA can be recast as

$$E \cong \sum_i \varepsilon_i - \frac{1}{2}\int d\mathbf{r}\, d\mathbf{r}' \frac{\rho(\mathbf{r})\rho(\mathbf{r}')}{|\mathbf{r}-\mathbf{r}'|} + \int d\mathbf{r}\, \rho(\mathbf{r})\{\varepsilon_{exc}(\rho) - \mathrm{v}_{exc}(\rho)\}. \tag{1.191}$$

Thus by a local approximation the problem of exchange and correlation in an inhomogeneous system is reduced to calculating the exchange-correlation energy density $\varepsilon_{exc}(\rho,\varphi)$ of a homogeneous electron gas that has already been examined in the preceding section of this study.

Potential (1.180) is simplified for relativistic calculations in the LDA and takes the form

$$\mathrm{v}_{eff}^{LDA}(\mathbf{r}) = -\left[\mathrm{v}(\mathbf{r}) + 2\int d\mathbf{r}' \frac{\rho(\mathbf{r})}{|\mathbf{r}-\mathbf{r}'|} + \frac{d}{d\rho}(\varepsilon_{exc}(\rho)\rho)\right] - \alpha\cdot\mathbf{A}. \tag{1.192}$$

Equation (1.179) with potential (1.192) in the majority of practical cases are solved in a central field. Then we have a first order system of differential equations:

$$\frac{d}{dr}\begin{pmatrix} P_\varkappa(r) \\ Q_\varkappa(r) \end{pmatrix} = \begin{pmatrix} -\varkappa/r & 2M_c \\ -\frac{1}{c}[\varepsilon - \mathrm{v}_{eff}^{LDA}(r)] & \varkappa/r \end{pmatrix}\begin{pmatrix} P_\varkappa(r) \\ Q_\varkappa(r) \end{pmatrix}, \tag{1.193}$$

where $\varkappa = -\sigma(J + 1/2)$; $2M_c \equiv c[1 + (\varepsilon - \mathrm{v}_{eff}^{LDA})/c^2]$; and $P_\varkappa$ and $Q_\varkappa$ are the large and small components multiplied by r.

Solution (1.193) does not yield the desired result for a certain class of condensed matter. The problem here is that, for example, an investigation of the properties of f metals and their various compounds is desirable so that the spin remains a "good" quantum number. A situation arises where relativistic effects are important on the one hand yet exchange interaction between the electrons dominates spin–orbit interaction on the other. In other words, spin-polarized calculations of the electronic structure are very important for a certain class of condensed matter consisting of heavy atoms. These circumstances have led to a review of traditional methods of self-consistent relativistic calculations of the electronic structure. It is generally well known that only such relativistic effects as the velocity dependence of the mass, the Darwin correction, and other kinetic effects have an influence on the energy position.

Incorporating spin–orbit interaction will only cause splitting of the levels or bands. It therefore would be desirable to formulate wave functions so that the spin would remain a quantum number although the principal relativistic effects would still be accounted for in this case: the velocity dependence of the mass and the Darwin correction. One simple method for solving this problem is to use the Pauli equation and then neglect spin–orbit interaction. However, in this case we may obtain an incorrect result for small values of r, since the Pauli equation ignores one of the terms of the series $v_{\text{eff}}^{\text{LDA}}(r)/mc^2$. For small r this term may be ~ 1 and the approximation becomes unsatisfactory. One alternative that eliminates this error is the scalar relativistic approximation (SRA).[55]

The SRA makes it possible to exclude spin–orbit interaction while also carrying out the calculation in the nonrelativistic case with quantum numbers l and S, thereby exactly accounting for all remaining relativistic effects. The idea of the SRA lies in averaging the large and small components of the wave function over J for the given l and neglecting spin–orbit splitting in the large component only. Then equation system (1.193) takes the following form:

$$\frac{d}{dr}\begin{pmatrix} P_l^\sigma(r) \\ Q_l^\sigma(r) \end{pmatrix} = \begin{pmatrix} 1/r & 2M_\sigma c \\ -\frac{1}{c}(\varepsilon_\sigma - v_{\text{eff}}^{LDA\sigma}) + \frac{l(l+1)}{2M_\sigma c r^2} & -1/r \end{pmatrix} \begin{pmatrix} P_l^\sigma(r) \\ Q_l^\sigma(r) \end{pmatrix}. \tag{1.194}$$

Solving system (1.194) we rigorously account for relativistic effects with the exception of spin–orbit interaction. The functions $P_l^\sigma(r)$ and $Q_l^\sigma(r)$ are normalized as follows:

$$\int_0^\infty dr \left\{ [Q_l^\sigma(r)]^2 + \left[\frac{l(l+1)}{(2M_\sigma c r)^2} + 1 \right] [P_l^\sigma(r)]^2 \right\} = 1. \tag{1.195}$$

It is clear from a comparison of Eqs. (1.193) and (1.194) that there are approximately half the number of calculations in the SRA compared to a complete relativistic calculation. The wave functions in the SRA are dependent on l and S, that is, spin-polarization calculations can be carried out. This method is particularly effective in self-consistent relativistic calculations, which can now be divided into two stages. The SRA can be used in the first stage to obtain the self-consistent potential and when necessary equation system (1.193) which already has the self-consistent potential can be solved in the second stage. This two-stage scheme was implemented for atomic and band structure calculations.[56,57]

The LDA is by virtue of its form exact in the limit of slowly varying spin densities and hence as we would expect this approximation will be correct only in systems where the density does not change appreciably at distances $\sim k_F^{-1}$. Unfortunately, the density at such distances varies quite substantially in the majority of real systems. Yet practical calculations have suggested that the LDA results for atoms, molecules, and a variety of condensed matter have been substantially more exact than we would have reason to expect. Extensive examples of such calculations are provided in surveys.[19,52] Hence here we will only discuss in greater detail the physical aspects responsible for the success of the LDA.

(a) Gradient corrections. The natural route to explain why theory in the zeroth approximation works is to estimate the, say, first order corrections in the hope that they will turn out to be small, since they are based on the limit of the slowly varying densities and in this case the corrections are gradient corrections such that[19]

$$E_{\text{exc}}[\rho]=\int d\mathbf{r}\,\varepsilon_{\text{exc}}(\rho)\rho(\mathbf{r})+\int d\mathbf{r}\,B_{\text{exc}}(\rho)|\nabla\rho(\mathbf{r})|^2. \qquad (1.196)$$

The function $B_{\text{exc}}(\rho)$ is given by a density response function for the uniform gas density and was obtained from a variety of approximations.

It is clear[58] that there are very strong arguments against using gradient corrections as a method of enhancing the LDA. However, there are certain situations in which a gradient expansion may be suitable. For example, when there is a possibility for error compensation in the case of severe, strong inhomogeneities by the killing each other the erroring terms. In these cases, the LDA yields quite accurate results even when gradient corrections are unsuitable. This means that the LDA cannot be substantiated as a zeroth order approximation in a rapidly diverging expansion. This is more likely an independent approximation and an explanation for its success should be sought in a different area.

(b) Pair correlation function and the sum rules. The exchange-correlation energy $E_{\text{exc}}[\rho]$ can be expressed through the pair correlation energy g_λ (Ref. 31):

$$E_{\text{exc}}[\rho]=\frac{1}{2}\int d\mathbf{r}\,d\mathbf{r}'\frac{\rho(\mathbf{r})\rho(\mathbf{r}')}{|\mathbf{r}-\mathbf{r}'|}\int_0^2 d\lambda[g_\lambda(\mathbf{r},\mathbf{r}')-1]. \qquad (1.197)$$

Here λ is the coupling constant [see formula (1.75)] which varies from 0 to e^2 (in our units of $e^2=2$). If the exchange-correlation hole is given as

$$\rho_{\text{exc}}(\mathbf{r},\mathbf{r}')=\rho(\mathbf{r}')\int_0^2 d\lambda[g_\lambda(\mathbf{r},\mathbf{r}')-1], \qquad (1.198)$$

the exchange-correlation energy is written as

$$E_{\text{exc}}[\rho]=\frac{1}{2}\int d\mathbf{r}\,d\mathbf{r}'\frac{\rho(\mathbf{r})\rho_{\text{exc}}(\mathbf{r},\mathbf{r}')}{|\mathbf{r}-\mathbf{r}'|}. \qquad (1.199)$$

The exchange-correlation energy in accordance with Eq. (1.199) arises due to Coulomb interaction of each electron with a charge distribution $\rho_{\text{exc}}(\mathbf{r},\mathbf{r}')$, that is, an exchange-correlation hole surrounding this electron. The hole is the result of exchange and Coulomb interaction that may reduce the charge in the vicinity of each electron. The fundamental requirement for charge concentration yields the sum rule

$$\int d\mathbf{r}'\,\rho_{\text{exc}}(\mathbf{r},\mathbf{r}')=-1, \qquad (1.200)$$

which is valid for all $\mathbf{r}$.

A different definition of the LDA that is entirely equivalent to Eq. (1.189) can be given in terms of the pair correlation function and the exchange-correlation hole. The transition to the LDA means that the exact functions $g_\lambda(\mathbf{r},\mathbf{r}')$ in Eq. (1.197) are replaced by pair correlation functions for the uniform electron gas of local density, that is,

$$g_\lambda(\mathbf{r},\mathbf{r}')\to g_h(|\mathbf{r}-\mathbf{r}'|;\rho(\mathbf{r})),$$

$$\rho_{exc}^{LD}(\mathbf{r},\mathbf{r}') = \rho(\mathbf{r}) \int_0^2 d\lambda [g_h(|\mathbf{r}-\mathbf{r}'|;\rho(\mathbf{r}) - 1)], \tag{1.201}$$

where $g_\lambda(\mathbf{r},\rho)$ is the pair correlation function of the uniform electron gas of density ρ. We note that the sum rule holds for the exchange-correlation hole in the LDA, that is, this rule will hold even for a uniform electron gas:

$$\int d\mathbf{r}' \, \rho_{exc}^{LD}(\mathbf{r},\mathbf{r}') = -1. \tag{1.202}$$

It follows from Eq. (1.201) that the LDA uses a spherically symmetric hole. The long-range nature of the Coulomb interaction will cause the interaction energy between two charge distributions to be weakly dependent on their distribution functions. A more important factor is the total charge of each distribution. In other words an incorrect representation of the exchange-correlation hole will not produce significant errors in the definition of E_{exc} as long as sum rule (1.200) holds. Moreover, if we define the spherically averaged $\bar{\rho}_{exc}(\mathbf{r},\mathbf{R})$ at point $\mathbf{r}$:

$$\rho_{exc}(\mathbf{r},\mathbf{R}) = \frac{1}{4\pi} \int d\Omega \, \rho_{exc}(\mathbf{r},\mathbf{r}+\mathbf{R}) \tag{1.203}$$

($d\Omega = d\varphi \, d\theta \sin\theta$), we can easily see that

$$E_{exc}[\rho] = \frac{1}{2} \int d\mathbf{r} \, \rho(\mathbf{r}) \int d\mathbf{R} \frac{\rho_{exc}(\mathbf{r},\mathbf{R})}{|\mathbf{r}-\mathbf{R}|}. \tag{1.204}$$

Equation (1.204) means that the exchange-correlation energy is dependent only on the spherically averaged exchange-correlation hole.

Therefore, the primary factors responsible for the success of the LDA can be formulated as follows: (1) the exchange-correlation energy is determined by the exchange-correlation hole averaged over the sphere; (2) the LDA yields a relatively exact description of this spherically averaged hole; and (3) systematic error compensation exists due to the sum rule.[51]

The arguments given above represent one possible explanation for the success of the LDA. However, this explanation cannot be treated as the only possible explanation. A discussion of this issue can be found, for example, in von Barth's survey.[52]

(c) Nonlocal functionals in real space. Two nonlocal schemes for improving the LDA were proposed.[59,60] These schemes are nonlocal in the sense that the local exchange-correlation energy density is made dependent on a certain average in the neighborhood of the examined point and not simply dependent on the local density as in the LDA case. Since the exchange and correlation effects are by their very nature nonlocal, we cannot preclude the possibility that the new schemes will better describe these effects. These nonlocal schemes have given rise to a local effective potential as does any other density functional scheme. As in the LDA the methods proposed by Gunnarson *et al.*[59,60] utilize a pair correlation function of a uniform electron gas to model an exchange-correlation hole.

The density average determined through the weight multiplier $W(\mathbf{r},\rho)$ which is dependent on the density, is taken in a first approximation which was called the average density approximation (ADA) (Ref. 59):

$$\rho(\mathbf{r}) = \int d\mathbf{r}'\, \rho(\mathbf{r}')\, W(\mathbf{r}-\mathbf{r}';\bar{\rho}(\mathbf{r})). \tag{1.205}$$

The exchange-correlation hole in the ADA takes the form

$$\rho^{\rm AD}_{\rm exc}(\mathbf{r},\mathbf{r}') = \bar{\rho}(\mathbf{r}) \int_0^{2} d\lambda [g_h(\mathbf{r}-\mathbf{r}';\bar{\rho}(\mathbf{r})) - 1]. \tag{1.206}$$

Like the LD hole the exchange-correlation hole is spherically symmetrical. We can easily obtain an equation for the exchange-correlation energy in the ADA:

$$E^{\rm AD}_{\rm exc}[\rho] = \int d\mathbf{r}\, \rho(\mathbf{r})\varepsilon_{\rm exc}(\bar{\rho}(\mathbf{r})). \tag{1.207}$$

We note that the ADA obeys sum rule (1.200) regardless of the selection of the weight function W by virtue of the construction of the ADA. Therefore, we can use the weight function to enhance the description of the exchange-correlation hole. Gunnarson *et al.*[59] sets up the calculation so that the ADA was exact when the density is near-constant. It turns out that this requirement unambiguously defines the weight function which we will find by solving the following equation[59]:

$$[W(q,r_s)]^2 + 2A(r_s)W(q,r_s)\left(1 - \frac{r_s}{3}\frac{\partial W(q,r_s)}{\partial r_s}\right) - [1+2A(r_s)]f(q,r_s) = 0, \tag{1.208}$$

where

$$f(q,r_s) = K_{\rm exc}(q,r_s)/K_{\rm exc}(0,r_s),$$

$$A(r_s) = \frac{\partial \varepsilon_{\rm exc}(\rho)}{\partial \rho} \Big| \rho \frac{\partial^2 \varepsilon_{\rm exc}(\rho)}{\partial \rho^2}\Big|_{\rho=\rho_0}$$

and $q = k/k_F(r_s)$. The quantities $K_{\rm exc}(q,r_s)$, etc. can be determined if we know the dielectric function of the uniform electron gas [see Eq. (1.186)].

The second scheme proposed[60] is called the weighted density approximation (WDA). In this scheme the exchange-correlation hole

$$\rho^{\rm WD}_{\rm exc}(\mathbf{r},\mathbf{r}') = \rho(\mathbf{r}') \int_0^{2} d\lambda [g^h_{\tilde{\rho}(r)}(\mathbf{r}-\mathbf{r}';\lambda) - 1]. \tag{1.209}$$

Here $\tilde{\rho}(\mathbf{r})$ is found from the relation for the sum rule:

$$\int d\mathbf{r}' \int_0^{2} d\lambda [g^h_{\tilde{\rho}(r)}(\mathbf{r}-\mathbf{r}';\lambda) - 1]\rho(\mathbf{r}') = -1, \tag{1.210}$$

that is, the sum rule is satisfied for each $\mathbf{r}$ by corresponding selection of $\tilde{\rho}(\mathbf{r})$ in the pair correlation function of a uniform electron gas. Therefore, the exchange-correlation hole in the WDA is obtained by "dressing" the hole with the true density. Consequently this hole, like a real hole, is not spherically symmetrical in the neighborhood of the electron unlike the LDA or ADA. Moreover, the WDA yields the exact exchange energy for a two-electron system.

If we denote the integral in Eq. (1.209) by $G^h[\mathbf{r} - \mathbf{r}';\rho(\mathbf{r})]$, the exchange-correlation energy in the WDA will take the following form:

$$\varepsilon_{\rm exc}^{\rm WD}(\mathbf{r})\,| = \int d\mathbf{r}' \frac{\rho(\mathbf{r}')}{|\mathbf{r}-\mathbf{r}'|} G^h(\mathbf{r}-\mathbf{r}';\tilde{\rho}(\mathbf{r})). \tag{1.211}$$

The function G^h can be calculated in the RPA. Its Fourier components in this approximation were obtained[30] as

$$G_{\rm RPA}^h(q,\rho) = \frac{1}{\rho}\left(\frac{1}{\rho}\int_0^\infty \frac{d\omega}{\pi}\frac{1}{\mathrm{v}(q)}\ln[1-\mathrm{v}(q)\chi_0(q,i\omega)] + 1\right)_{\rho=\tilde{\rho}}, \tag{1.212}$$

where $\mathrm{v}(q) = 8\pi/q^2$ and $q \equiv k/k_F$.

The test of the ADA and WDA on atomic structure calculations was carried out,[61] and demonstrated that the ADA is superior to the WDA when the opposite situation would be expected based on general theoretical considerations. Evidently one of the reasons is that the LDA, ADA, and WDA schemes attempt to simulate the exchange-correlation hole through the pair correlation function of a uniform electron gas although, for example, the atoms are highly nonuniform systems. Hence a modification of the WDA scheme was proposed,[52] in which the true pair correlation function is simulated by an analytical function of the type

$$g_m(r,\rho) - 1 = A(\rho)\{1 - \exp[-B^5(\rho)/r^5]\}. \tag{1.213}$$

Here the density-dependent parameters A and B are selected so that g_m satisfies the sum rules. The advantage of this model lies in the fact that the long-range oscillating tail of the pair correlation function of the electron gas which of course has no relation to such localized systems as atoms is replaced by a rapidly decreasing tail (r^{-5}) without violating the validity of the theory in the slowly varying density limit.

ADA and WDA schemes were tested[61] for calculating the exchange-correlation parts of the surface energy in a semi-infinite jellium model. The results were rather disappointing. It is not clear whether or not the ADA yields a converging result for this part of the energy. In any case the surface energy is too high. The WDA yields a diverging exchange energy although the total exchange-correlation energy converges, yet it is one-half the value of the exact result obtained in the RPA. Consequently, the applicability of the ADA and the WDA for this individual problem remains doubtful. The most likely reason is the incorrect volumetric representation of the exchange-correlation hole. Since the surface energy represents the difference in energies between the finite and semi-infinite system energies, minor errors in this case will have a tendency to accumulate rather than compensate. It is possible that this problem is not as severe in modified scheme (1.213) since the exchange-correlation hole in this case is less extended.

Of course calculating the characteristics of atoms and surfaces is rather complex for any approximation in a DFT based on results for a uniform electron gas. Hence the ADA and WDA represent an improvement over the LDA in systems where the density varies rather slowly.

One additional significant drawback of the schemes described above is the irregular asymptotic behavior of the exchange-correlation potential at large distances. In the LDA it decays exponentially and hence it is necessary to artificially introduce the so-called Letter's correction in atomic structure calculations.[63] The exchange-correlation potential decays as $-0.55/r$ in the ADA and WDA and qualitatively the behavior of v_{exc} is correct although the coefficient is not -1 but rather is -0.55. In the WDA this effect occurs due to violation of symmetry for the pair correlation function $g(\mathbf{r},\mathbf{r}') = g(\mathbf{r}',\mathbf{r})$. In the ADA the erroneous coefficient derives from the more subtle numerical effects.[61]

(d) Nonlocal functionals in reciprocal space. So far we have considered the LDA and its modifications based on a description of an exchange-correlation hole in coordinate space. The exchange-correlation energy has been proposed describing as the sum of contributions from different wave vectors in reciprocal space.[58,64,65] Defining $\widetilde{S}(\mathbf{q})$ as

$$\widetilde{S}(\mathbf{q}) = 1 + \frac{1}{N} \int d\mathbf{r}\, d\mathbf{r}'\, \rho(\mathbf{r})\rho(\mathbf{r}')[g(\mathbf{r},\mathbf{r}') - 1]e^{i\mathbf{q}(\mathbf{r}-\mathbf{r}')}, \tag{1.214}$$

we can represent Eq. (1.197) as

$$E_{exc}[\rho] = \frac{N}{2} \int \frac{d\mathbf{q}}{(2\pi)^3} v(\mathbf{q})[\widetilde{S}(\mathbf{q}) - 1], \tag{1.215}$$

where N is the number of particles in the system, $v(\mathbf{q}) = 8\pi/q^2$. $\widetilde{S}(\mathbf{q})$ is the regular structure factor integrated with respect to λ and this tends to unity with large $\mathbf{q}$, while at small $\mathbf{q}$: $S(\mathbf{q}) \to 0$ due to the sum rule. Since Eq. (1.215) contains a Coulomb potential, large wave vectors will make a relatively small contribution to E_{exc} at the same time that small and average $\mathbf{q}$ will make a significant contribution.

Langreth and Perdew[64] investigated the part of the structure factor $\delta S(\mathbf{q})$ attributable to purely surface effects in the RPA for a surface in the jellium model. It was demonstrated that $\delta S(\mathbf{q})$ behaves as $\alpha \cdot \mathbf{q}$ for small $\mathbf{q}$. The proportionality factor α was calculated in all interelectron interaction orders.[64] As demonstrated above the LDA consists of substituting the pair correlation function with the corresponding quantity for a uniform electron gas and substituting $\mathbf{r}'$ by $\mathbf{r}$ in the second density in Eq. (1.214). Ceperley[34] determined that the corresponding quantity $\delta\widetilde{S}^{LDA}(\mathbf{q})$ manifests irregular behavior with small $\mathbf{q}$ (it is proportional to $\mathbf{q}^2$). The LDA yields a satisfactory description of $\delta S(\mathbf{q})$ for large wave vectors. Hence after formulating a simple interpolation scheme between the known exact result for small $\mathbf{q}$ and the LDA results for medium and large $\mathbf{q}$, Langreth and Perdew were able to produce the results[64] of a complete RPA for the exchange-correlation surface energy in the jellium model examined within the framework of the infinite barrier approximation. Then this interpolation scheme was applied to a surface with a more realistic electron density distribution[56] and yielded a correction to the LDA results of approximately 5%–10% depending on the value of $\rho(\mathbf{r})$. Therefore, the LDA is surprisingly accurate for such nonuniform systems and hence a subsequent study[58] selected a different approach. The LDA was left untouched, but a gradient correction was added. However, as noted above the low order gradient correction does not enhance the LDA. Langreth and Perdew[58] carried out the analysis and identified the difficulties with the gradient correction. In the case of a surface the gradient term tends toward a large constant in the small wave vector limit. This results in a strong overestimation of the correction to the LDA which in turn yields excessively small exchange-correlation energy. It was demonstrated[58] that use of the local density curvature in the gradient term is nonphysical in the case of fluctuations with a long wavelength which clearly do not sense subtle details in density behavior. In place of this arrangement it was proposed[58] using a curvature averaged over the fluctuation "dimensions" ($\sim 1/q$). The resulting gradient correction manifests proper analytical behavior ($\beta \cdot \mathbf{q}$) with small wave vectors. The new scheme yields surface exchange-correlation energies very similar to those obtained from the interpolation scheme described above although unlike the interpolation scheme the average gradient scheme permits easy parametrization so that this technique can be applied to arbitrary nonuniform systems.[67] The idea is to simply ignore the correlation part of the gradient correction for wave vectors below a certain boundary wave vector proportional to the local density gradient and conserve the total correction with large wave vectors. The resulting correction for the exchange-correlation energy takes the form

$$E_{\text{exc}}[\rho] = E_{\text{exc}}^{\text{LD}}[\rho] + c \int dr |\nabla \rho|^2 \rho^{-4/3} \left\{ 2e^{-F} - \frac{7}{9} \right\}, \qquad (1.216)$$

where $C = (3/\pi)^{2/3}/(7/2\pi)$; $F = 0.262|\nabla\rho|\rho^{-7/6}$. The calculations for the total energy of certain atoms demonstrated[65,67] that the parametrized gradient scheme reduces the LDA error by a factor of approximately 3. In this case the self-consistent atomic charge densities were superior to the Hartree–Fock densities obtained accounting for the overlying configuration.

It was also demonstrated[65,67] that the greatest error in their energy functional (1.216) is due to manipulations with the exchange energy. After formulating the scheme to separate the exchange and correlation effects they obtained the correlation energy functional which can be added to the energy of the exact Hartree–Fock calculation. Therefore, they succeeded in calculating the atomic correlation energies accurate to better than 0.01 Ry.

The approximation developed by Langreth and Mehl[65] was tested[52] in calculating the ground state properties of bulk silicon. According to its preliminary results the binding energy is superior to that obtained in LDA while the equilibrium volume and bulk modulus within numerical accuracy are equal to the corresponding experimental results. However, at the same time the band gap in silicon was one-half the value in this scheme as was the case in the LDA.

Therefore, the parametrized gradient scheme significantly reduces LDA errors in such similar systems as atoms and surfaces for which errors have opposite signs. This indicates the obvious advantage of this scheme over all current schemes presently used to improve the LDA. The primary drawback of the new scheme is the fact that it cannot be applied to spin-polarized systems.

(e) Self-interaction corrections. The transition to the LDA in density-functional theory eliminates compensation of the self-action energy in the Coulomb interaction functional $U[\rho]$ and the functional $E_{\text{exc}}[\rho]$.[68] Hence one alternative approach to enhancing the LDA is introducing direct self-interaction corrections (SIC) to the total energy functional in the LDA (SIC-LDA) in the calculation. The corrected exchange-correlated energy functional can be given as[68]

$$E_{\text{exc}}^{\text{AD}}[\rho_\uparrow,\rho_\downarrow] = E_{\text{exc}}^{\text{LD}}[\rho_\uparrow,\rho_\downarrow] - \sum_{nl\sigma} \delta_{nl\sigma}, \qquad (1.217)$$

where

$$\delta_{nl\sigma} = U[\rho_{nl\sigma}] + E_{\text{exc}}^{\text{LD}}[\rho_{nls}, 0]. \tag{1.218}$$

The variational principle yields Kohn–Sham equations with the effective orbital-dependent potential

$$\text{v}_{\text{eff}}^{nl,\sigma}(\mathbf{r}) = \text{v}(\mathbf{r}) + \int d\mathbf{r}' \frac{\rho(\mathbf{r}')}{|\mathbf{r} - \mathbf{r}'|} + \text{v}_{\text{exc}}^{nl,\sigma}(\mathbf{r}), \tag{1.219}$$

$$\text{v}_{\text{exc}}^{nl,\sigma}(\mathbf{r}) = \text{v}_{\text{exc}}^{\text{LD},\sigma}(\mathbf{r}) - \int d\mathbf{r}' \frac{\rho_{nl\sigma}(\mathbf{r}')}{|\mathbf{r} - \mathbf{r}'|} - \text{v}_{\text{exc}}^{\text{LD},\sigma}[\rho_{nl\sigma}]. \tag{1.220}$$

The SIC-LDA enhances the representation of an exchange-correlation hole in atoms and yields the correct asymptotic potential far from the nucleus. Superior total energies and electron charge densities are obtained. Unlike the LDA in which the instability of negative ions is erroneously predicted, in the SIC-LDA calculation of the negative ion bonding energies yields good results.

The SIC-LDA has certain fundamental limits. This approximation is derived in orbital-functional form which yields orbital-dependent exchange-correlated potential [Eq. (1.220)] and nonorthogonality of the spin orbitals. At the same time the exact potential is local and yields the correct total energy and density. This drawback of the SIC-LDA is due to the zero value of the self-interaction effect for Bloch functions in the extended system. Hence the SIC-LDA scheme can be applied to extended periodic systems only after a unitary transformation of the occupied orbitals into localized orbitals such as Wannier functions in the case of nonmetals. All this substantially complicates the application of the SIC-LDA scheme to electronic structure calculations of condensed matter. To date there are only certain integrated studies that utilize the SIC-LDA technique for electronic spectral calculations in ionic insulators and semiconductors.[52]

1.4. The DFT and a description of excited states

1.4.1. The Δ SCTE method

The density-functional theory examined above is a theory that examines the physical properties of an electronic system in the ground state. However, there are several obvious cases in which the DFT can be used to obtain the excitation energies. The ionization potential (I) represents the difference of the two ground state energies: the energy of the initial N-particle system $E_0(N)$ and the energy of the corresponding ion $E_0(N - 1)$:

$$I = E_0(N-1) - E_0(N). \tag{1.221}$$

Analogously for the electron affinity A we have

$$A = E_0(N) - E_0(N+1). \tag{1.222}$$

This method of obtaining the excitation energies includes a self-consistent calculation of the total energies and is called the Δ SCTE method.

In the case of metallic systems the ionization potential and the electron affinity are the same quantity; specifically, the chemical potential relative to the vacuum level which is exactly determined by the Δ SCTE method. In the case of semiconductors and dielectrics the difference $I - A = E_0(N - 1) + E_0(N + 1) - 2E_0(N)$ neglecting the electron and hole interaction effects is equal to the energy gap ε_g describing the minimum excitation energy in semiconductors.

The density-functional theory was formally generalized[69] to describe excited states. This study demonstrated that there exist subspaces of the eigenstates of the Hamiltonian such that the corresponding subspace energies are density functionals of the subspaces. We consider, for example, a subspace filled with all eigenstates belonging to the two lower energy states. The subspace energy defined as the trace of the matrix of the Hamiltonian relative to the subspace will in this case simply be equal to the sum of the two low energy eigenvalues with weights equal to their degrees of degeneracy. The subspace density is defined analogously as the trace of the operator of the subspace density matrix, that is, it is equal to the sum of all particle densities in the individual states of the subspace. It was also demonstrated[69] that the functional for the subspace energy is minimized by the exact density of the subspace. This theory makes it possible in principle to calculate the subspace energy and, consequently, by calculation determine the energy of the first excited state when the ground state energy is known.

The difficulties of using the subspace theory developed by Theophilou[69] lie in our insufficient knowledge of the exchange-correlated part of the functional for the energy subspace. The functional E_{exc} must be explicitly dependent on the selected subspace and the LDA cannot be used to describe the exchange-correlation subspace energy. Hence in order for the subspace theory to be used in practical calculations it is necessary to find a certain method of including the subspace dependence in E_{exc}.

1.4.2. Multiplet structure

We consider the case where the excited state has symmetry S that differs from the ground state symmetry and this excited state is the lowest state in the subspace S. We can then obtain the energy of the excited state by minimizing the density functional for the total energy in this subspace.[52] Clearly, in principle, the total energy functional is dependent on symmetry S and consequently the exchange-correlation energy functional is also dependent on S, that is, $E^S_{exc}[\rho]$. Unfortunately, there has yet to be a theory to put this symmetry into $E^S_{exc}[\rho]$.

We consider a consequence of using the LDA for $E^S_{exc}[\rho]$. As an example we use the excited configuration $1S2S$ of the helium atom[52] which generates four states: ^1S designated $|0, 0\rangle$ and three states ^3S denoted as $|1,M_S\rangle (M_S = -1,0,1)$. We can easily see that the density matrices corresponding to these states are diagonal and spherically symmetrical. They take the following form:

$$\begin{pmatrix} \frac{1}{2}\rho(r) & 0 \\ 0 & \frac{1}{2}\rho(r) \end{pmatrix}$$

for the $|0,0\rangle$ and $|1,0\rangle$ states and

$$\begin{pmatrix} \rho(r) & 0 \\ 0 & 0 \end{pmatrix}$$

for the $|1,1\rangle$ state. Since there is no explicit dependence on the symmetry of $E^S_{exc}[\rho]$ in the LDA but rather only an implicit dependence through the density matrix, the states with an identical form of the density matrix in the minimization process will acquire the same energy. Consequently, the scheme predicts identical energies for the $|0,0\rangle$ and $|1,0\rangle$ states, as well as for the $|1,0\rangle$ and $|1,1\rangle$ states, yet $|1,1\rangle$ and $|0,0\rangle$ must be different. This has no physical meaning. These difficulties derive from the fact that the LDA is invariant with respect to rotations in the spin space, which yields different energies for the different quantum numbers M_S. Moreover, there may be errors associated with the invariance of the LDA relative to rotations in the coordinate space.[52] Therefore, the DFT cannot be used as a practical tool for calculations if the drawbacks of the LDA outlined above are not eliminated.

At present there is no description of $E^S_{exc}[\rho]$ dependent on symmetry S. An alternative procedure was proposed[70] based on the concept of states with mixed symmetry that conserves the LDA for the exchange-correlation functional. We know that the total energies for the individual asymmetrical states that represent individual Slater determinants in the absence of Coulomb interaction between electrons are obtained in the LDA. At the same time the energies of states described by several Slater determinants are often poor. This can be understood from the equation for $E_{exc}[\rho]$ obtained using Hartree–Fock pair correlation function (1.88). Since $g^{\sigma\sigma'}_{HF}$ has four spin components ($\sigma\sigma' = \uparrow,\downarrow$), we can treat the exchange energy as the sum of four components of $E^{\sigma\sigma'}_x$. Here $g^{\uparrow\downarrow}_{HF} = g^{\downarrow\uparrow}_{HF} = 1$ in single-determinant states, and consequently, the exchange energies for different spins $E^{\uparrow\downarrow}_x$ and $E^{\downarrow\uparrow}_x$ are equal to zero. The true pair correlation function in the LDA is replaced by a function for a homogeneous electron gas which in the Hartree–Fock approximation is obtained from a single Slater determinant. Hence the exchange energies will always be equal to zero in the LDA for four different spins. This part of the exchange energies will therefore be determined exactly in the LDA for states described by single determinants in the absence of interaction. However, states characterized by several determinants may have large exchange energies for different spins and then the LDA becomes unsuitable.

In order to overcome these difficulties a rigorous density functional theory was formulated[70] for states with a mixed symmetry. These states are defined as linear combinations of the states with pure symmetry that have the lowest energy:

$$|D_i\rangle = \sum_j \alpha_{ji}|S_j,0\rangle. \tag{1.223}$$

Here $|S_j,0\rangle$ is the ground state of S symmetry. The coefficients $\alpha_{j,i}$ in Eq. (1.223) are selected so that the state with mixed symmetry $|D_i\rangle$ tended toward a single Slater determinant in the case where interaction was absent. The average energy value of this state is given by the equation

$$E(D_i) = \sum_j |\alpha_{ji}|^2 E_0(S_j), \tag{1.224}$$

where $E_0(S_j)$ is the lowest energy of S symmetry. The quantity $E(D_i)$ can be estimated in the LDA by minimizing the total energy functional with respect to the density matrices with a form corresponding to the state $|D_i\rangle$. If we can find the same number of states with mixed symmetry as energies of the pure states, we can invert the linear equation system (1.224) and determine the energies of the pure states $E_0(S_j)$.

A simpler method of calculating the multiplet structure of the excited states was proposed by Farberovich.[71] The complete Hamiltonian of the system is given as

$$H = H_{\text{eff}} + H'_{ll} + H_{\text{so}}. \tag{1.225}$$

Here $H_{\text{eff}} = -\nabla^2 + v_{\text{eff}}^{\text{DA}}(\mathbf{r})$ is calculated within the framework of the LDA and determines the mean weighted energy of the multiplet group; H'_{ll} is the Hamiltonian of the off-diagonal section of the multiplet expansion in Legendre polynomials; H_{so} is the Hamiltonian accounting for spin–orbit interaction. Hamiltonians H'_{ll} and H_{so} determine splitting between the multiplet terms. The subsequent calculation is based on perturbation theory where the LDA calculations are used as the zeroth approximation. The matrix perturbation elements of electrostatic and spin–orbit interaction are calculated utilizing atom-like functions derived from the Kohn–Sham equations with a self-consistent $v_{\text{eff}}^{\text{LD}}(\mathbf{r})$. A secular equation is solved in the quantitative analysis of intermediate-type coupling in the one-configuration approximation in order to determine the structure of the multiplet terms; this equation is written for each total angular momentum J, while the intermediate coupling state is expressed by the linear combination of states

$$|\alpha LSJM\rangle = \sum_{\alpha_i L_i S_i} c(\alpha_i L_i S_i; \alpha LSJ)\,|\alpha_i L_i S_i; \alpha LSJM\rangle, \tag{1.226}$$

where $c(\alpha_i L_i S_i; \alpha LSJM)$ are the eigenvectors of the matrix

$$\begin{vmatrix} \langle\beta_1|H'_{ll}+H_{\text{so}}|\beta_1\rangle - \varepsilon^{(J)} & \langle\beta_1|H_{\text{so}}|\beta_2\rangle & \cdots \\ \langle\beta_2|H_{\text{so}}|\beta_1\rangle & \langle\beta_2|H'_{ll}+H_{\text{so}}|\beta_2\rangle - \varepsilon^{(J)} & \cdots \\ \cdots & \cdots & \cdots \end{vmatrix}. \tag{1.227}$$

Here $\beta_i \equiv \alpha_i L_i S_i \alpha LSJM$, $\varepsilon^{(J)}$ are the term energies for the given value of J.

1.4.3. *Eigenvalues of the excited states in the DFT*

The exact eigenvalues $\varepsilon_i^{\text{DF}}$ in the DFT are defined as solutions of the effective single-particle Kohn–Sham equation (1.175) with the exchange-correlation potential (1.176). In the initial formulation of the DFT these eigenvalues were treated as auxiliary quantities for calculating the total energy (1.178) and the charge density (1.177). It was assumed that these quantities had no physical meaning aside from the fact that the eigenvalue of the highest occupied state yields the Fermi level in the metal. Nonetheless, these eigenstates obtained in the LDA have been extensively and rather successfully utilized to interpret excitation spectra (e.g., the photoelectron and x-ray spectra of metals and their compounds). However, in certain cases the LDA eigenvalues were incompatible with the experimental data.[52] Hence, it is very important to explain the relation between the one-electron excitation energies and the exact $\varepsilon_i^{\text{DF}}$ as well as how the latter are distorted in the LDA.

In the DFT the eigenvalues are given from[52]

$$\varepsilon_i^{\mathrm{DF}} = \frac{\partial E[\rho]}{\partial n_i}. \tag{1.228}$$

The subsequent arguments are entirely analogous to those above for the X_α method. Equation (1.60) relates ΔSCTE to the eigenvalues $\varepsilon_i^{\mathrm{DF}}$. An exact relation for U_{ij} in Eq. (1.60) exists[52]:

$$U_l = \frac{\partial \varepsilon_i}{\partial n_j} = \langle ij | (\mathrm{v} + k_{\mathrm{exc}}) \tilde{\varepsilon}^{-1} | ij \rangle, \tag{1.229}$$

where $\tilde{\varepsilon}^{-1}$ is the inverse dielectric function. The normalization coefficients of the orbitals together with the two volume integrations of Ω and Coulomb interaction cause the right-hand side of Eq. (1.229) to decrease as $\Omega^{-1/3}$. In the case of common systems such as condensed matter $\Omega \to \infty$ and then $u_{ij} \to 0$, that is, $\partial\varepsilon_i/\partial n_j = 0$ and we obtain from Eq. (1.60),

$$\varepsilon_k^{\mathrm{DF}} = E(n_k=1) - E(n_k=0). \tag{1.230}$$

If k denotes the highest occupied state in the metal, then the right-hand side of Eq. (1.230) represents the difference between the two ground state energies which corresponds to the escape energy of a single electron, that is, the Fermi energy which, consequently, is exactly determined in the DFT. Analogously if k denotes the state at the top of the valence band of the semiconductor and dielectric, the right-hand side of Eq. (1.230) is the ΔSCTE result for the ionization energy of the system. Equation (1.230) shows that this excitation energy is also correctly given by the corresponding DF eigenvalues. The same situation occurs for the energy acquired by the system when a particle is added. Consequently, the DFT can be used to determine the band gap ε_g in semiconductors and dielectrics.

The arguments given above in deriving Eq. (1.230) can also be shown to have a certain empirical nature. In this connection a theoretically rigorous proof of relation (1.230) was offered[72] not only for common solid state systems but also for such finite systems as atoms.

Now, if we move away from the Fermi level in metals or from the band gap in semiconductors and dielectrics, the right-hand side of Eq. (1.230) will no longer represent the difference between the total energies of the ground state and it then becomes doubtful whether the exact $\varepsilon_i^{\mathrm{DF}}$ will represent the excitation energies. However, the formal affinity between Eq. (1.218) and the corresponding equation from the exact Landau–Fermi liquid theory,

$$\frac{\delta E}{\delta n(k)} = \varepsilon(k) \tag{1.231}$$

[$n(k)$ is the distribution function] assumes that they in fact may represent the excitation energies. This is a seductive conclusion based on the fact that the experimental structure data obtained from the photoemission spectra of simple metals recorded with angular resolution coincide with calculation results in the LDA.

There is a system for which the correlation between the eigenvalues from the ground state theory and the quasiparticle spectrum can be investigated in detail without the uncertainties ordinarily induced by the LDA. This system is a uniform interacting electron gas.

The exact spectrum of quasiparticle excitation $E(k)$ is related to the self-energy $\Sigma(k, \omega)$ which is dependent on the momentum and frequency by the following implicit relation[73]:

$$E(k)=\frac{k^2}{2m}+\Sigma(k,E(k)). \tag{1.232}$$

On the other hand, the single-particle equations of ground state theory contain a translationally invariant potential independent of k. Then the variance of the self-energies of the ground state is proportional to k^2. However, as stated above the self-energies of the ground state correctly describe the Fermi energy, such that

$$\varepsilon(k)=\frac{k^2}{2m}+\Sigma(k_F,E(k_F)). \tag{1.233}$$

It is clear from a comparison of Eqs. (1.232) and (1.233) that eigenvalues in the ground state $\varepsilon(k)$ and the quasiparticle spectrum $E(k)$ differ increasingly as we move away from the Fermi energy although only due to the change in the self-energy part as k and ω change. The quasiparticle energies and $\varepsilon_i^{\mathrm{DF}}$ can be expanded into a Taylor series in the immediate vicinity of the Fermi level. Limiting the analysis to first order terms we obtain

$$E(k)=E_F+\frac{k_F}{m^*}(k-k_F), \tag{1.234}$$

$$\varepsilon(k)=E_F+\frac{k_F}{m}(k-k_F), \tag{1.235}$$

where

$$\frac{m^*}{m}=\left(1-\frac{\partial\Sigma}{\partial\omega}\right)\left(1+\frac{m}{k_F}\frac{\partial\Sigma}{\partial k}\right)^{-1}. \tag{1.236}$$

It is clear from Eqs. (1.234) and (1.235) that the differences between the eigenvalues from the DFT and the quasiparticle spectrum near E_F are described by the ratio of the effective mass m^* to the free electron mass m.

The difference between m and m^* clearly reveals that the eigenvalues from the DFT are formally equivalent to the excitation energies. Therefore, the total energy functional in the DFT is different from the corresponding functional in Fermi liquid theory while the state obtained by adding an orbital with wave vector k generally speaking is not equivalent to the multiparticle state corresponding to a quasiparticle with the same wave vector. Hence although the DFT in principle yields the correct Fermi energy the issue as to whether or not it yields the correct Fermi surface remains open. Moreover, the deviation of m^* from m suggests that the density of states on the Fermi level, where the lifetime effects of the states are not significant, is not given by the exact distribution of the self-energies obtained from the DFT. This leads to the question of what the difference is in practice between m^* and m. It turns out that in the majority of practical cases this difference is approximately 5% in the band energy range. Hence, the following simple approximation can be used for moderate excitation energies:

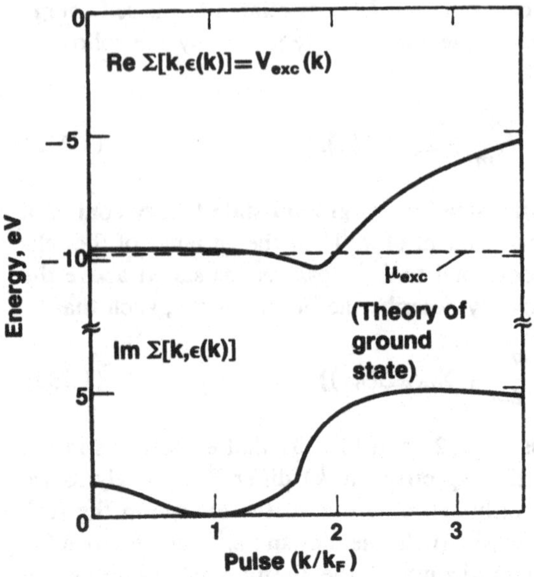

FIG. 1.2. Self-energy part plotted as a function of **k**.

$$\Sigma(k,\omega) \approx v_{exc}. \tag{1.237}$$

The difference between the eigenvalues of the ground state and the excitation spectrum far from the Fermi surface can be expressed graphically as a plot of the self-energy part as a function of k (see Fig. 1.2). The most significant qualitative change in self-energy moving away from the Fermi surface is the appearance of an imaginary part corresponding to the finite quasiparticle excitation time. Moreover, it is clear that the eigenvalues from the DFT for $k \sim 2k_F$ do not reflect a rapid change in $\Sigma(k,\omega)$ (this is due to a decay of the quasiparticles attributable to plasmon creation) nor a gradual degradation in exchange-correlated energy with an increasing quasiparticle velocity. On the other hand, we can conclude that the self-energies of the ground state accurately describe the spectra of the valence band of simple metals since, as we see from the figure, the variations in Σ are relatively small in the vicinity of the valence band.

We now consider localized systems. It follows from Eq. (1.60) that the following relation will hold for localized systems:

$$E(n_i=1) - E(n_i=0) = \varepsilon_i + \tfrac{1}{2}u_{ii}, \tag{1.238}$$

This relation differs from Eq. (1.230) by $u_{ii}/2 = 1/2(\partial\varepsilon_i/\partial n_i)$. As discussed above relation (1.230) holds for the highest level, that is, $\partial\varepsilon_N/\partial n_N = 0$. Here Eq. (1.238) holds for the lower eigenvalues, that is, $\partial\varepsilon_i/\partial n_i > 0$, and they lie above the corresponding excitation energies.

We now consider how the eigenvalues become distorted within the framework of the LDP. There is a wide range of contradictory results obtained in the LDA. Some of these can be formulated as follows: (a) situations arise in the atoms of transition metals where the s and d levels are ordered opposite to the configuration predicted by Hartree–Fock theory; (b) the spin splittings of the d bands in ferromagnetic metals calculated in the LDA are often much greater than that observed in experiment[68]; (c) self-consistent calculations of the band structure of semiconductors and insulators in the LDA systematically underestimate the one-electron width of the forbidden bands by more than 40% (Ref. 68); (d) a significant deviation from experiments is often observed when the eigenvalues $\varepsilon_i^{\text{LD}}$ are identified with the positions of the surface states, the deep levels of defects in crystals, and the core bands; and (e) the experimentally observed stable negative ions (H^-,O^-,F^-) are unstable in the LDA.[68]

The majority of these drawbacks of the LDA technique can be corrected if self-interaction corrections are introduced by the technique described above. We estimate the self-interaction correction $\delta_{nl\sigma}$ (Eq. 1.218). It was demonstrated[68] that

$$U[\rho] \leqslant 1.092 N^{2/3} \int dr\, \rho^{4/3}(r), \tag{1.239}$$

and so the LDA for exchange is calculated by the following expression:

$$E_x^{\text{LD}}[\rho_\uparrow, \rho_\downarrow] = -0.9305 \int d\mathbf{r} [\rho_\uparrow(\mathbf{r})^{4/3} + \rho_\downarrow(\mathbf{r})^{4/3}]. \tag{1.240}$$

The poor compensation between $U[\rho_{nl\sigma}]$ and $E_x^{\text{LD}}[\rho_{nl\sigma}, 0]$ for the $nl\sigma$ orbital is obvious. The remainder from incomplete compensation

$$\delta_{nl\sigma} \leqslant 0.16 \int d\mathbf{r}\, \rho_{nl\sigma}(\mathbf{r})^{4/3} \tag{1.241}$$

may be significant. Since Eq. (1.239) is normally close to an equality, Eq. (1.241) can also be replaced with an equality. Clearly the correction becomes more significant when the orbital localization rises. For example, for a hydrogen-like $1S$ orbital we have $\delta_{1S} < 0.047Z$. On the other hand, for a delocalized electron in volume $\Omega \rho_{nl\sigma} \sim \Omega^{-1}$ and $\delta_{nl\sigma} \sim \Omega^{-1/3}$, that is, it vanishes as $\Omega \to \infty$. Consequently the LDA and SIC-LDA both yield the exact energy per electron for densities that have little variability.

If we neglect the difference in the orbitals $\Psi_{nl\sigma}$ obtained in the LDA and SIC-LDA, that is, the self-consistency effect, we can show[68] that

$$\varepsilon_{nl\sigma}^{\text{SI-LD}} \approx \varepsilon_{nl\sigma}^{\text{LD}} - 0.94 \int d\mathbf{r}\, \rho_{nl\sigma}^{4/3}(\mathbf{r}). \tag{1.242}$$

It is clear from Eq. (1.242) how significant the eigenvalues obtained in the LDA are distorted due to the neglect of the self-interaction effect.

Koopman's theorem in Hartree–Fock theory [see formula (1.47)] yields an explicit physical interpretation of the eigenvalues

$$I \equiv E(n_i = 0) - E(n_i = 1) = -\varepsilon_i^{\text{HF}}. \tag{1.243}$$

The relaxation effects can be calculated in Hartree–Fock theory by means of two independent self-consistent calculations for the initial and final states (the ΔSCTE method):

$$I^* = I + \Delta. \tag{1.244}$$

For a specific orbital, Eq. (1.244) can be given as

$$I^*_{nl\sigma}=I_{nl\sigma}+\Delta_{nl\sigma}. \tag{1.245}$$

Here $\Delta_{nl\sigma}$ is the relaxation energy of the $nl\sigma$ orbital. It is clear from Eq. (1.238) that Koopman's theorem is not satisfied in the LDA. However, if we use the Slater transition state method, as demonstrated above, we can write

$$I^*_{nl\sigma}\approx-\varepsilon_{nl\sigma}(\tfrac{1}{2}). \tag{1.246}$$

In order to obtain an analog of Koopman's theorem it is necessary to have $\varepsilon_{i\sigma}(1)$ in place of $\varepsilon_{i\sigma}$ in Eq. (1.246). However, such an analog of Koopman's theorem is not possible in the LDA.

We now consider the situation with the calculation of I^* in the SIC-LDA. Denoting $P_{i\sigma}=U_{ii}/2$ in Eq. (1.238) we have

$$I^{\text{LD}}_{i\sigma}=-\varepsilon^{\text{LD}}_{i\sigma}+P^{\text{LD}}_{i\sigma}, \tag{1.247}$$

$$I^{\text{SI-LD}}_{i\sigma}=-\varepsilon^{\text{SI-LD}}_{i\sigma}+P^{\text{SI-LD}}_{i\sigma}, \tag{1.248}$$

where

$$\begin{aligned} P^{\text{LD}}_{i\sigma}&=U[\rho_{i\sigma}]+S_{i\sigma},\\ P^{\text{SI-LD}}_{i\sigma}&=-\frac{1}{3}\int d\mathbf{r}\,\rho_{i\sigma}\varepsilon_x(\rho_{i\sigma},0)+S_{i\sigma}, \end{aligned} \tag{1.249}$$

$$S_{i\sigma}=\int d\mathbf{r}\left[\varepsilon_x(\rho_\sigma-\rho_{i\sigma};0)\varepsilon_x(\rho_\sigma-\rho_{i\sigma};0)-\rho_\sigma\varepsilon_x(\rho_\sigma;0)+\frac{4}{3}\rho_{i\sigma}\varepsilon_x(\rho_\sigma;0)\right]. \tag{1.250}$$

Using Eq. (1.239) for the estimates we find

$$P^{\text{LD}}_{i\sigma}\approx1.09\int d\mathbf{r}\,\rho^{4/3}_{i\sigma}(\mathbf{r})+S_{i\sigma}, \tag{1.251}$$

$$P^{\text{SI-LD}}_{i\sigma}\approx0.31\int d\mathbf{r}\,\rho^{4/3}_{i\sigma}(\mathbf{r})+S_{i\sigma}. \tag{1.252}$$

As in Hartree–Fock theory we can write

$$I^{*\text{LD}}_{i\sigma}=I^{\text{LD}}_{i\sigma}+\Delta^{\text{LD}}_{i\sigma}=-\varepsilon^{\text{LD}}_{i\sigma}+P^{\text{LD}}_{i\sigma}+\Delta^{\text{LD}}_{i\sigma}, \tag{1.253}$$

$$I^{*\text{SI-LD}}_{i\sigma}=I^{\text{SI-LD}}_{i\sigma}+\Delta^{\text{SI-LD}}_{i\sigma}=-\varepsilon^{\text{SI-LD}}_{i\sigma}+P^{\text{SI-LD}}_{i\sigma}+\Delta^{\text{SI-LD}}_{i\sigma}. \tag{1.254}$$

It was demonstrated[68] that

$$\Delta^{\text{LD}}_{i\sigma}\approx k_1U[\rho_{i\sigma}], \tag{1.255}$$

specifically for atoms $k_1=-0.44$. Using Eq. (1.239), Eq. (1.255) can be recast as

$$\Delta^{\text{LD}}_{i\sigma}\approx-0.48\int d\mathbf{r}\,\rho^{4/3}_{i\sigma}(\mathbf{r}). \tag{1.256}$$

Comparing Eq. (1.256) to Eq. (1.251) we see that $\Delta^{\text{LD}}_{i\sigma}$ can compensate only less than one-half of $P^{\text{LD}}_{i\sigma}$ since the electron escape energy $I^*_{i\sigma}$ is poorly approximated by the self-energy $\varepsilon_{i\sigma}^{\text{LD}}$. At the same time it is clear from a comparison of Eqs. (1.256) and (1.252) that $\Delta^{\text{SI-LD}}$ and $P_{i\sigma}^{\text{SI-LD}}$ effectively cancel. We note that the $S_{i\sigma}$ term can always be dropped if the orbital densities ρ_σ make a small contribution to the total density ρ_σ, that is, $\rho_{i\sigma}/\rho_\sigma \ll 1$. Therefore, the eigenvalue $-\varepsilon_{i\sigma}^{\text{SI-LD}}$ in the SIC-LDA accurately describes the relaxation ionization energy I_{ij}^*.

In summary we conclude that in many cases the DFT eigenvalues may be a useful tool for interpreting single-particle excitation spectra. It is, however, necessary to recall at all times that the inaccuracy associated with the LDA often exceeds the inaccuracy that occurs due to the application of ground state theory to perturbed systems. Hence, an approach based on the SIC-LDA is often useful here. At the same time although the self-energies are a useful tool for interpreting the excitation spectrum one may encounter a situation where one or several of the assumptions underlying the applications prove to be invalid for certain specific systems.

1.4.4. The self-energy approach to calculating excitation energies

As we see from the preceding section the orbital eigenvalues $\varepsilon_i^{\text{DF}}$ of the ground state, generally speaking, do not yield exact single-particle excitations for multi-electron systems. However, we have also seen that these eigenvalues reproduce with sufficient accuracy the excitation energies of the quasiparticles, particularly when they are not distorted by self-action associated with the LDA. The primary drawback of this description is that it is not a well-defined approximation of a certain description that is in principle exact. Hence the advantage of the self-energy approach lies in the fact that it provides a description of the spectrum of quasiparticle excitations which in principle is exact.

We define the one-particle Green's function as[74]

$$G(1,2) = -i\langle T(\Psi(1)\Psi^+(2))\rangle. \tag{1.257}$$

Here 1 and 2 denote the set of electron coordinates: $(1)\equiv(r_1,\sigma_1,t_1) = (x_1,t_1)$; T is the chronological ordering operator of the multipliers in the operator product; ψ is the field operating in the Heisenberg representation. The $\langle\cdots\rangle$ denote averaging over the ground state.

Green's function (1.257) satisfies equation (1.74):

$$[\varepsilon - h(\mathbf{x}) - v^H(\mathbf{x})]G(\mathbf{x},\mathbf{x}';\varepsilon) - \int d\mathbf{x}'' \sum(\mathbf{x},\mathbf{x}'',\varepsilon)G(\mathbf{x}'',\mathbf{x}';\varepsilon) = \delta(\mathbf{x}-\mathbf{x}'), \tag{1.258}$$

where

$$h(\mathbf{x}) = -\nabla^2 - \sum_n \frac{2Z_n}{|\mathbf{x}-\mathbf{R}_n|} = -\nabla^2 + v(\mathbf{x}),$$

$$v^H(\mathbf{x}) = \int d\mathbf{x}' \frac{\rho(\mathbf{x}')}{|\mathbf{x}-\mathbf{x}'|}.$$

Z_n and R_n are the nuclear charge and its radius vector,

$$\rho(\mathbf{x}) = \langle\Psi^+(\mathbf{x})\Psi(\mathbf{x})\rangle,$$

$$G(\mathbf{x},\mathbf{x}';\varepsilon)=\int d(t-t')G(\mathbf{x},t;\mathbf{x}',t')\exp[i\varepsilon(t-t')].$$

$\Sigma(x,x',\varepsilon)$ is the mass operator (the self-energy part) describing all correlation effects in the multielectron system.

We now redefine the Green's function in the following manner[74]:

$$G(\mathbf{x},\mathbf{x}';\varepsilon)=\sum_s f_s(\mathbf{x})f_s^*(\mathbf{x}')/(\varepsilon-\varepsilon_s), \tag{1.259}$$

where

$$f_s(\mathbf{x})=\langle N,0|\Psi(\mathbf{x})|N+1,S\rangle,$$

$$\varepsilon_s=E(N+1,s)-E(N,0)-i\Delta \quad \text{if } \varepsilon_s\gg\mu$$

or

$$f_s(\mathbf{x})=\langle N-1,s|\Psi(\mathbf{x})|N,0\rangle,$$

$$\varepsilon_s=E(N,0)-E(N-1,S)+i\Delta \quad \text{if } \varepsilon_s<\mu$$

and $\mu=E(N+1,0)-E(N,0)$ is the chempotential or the electron affinity; $|N,\,0\rangle$ is the ground state of the N-particle system; S runs over all states of the $(N+1)$ and $(N-1)$ particle system. The amplitudes $f_S(x)$ and excitation energies ε_S are solutions of the Schwinger equation[75]:

$$[\varepsilon-h(\mathbf{x})-\mathrm{v}^H(\mathbf{x})]f_s(\mathbf{x})-\int d\mathbf{x}''\sum(\mathbf{x}',\mathbf{x}'',\varepsilon_s)f_s(\mathbf{x}'')=0 \tag{1.260}$$

for the case of a discrete excitation energy spectrum ε_S. For a continuous excitation energy spectrum the solutions of Eq. (1.260) will generally be complex.

It is possible to impart a Schrödinger equation form* to Eq. (1.260):

$$[-\nabla^2+\mathrm{v}(\mathbf{r})+\mathrm{v}^H(\mathbf{r})]f_s(\mathbf{r})+\int d\mathbf{r}''\sum(\mathbf{r},\mathbf{r}'';\varepsilon_s)f_s(\mathbf{r}'')=\varepsilon_s f_s(\mathbf{r}). \tag{1.261}$$

If we compare Eq. (1.261) to the single-particle Kohn–Sham equation (1.175), it is clear that the exchange-correlation potential $\mathrm{v}_{\mathrm{exc}}(\mathbf{r})$ figures into Eq. (1.175) in place of the self-energy part. We can hence conclude that the degree of proximity of $\varepsilon_i^{\mathrm{DF}}$ to the excitation energies ε_S is given by the correlation between ε and $\mathrm{v}_{\mathrm{exc}}$ which we discussed in the preceding section.

We now consider methods of calculating Σ and the excitation spectrum in actual systems. The first approximation for calculating Σ was proposed by Hedin.[74] This technique was based on the application of Green's function technique to the electron gas problem and was called the GW approximation. Formally Σ can be expressed through the renormalized Green's function G and the screened Coulomb interaction W. For the Fourier transform of $\Sigma_h(\mathbf{k},\varepsilon)$ we have[74]

*Here and henceforth we neglect the spin.

$$\Sigma_h(\mathbf{k},E)=\frac{i}{(2\pi)^4}\int_{-\infty}^{\infty}d\omega'\int d\mathbf{k}'\,G(\mathbf{k}-\mathbf{k}',E-\omega')W(\mathbf{k}',\omega')e^{i\omega'0^+},\tag{1.262}$$

$$W(\mathbf{k},\omega)=\frac{8\pi}{\Omega k^2\varepsilon(\mathbf{k},\omega)}.\tag{1.263}$$

$\varepsilon(\mathbf{k},\omega)$ is the dielectric function.

Using for the Green's function the equation

$$G(\mathbf{k},E)=1/[E-E(\mathbf{k})+i\Delta\,\mathrm{sign}(k_F|\mathbf{k}|)]\tag{1.264}$$

we obtain

$$\Sigma_h(\mathbf{k},E)=\frac{i}{(2\pi)^4}\int_{-\infty}^{\infty}d\omega'\int d\mathbf{k}'\frac{8\pi}{\Omega k^2\varepsilon(\mathbf{k}',\omega)}\times\frac{e^{-i\omega'0^+}}{E-\omega'-E(\mathbf{k}-\mathbf{k}')+i\Delta\,\mathrm{sign}(k_F-|\mathbf{k}-\mathbf{k}'|)}.\tag{1.265}$$

This equation for Σ_h was divided into two terms[74]: Σ_h is the "Coulomb hole + screened exchange" term and Σ_h^d is the term including dynamical screening. These two terms are calculated by the following equations[74]:

$$\Sigma_h^c(k)=\frac{4}{\pi a r_s}\left\{\int_0^{\infty}d\mathbf{k}'[\varepsilon^{-1}(\mathbf{k}',0)-1]-\int_{-1}^{1}d\xi\int_0^{\infty}d\mathbf{k}'\frac{\theta(0.25-k^2-k'^2-2\mathbf{k}\mathbf{k}'\xi)}{\varepsilon(\mathbf{k}',0)}\right\},\tag{1.266}$$

$$\begin{aligned}\Sigma_h^d(\mathbf{k},E)=&\frac{4}{\pi a r_s}\int_{-1}^{1}d\xi\int_0^{\infty}d\mathbf{k}'\left(\frac{1}{\varepsilon(\mathbf{k}',\omega-E(\mathbf{k}-\mathbf{k}'))}-\frac{1}{\varepsilon(\mathbf{k}'0)}\right)\\&\times[\theta(\omega-E(\mathbf{k}-\mathbf{k}'))-\theta(0.25-E(\mathbf{k}-\mathbf{k}'))]\\&+\frac{1}{\pi^2 a r_s}\int_0^{\infty}d\omega'\int_0^{\infty}d\mathbf{k}'\left(\frac{1}{\varepsilon(\mathbf{k}',i\omega')}-\frac{1}{\varepsilon(\mathbf{k}',0)}\right)\\&\times\frac{1}{kk'}\ln\frac{(\omega-(\mathbf{k}+\mathbf{k}')^2)^2+\omega'^2}{(\omega-(\mathbf{k}-\mathbf{k}')^2)^2+\omega'^2},\end{aligned}\tag{1.267}$$

where

$$a=\left(\frac{4}{9\pi}\right)^{1/3},\quad \xi=\mathbf{k}\cdot\mathbf{k}'/(|\mathbf{k}|\cdot|\mathbf{k}'|),$$

θ is the step or Heaviside function. The integrals in Eqs. (1.266) and (1.267) can be evaluated numerically.

Systems with a slowly varying density $\rho(\mathbf{r})$ were examined by Sham and Kohn[76] in the spirit of the LDA. Schwinger substituted[75] the sum Σ with the mass operator for a uniform electron gas:

$$\sum(\mathbf{r},\mathbf{r}';E) \rightarrow E_h(\mathbf{r}-\mathbf{r}',E-E_F+\mu_h(\bar{\rho});\rho), \tag{1.268}$$

where $E_h(\mathbf{r} - \mathbf{r}',E;\rho)$ is the mass operator for the uniform electron gas of density ρ and $\bar{\rho}=\rho[(\mathbf{r}+\mathbf{r}')/2]$. In the ADA the amplitude $f_S(\mathbf{r})$ can be represented as a plane wave, that is,

$$f_s(\mathbf{r}) \approx \exp[ik(\mathbf{r},E)\cdot\mathbf{r}]. \tag{1.269}$$

Substituting Eq. (1.269) into Eq. (1.261) and neglecting nonlocal terms we obtain

$$-E+k_{\mathrm{LD}}^2+\mathrm{v}^H(\mathbf{r})+\Sigma_h(k_{\mathrm{LD}},E-E_F+\mu_h(\rho);\rho)=0. \tag{1.270}$$

For a slowly varying density in the spirit of the Fermi–Thomas method we can assume that

$$\mathrm{v}^H(\mathbf{r})=E_F-\mu_h(\rho(\mathbf{r})). \tag{1.271}$$

Defining

$$\mu_h(\rho)=k_F^2(\rho)+\mathrm{v}_{\mathrm{exc}}(\rho)=k_F^2(\rho)+\Sigma_h(k_F,\mu_h(\rho);\rho), \tag{1.272}$$

we finally obtain

$$k_{\mathrm{LD}}^2-k_F^2=(E-E_F)-\left|\Sigma_h(k_{\mathrm{LD}},E-E_F+\mu_h(\rho);\rho)-\Sigma_h(k_F,\mu_h(\rho);\rho)\right|. \tag{1.273}$$

Thus substitution of the amplitude $f_s(\mathbf{r})$ with a plane wave takes us from a nonlocal mass operator to a local mass operator, that is,

$$\sum(\mathbf{r}-\mathbf{r}';E) \rightarrow E_h(k_{\mathrm{LD}},E;\rho(r))\delta(\mathbf{r}-\mathbf{r}'). \tag{1.274}$$

The local approximation for the mass operator is called the time-dependent local density approximation (TDLDA). On the Fermi level ($E=E_F$) $k_{\mathrm{LD}}=k_F$ and then the mass operator (1.274) reduces to the exchange-correlation potential of the ground state theory in the LDA, that is,

$$\mathrm{v}_{\mathrm{exc}}=\Sigma_h(k_F,\mu_h;\rho). \tag{1.275}$$

Subject to Eq. (1.275) we can also write

$$\Sigma_h(k,E;\rho)=\mathrm{v}_{\mathrm{exc}}(\rho)+\left|\Sigma_h(k,E;\rho)-\Sigma_h(k_F,\mu_h;\rho)\right|\equiv\mathrm{v}_{\mathrm{exc}}(\rho)+\Delta(k,E;\rho). \tag{1.276}$$

Numerical calculations of $\Delta(k,E;\rho)$ were first carried out by Albman and Barth[77] and subsequently by others.[78–80] It turns out that very minor additions to the energy bands obtained in the ground state were derived for actual metals. For example, for Ni the addition is ~0.1 eV.[78]

The advantage of calculating the difference Δ between Σ and $\mathrm{v}_{\mathrm{exc}}$ lies in the fact that a direct calculation of the mass operator in the GW approximation does not include the important short-range correlations that arise in nonuniform systems. At the same time in the LDA v_{exc} is known exactly, including the short-range correlations. Hence the GW approximation will work substantially better for Δ than for Σ. We also note that at the Fermi level $\Delta(k_F,E_F;\rho) = 0$, that is, the Kohn–Sham eigenvalues and wave functions represent a different zeroth approximation of the quasiparticle excitation energies and amplitudes. It is obvious that the difference $\Delta(k,E,\rho)$ can be interpreted as the difference in the exchange-correlated processes that "sense" the quasiparticle state f_S described by Σ_h and the Kohn–Sham state φ_i given by $v_{exc}(\rho)$.

1.4.5. Time-dependent density-functional theory

In quantum mechanics the problem of finding excited states is closely related to the problem of system response to an external time-dependent excitation. The difficulties one encounters when using the DFT to describe excited states occur primarily to the fact that essentially this theory is static. Hence the DFT can be used only to analyze stationary time-independent processes. We can, therefore, calculate the static atomic polarizabilities $\alpha(0)$ (Ref. 81) within the framework of the DFT at the same time that it is not possible to calculate the dynamical polarizabilities.

A time-dependent generalization of the DFT was proposed by Peuckert.[82] The structure of the new theory is very similar to that of the DFT. The problem of N interacting electrons, as in the DFT case, reduces to a problem of N noninteracting electrons traveling in an effective time-dependent potential which is the sum of the external potential $v(\mathbf{r},t)$, the Hartree potential $v^H(\mathbf{r},t)$, and the time-dependent exchange-correlated potential:

$$\left[-\nabla^2 + v(\mathbf{r},t) + \int d\mathbf{r}' \frac{\rho(\mathbf{r}',t)}{|\mathbf{r}-\mathbf{r}'|} + v_{exc}(\mathbf{r},t)\right]\Psi_i(\mathbf{r},t) = i\frac{\partial}{\partial t}\Psi_i(\mathbf{r},t), \tag{1.277}$$

$$\rho(\mathbf{r},t) = \sum_{i=1}^{N} |\Psi_i(\mathbf{r},t)|^2, \tag{1.278}$$

$$v_{exc}(\mathbf{r},t) = \frac{\partial}{\partial\rho(\mathbf{r},t)} \int dt \int \frac{d\lambda}{2\lambda} d\mathbf{r}\, d\mathbf{r}' \times\{\langle\Psi_\lambda(t)|\Psi^+(\mathbf{r})\Psi^+(\mathbf{r}')\Psi(\mathbf{r}')\Psi(\mathbf{r})|\Psi_\lambda(t)\rangle - \rho(\mathbf{r},t)\rho(\mathbf{r}',t)\}\frac{\lambda}{|\mathbf{r}-\mathbf{r}'|}. \tag{1.279}$$

Here $|\Psi(t)\rangle$ is the state vector of the interacting system at time t; $\Psi^+(\mathbf{r})$ and $\Psi(\mathbf{r})$ are the field electron creation and destruction operators at point $\mathbf{r}$; λ is the interaction constant.

It is well known (see Sec. 1.4.2 of this study) that the interaction energy of a static system can be expressed through the frequency-dependent density–density response function. For example, as we can demonstrate from an analysis of Feynman diagrams[73] this is also true of interacting electrons in a time-dependent potential. We then have

$$\mathrm{v}_{exc}(\mathbf{r},t)=\frac{\partial}{\partial\rho(\mathbf{r},t)}\int dt\int\frac{d\lambda}{2\lambda}\,d\mathbf{r}\,d\mathbf{r}'$$

$$\times\{i\Pi(\mathbf{r},t;\mathbf{r}',t')-\rho(\mathbf{r},t)\delta(\mathbf{r}-\mathbf{r}')\}\frac{\lambda}{|\mathbf{r}-\mathbf{r}'|}, \qquad (1.280)$$

where Π is the time-dependent response function (the polarization operator). In principle, Eq. (1.280) provides us with the capacity to exactly calculate $\mathrm{v}_{exc}(\mathbf{r},t)$. If we use a certain approximation for v_{exc} we can calculate $\Pi(\mathbf{r},t;\mathbf{r}',t')$ using an equivalent single-particle scheme. The derived function Π can again be substituted into Eq. (1.280) to find the enhanced potential v_{exc} which in turn can be used to formulate the enhanced function Π. An exact v_{exc} can in principle be formulated in this manner although in practice insurmountable difficulties are encountered as early as the second step. Therefore, at present only the first step is possible, that is, formulation of Π from a certain known approximation for v_{exc} which was in fact carried out in Ref. 33.

Writing Eq. (1.180) in coordinate space we have

$$\delta\rho_{ind}(\mathbf{r},\omega)=2\int d\mathbf{r}'\,\chi_0(\mathbf{r},\mathbf{r}';\omega)\mathrm{v}_{tot}^{eff}(\mathbf{r}',\omega), \qquad (1.281)$$

where

$$\mathrm{v}_{tot}^{eff}(\mathbf{r},\omega)=\mathrm{v}_{ext}(\mathbf{r},\omega)+\delta\mathrm{v}_{ind}^{eff}(\mathbf{r},\omega) \qquad (1.282)$$

$\mathrm{v}_{ext}(\mathbf{r},\omega) = -rE_0P_1(\cos\theta)$ is the external electromagnetic field; the direction of the external electrical field E_0 is used as the polar axis; $P_1(\cos\theta)$ is a first order Legendre polynomial:

$$\delta\mathrm{v}_{ind}^{eff}(\mathbf{r},\omega)=\delta\mathrm{v}_{ind}^{Coul}(\mathbf{r},\omega)+\delta\mathrm{v}_{ind}^{exc}(\mathbf{r},\omega), \qquad (1.283)$$

$$\delta\mathrm{v}_{ind}^{Coul}(\mathbf{r},\omega)=\int d\mathbf{r}'\,\frac{\delta\rho_{ind}(\mathbf{r}',\omega)}{|r_1'-\mathbf{r}'|}, \qquad (1.284)$$

$$\delta\mathrm{v}_{ind}^{exc}(\mathbf{r},\omega)=\frac{1}{2}\frac{\partial\mathrm{v}_{exc}}{\partial\rho(\mathbf{r})}\bigg|_{\rho=\rho_0}\delta\rho_{ind}(\mathbf{r},\omega) \qquad (1.285)$$

in the linear response:

$$\chi_0(\mathbf{r},\mathbf{r}';\omega)=2\sum_{i,j}\varphi_i(\mathbf{r})\varphi_i^*(\mathbf{r}')\varphi_j(\mathbf{r}')\varphi_j^*(\mathbf{r})\frac{f(\varepsilon_i)-f(\varepsilon_j)}{\varepsilon_i-\varepsilon_j\pm h\omega+i\delta}, \qquad (1.286)$$

$f(\varepsilon)$ is the Fermi–Dirac distribution function. Equation (1.286) contains the sum over the entire spectrum of the effective single-particle potential $\mathrm{v}_{eff}(\mathbf{r})$. Hence it is necessary to carry out summation not only over the occupied orbitals in Eq. (1.286) but also over the continuous spectrum. At the same time if Eq. (1.286) is expressed through the single-particle Green's function the problem of summation over the entire spectrum can be avoided. We will calculate the Green's function which satisfies the following differential equation:

$$[-\nabla^2+\mathrm{v}_{eff}(\mathbf{r})-\varepsilon]G(\mathbf{r},\mathbf{r}';\varepsilon)=-\delta(\mathbf{r}-\mathbf{r}'). \qquad (1.287)$$

Using the spectral expansion for Green's function (1.259) and the symmetry relation $G(\mathbf{r},\mathbf{r}';\varepsilon) = G^*(\mathbf{r}',\mathbf{r};\varepsilon)$ we find

$$\chi_0(\mathbf{r},\mathbf{r}';\omega) = \sum_{i<i_{\max}} \varphi_i(\mathbf{r})\varphi_i^*(\mathbf{r}')G(\mathbf{r},\mathbf{r}';\varepsilon_i+\omega)$$

$$+ \sum_{i<i_{\max}} \varphi_i^*(\mathbf{r})\varphi_i(\mathbf{r}')G^*(\mathbf{r},\mathbf{r}';\varepsilon_i-\omega). \tag{1.288}$$

The summation in Eq. (1.288) is now carried out over only the occupied states.

Therefore, the problem has been reduced to formulating Green's function. We consider one of the simplest cases: an atom with completely filled shells. Then

$$G(\mathbf{r},\mathbf{r}';\varepsilon) = \sum_{lm} Y_{lm}(\hat{\mathbf{r}})Y_{lm}^*(\hat{\mathbf{r}}')G_l(\mathbf{r},\mathbf{r}';\varepsilon), \tag{1.289}$$

$$\chi_0(\mathbf{r},\mathbf{r}';\omega) = \frac{1}{4\pi}\sum_{nl} W_{nl}\varphi_{nl}(\mathbf{r})\varphi_{nl}(\mathbf{r}') \sum_{l'=|l-1|}^{l+1} [C_{l010}^{l'0}]^2\{G_{l'}(\mathbf{r},\mathbf{r}';\varepsilon_{nl}+\omega)$$

$$+ G_{l'}^*(\mathbf{r},\mathbf{r}';\varepsilon_{nl}-\omega)\}, \tag{1.290}$$

$$\mathrm{v}_{\mathrm{tot}}^{\mathrm{eff}}(\mathbf{r},\omega) = \mathrm{v}_{\mathrm{ext}}(\mathbf{r},\omega) + \frac{4\pi}{3}\left[\frac{1}{r^2}\int_0^r d\mathbf{r}'\,\delta\rho_{\mathrm{ind}}(\mathbf{r}',\omega)r' + r\int_0^\infty d\mathbf{r}'\,\delta\rho_{\mathrm{ind}}(\mathbf{r}',\omega)\right]$$

$$+ \frac{1}{2}\,\delta\rho_{\mathrm{ind}}(\mathbf{r},\omega)\left.\frac{\partial \mathrm{v}_{\mathrm{exc}}}{\partial\rho}\right|_{\rho=\rho_0}, \tag{1.291}$$

$$\delta\rho_{\mathrm{ind}}(\mathbf{r},\omega) = \frac{1}{2\pi}\sum_{nl} W_{nl}\varphi_{nl}(\mathbf{r}) \sum_{l'=|l-1|}^{l+1} [C_{l010}^{l'0}]^2$$

$$\times \int d\mathbf{r}'\,\varphi_{nl}(\mathbf{r}')[G_{l'}(\mathbf{r},\mathbf{r}'\varepsilon_{nl}+\omega)$$

$$+ G_{l'}^*(\mathbf{r},\mathbf{r}',\varepsilon_{nl}-\omega)]\mathrm{v}_{\mathrm{ind}}^{\mathrm{eff}}(\mathbf{r}',\omega)r'. \tag{1.292}$$

$C_{l010}^{l'0}$ is the Clebsch–Gordan coefficient; W_{nl} is the occupation number of the shell *nl*.

Carrying out a self-consistent solution of Eqs. (1.291) and (1.292) we obtain the charge density and the total effective potential.

The method described above is called the time-dependent local-density approximation (TDLDA) and has been used to investigate the optical properties of atoms, molecules, and small metallic particles.[84–86]

Chapter 2

Methods of calculating the electronic structure of condensed matter

2.1. The problem of calculating the electronic states of condensed matter

The development of quantum many-body theory (specifically, uniform electron gas theory) using computational techniques led to the development of modern methods of calculating the electronic structure of condensed matter which made it possible to calculate from first principles the thermodynamic, magnetic, spectral, and superconducting characteristics of metals and their compounds. Here the term "metals" refers to ideal single crystals, single crystals containing defects, films, liquids, and amorphous solids. Such a variety of objects of study has been responsible for the lack of a completely consistent theory of the electronic states of condensed matter from first principles. We can say that, at present, there is no general method of calculating the electronic structure of condensed matter consisting of a cluster of atoms distributed at random points in space. If we consider the particular case where these atoms are at the sites of a spatially periodic lattice, it is possible to use translational symmetry to formulate a structural theory of an ideal crystal (solid state band theory) which has recently found extensive applications. In principle, it would be desirable to have a theory in which an ideal crystal would be treated as a particular case of an amorphous solid.

We now examine this problem from the other end. We consider the so-called molecular models of condensed matter. Like crystals a molecule can be represented as a system of electrons traveling in the field of ions. Hence the approach where the crystal is simulated by a certain "large molecule" fragment with corresponding boundary conditions is quite valid. Such a large molecule has been called a cluster while the approach itself is called the cluster approach. The cluster approach utilizes a variety of techniques from quantum chemistry employed to describe the molecular electronic properties. Unlike an ideal crystal, a cluster only has point symmetry, which makes it quite flexible for describing the electronic properties of different types of "nonideal" condensed matter, that is, where translational symmetry is broken. The loss of translational symmetry of the crystal makes the band approach meaningless in the majority of cases at the same time that the loss of symmetry does not present any additional difficulties for clusters of comparatively moderate size. Hence, in analyzing the electronic structure of various types of nonideal condensed matter it is necessary to use an entirely different conceptual approach from that on which the study of the electronic structure of ideal crystals is based. Specifically, the concepts from Bloch's theorem applied to **k** space and used in ideal crystal theory are not suitable for describing nonideal bodies and in this case it is necessary to rely on the cluster approach which is related to real space.

Clusters are not suitable for describing physical phenomena associated with long-range order due to their finite size. Band theory techniques are more suitable for analyzing such processes. At the same time phenomena associated with short-range order can be successfully analyzed by means of the cluster approach. For example, point defects in crystals,[87] localized excitations, self-localized quasiparticles, chemisorption,[88] etc., have been successfully analyzed within the framework of the cluster approach.

The electronic structure of disordered systems is more complex than that of ideal crystals. Since these systems do not have translational symmetry, in addition to the "broad" band states distributed throughout the entire crystal there will also be localized states whose wave functions are concentrated near the lattice sites. Anderson was the first to identify localized states in disordered systems.[89] He formulated a fundamental theorem on the effect of disorder on the excitation spectrum: the wave functions are localized if the disorder "quantity" exceeds a certain defined value. The disorder quantity is given by the ratio W/B, where W is the maximum change in the random potential (the well depth) and B is the bandwidth that results from the overlapping of the wave functions of neighboring atoms. According to Anderson there exists a critical value $(W/B)_{cr}$ at which all states are localized. Anderson calls a degenerate electron gas with wave function localization due to a random field W "Fermi glass."

The density of electronic states $n(E)$ more completely characterizes the electronic structure of disordered systems. Hence in our treatment special attention will be devoted to certain methods of determining $n(E)$. A survey of the primary approaches to calculating the electronic structure of disordered systems has been provided[90] in a collection made by Zakis. In order to assure a comprehensive analysis of the issue we provide a summary table from this study (with some of our additions). Undoubtedly this table is not complete although it provides a comprehensive and accurate reflection of the current state of the problem. The random field concept accounts for the fact that the potential electron energy in a disordered system varies arbitrarily. This is the most common approach to a disordered system.

Below we will focus on certain techniques indicated in the table without, of course, being able to provide a comprehensive treatment of all approaches listed here. Descriptions and the primary conceptual bases of all theoretical approaches to analyzing disordered systems can be found in the references cited.

The method of moments and the recursive method have been widely used in recent years; the density of states in any condensed systems can be determined within the framework of these approaches.

We provide a brief discussion of these methods. The electron state density $n(E)$ is given by the familiar relation[87]

TABLE 2.1. Various methods of electronic structure calculations.

Authors	Fields	Source
Bonch-Burevich, V. L. Lifshits, I. M. Pastur, A. A. Mironov, A. G.	Random-field concept	Refs. 91–93
Anderson, R Mott, N. Weissman, M. Cohan, N. V., *et al.*	Random rectangular potential wells of different depths; development of the Kronig–Penney model	Refs. 93–96
Gubanov, A. I. Kramer, V. Maschke, K. Thomas, L. *et al.*	"Distorted" crystal model	LCAO, APW, OPWM and multiple scattering Green's function theory Refs. 93–100
Soven, P. Schwartz, L. Vedyaev, A. V.	Coherent potential approximation	Refs. 101–106
Edwards, S. F. Beeby, J. L. Ziman, J. Lloyd, D.	Liquid metal model MG approximation	Refs. 107–110
Weaire, D. Thorpe, M. F. Kelly, M. J. Bullett, D. V.	LCAO, random grid structure model, Bethe lattice	Refs. 111–117
Keller, J. McGill, T. C. Klima, J.	The cluster approach applied to amorphous matter. The Lloyd formula for $n(E)$	Refs. 118–120
Gubanov, A. I.	Formula of the cell method (CM) LCAO, and OPW for amorphous matter	CM, LCAD, OPWM Refs. 120–122
Slater, G. Johnson, K. Smith, F.	Molecular clusters	Ref. 123
Joannopoulos, J. D. Cohen, M. L. Schluter, M. Shimitsu, T., *et al.*	LCAO, random grid structure model, empirical pseudopotentials, and first-principle pseudopotentials	Refs. 124–128
Wilson, K. Sadovskiy, M. V. Migdal, A. B. Ma, Sh., Kadanov, L.	Renorm group method	Refs. 129 and 130
Klinger, M. I. Karpov, V. G.	Self-localization theory Critical potentials	Refs. 131 and 132
Stonham, A. M. Evarestov, R. A. Dederix, L. S.	Molecular clusters Electronic structure of point defects	Refs. 133 and 134
Egorushkin, V. E.	Alloys with an arbitrary degree of long-range order	Ref. 135
Ducastelle, F. Cyrot-Lackmann, F. Haydock, R., Heine, V. Kelly, M., Friedel, J.	Method of moments, recursive method	Refs. 136–140
Rahman, A.	Molecular dynamic method, equations of motion	Ref. 141
	Use of electron density functional theory in the molecular dynamics method	Ref. 143

$$n(E)=\frac{1}{N}\sum_i \delta(E-\varepsilon_i)=-\frac{1}{\pi N}\lim_{\eta\to+0}\operatorname{Im}\sum_\alpha G_{\alpha\alpha}(E+i\eta), \tag{2.1}$$

where $|\alpha\rangle$ are the basis orbitals over which the eigenfunctions of the Hamiltonian of the system $|\alpha\rangle\equiv\chi_\alpha(\mathbf{r}-\mathbf{R}_j)$ are summed:

$$H=T+\sum_i V(\mathbf{r}-\mathbf{R}_j). \tag{2.2}$$

$G_{\alpha\alpha}$ is the matrix of Green's function $G(z)$; $z=E+i\eta$ (η is a small positive quantity), and ε_i are the eigenvalues of H. The basis orbitals $\chi_\alpha(\mathbf{r}-\mathbf{R}_j)$ are the solution of the equation

$$[T+V(\mathbf{r}-\mathbf{R}_j)]\chi_\alpha(\mathbf{r}-\mathbf{R}_j)=\varepsilon_0\chi_\alpha(\mathbf{r}-\mathbf{R}_j). \tag{2.3}$$

We introduce the matrix

$$H_{\alpha\alpha'}(\mathbf{R}_i-\mathbf{R}_j)=H_{\alpha\alpha'}(\mathbf{R})=\int d^3\mathbf{r}\,\chi_\alpha^*(\mathbf{r})H\chi_{\alpha'}(\mathbf{r}-\mathbf{R}) \tag{2.4}$$

with the coordinate origin in atom j. We set $\varepsilon=0$ (the energy reference) and limit the analysis to two-center integrals, then

$$H_{\alpha\alpha'}(\mathbf{R})=\int d^3\mathbf{r}\,\chi_\alpha^*(\mathbf{r})V(\mathbf{r})\chi_{\alpha'}(\mathbf{r}-\mathbf{R}), \tag{2.5}$$

$$H_{\alpha\alpha'}(0)=\int d^3\mathbf{r}\,\chi_\alpha^*(\mathbf{r})\left[\sum_{i\neq j}V(\mathbf{r}-\mathbf{R}_j)\right]\chi_{\alpha'}(\mathbf{r}). \tag{2.6}$$

We determine the moments of the density of states:

$$\mu_n=\int_{-\infty}^{\infty} dE\,E^n n(E)=\frac{1}{N}\sum_i \varepsilon_i^n=\frac{1}{N}\operatorname{Sp}\{H^n\}. \tag{2.7}$$

We can obtain

$$\mu_n=\operatorname{Sp}\sum_{\mathbf{R}_1\ldots\mathbf{R}_n} H(\mathbf{R}_1)H(\mathbf{R}_2)\ldots H(\mathbf{R}_n) \tag{2.8}$$

when

$$\sum_{i=1}^{n}\mathbf{R}_i=0. \tag{2.9}$$

Equation (2.8) shows that μ_n is determined by the contribution of all closed configurations of interacting atoms to the sum. The calculation procedure is as follows: we determine the moments μ_n and select a certain parametrized function $f(E)$. We fit its parameters so that the moments calculated with this function are identical to the previously found values of μ_n. The resulting function $f(E)$ is identified with the density of states $n(E)$.

We represent the diagonal element of Green's function as

$$G_{\alpha\alpha}(z)=\langle\alpha|(z-H)^{-1}|\alpha\rangle=\frac{1}{z}\sum_{n=0}^{\infty}\frac{\mu_n(\alpha)}{z^n}, \tag{2.10}$$

where the latter equality is obtained by expanding $(z - H)^{-1}$ into a power series:

$$(z-H)^{-1}=\sum_{n=0}^{\infty}\frac{H^n}{z^{n+1}} \tag{2.11}$$

and by using Eq. (2.8). Carrying out the substitution $\mu_n(a) \to \mu_n(a)/\mu_0(a)$, that is, normalizing the local partial density of states to a single state, we recast Eq. (2.10) as

$$G_{\alpha\alpha}(z)=z^{-1}\left(1+\sum_{n=1}^{\infty}\frac{\mu_n(\alpha)}{z^n}\right). \tag{2.12}$$

The concept of irreducible chains, that is, chains in which the atom A is not an intermediate link is introduced in this technique. All other chains are assumed to be reducible. It can be shown[87] that Eq. (2.10) in terms of the moments of the irreducible chains can take the form

$$G_{\alpha\alpha}(z)=\frac{1}{z-\Sigma_{n=1}^{\infty}\bar{\mu}_n(\alpha)/z^{n-1}}, \tag{2.13}$$

where $\mu_n(\alpha)$ is the moment obtained by summing over only the irreducible chains. If the series in the denominator of Eq. (2.13) is broken off with arbitrary n, the resulting function $G_{\alpha\alpha}(z)$ will not satisfy the following requirements: $G_{\alpha\alpha}$ must be analytic over the entire complex plane z (aside from the poles on the real axis), $\mathrm{Im}\, G_{\alpha\alpha} < 0$ for $\mathrm{Im}\, z > 0$.[138]

These conditions are satisfied by the continued fraction

$$G_{\alpha\alpha}(z)=\cfrac{1}{z-a_0-\cfrac{b_1^2}{z-a_1-\cfrac{b_1^2}{z}\cdots}}. \tag{2.14}$$

The relations between the coefficients a_n and b_n are determined by the formulas[144]

$$a_n=\frac{1}{\Delta'_{n-1}}\left[\frac{\Delta_{n-1}\Delta'_n}{\Delta_n}+\frac{\Delta_n\Delta'_{n-2}}{\Delta_{n-1}}\right], \tag{2.15}$$

$$b_n^3=\frac{\Delta_n\Delta_{n-2}}{\Delta_{n-1}^2}, \tag{2.16}$$

$$\Delta_n=\begin{vmatrix} \mu_0 & \mu_1 & \cdots & \mu_n \\ \mu_1 & \mu_2 & \cdots & \mu_{n+1} \\ \cdots & \cdots & \cdots & \\ \mu_n & \mu_{n+1} & \cdots & \mu_{2n} \end{vmatrix}, \tag{2.17}$$

$$\Delta'_n = \begin{vmatrix} \mu_1 & \mu_2 & \cdots & \mu_{n+1} \\ \mu_2 & \mu_3 & \cdots & \mu_{n+2} \\ \cdots & \cdots & \cdots & \\ \mu_{n+1} & \mu_{n+2} & \cdots & \mu_{2n+1} \end{vmatrix}, \tag{2.18}$$

$$\Delta'_{-2}=0;\quad \Delta'_{-1}=\Delta_{-1}=1. \tag{2.19}$$

Often the recursion method proposed in Haydock's studies[138,145] is used in specific calculations. We will assume that each atom interacts solely with the nearest neighbors. A single function $|u_n\rangle$ from the orthogonal basis set corresponds to each lattice site. If we wish to find the density of states, then $|u_0\rangle$ must coincide with the orbital $|\alpha\rangle$. The recursion routine is carried out as follows. The orthogonal basis $\{u_n\}$ is formulated. In order to satisfy orthogonality of $|u_n\rangle$ to all preceding states the following condition must hold:

$$H|u_n\rangle = a_n|u_n\rangle + b_{n+1}|u_{n+1}\rangle + b_n|u_{n-1}\rangle. \tag{2.20}$$

From the recursion routine we have the sequence of states $\{|u_0\rangle,|u_1\rangle,\ldots,|u_n\rangle\}$ and two sequences of coefficients $\{a_0,a_1,\ldots\},\{b_0,b_1,\ldots\}$. For an arbitrary n for which all $\{a_0,a_1,\ldots,a_{n-1}\},\{b_0,b_1,\ldots,b_n\}$ are known we find from Eq. (2.20):

$$|u_{n+1}\rangle = \frac{1}{b_{n+1}}\,[H|u_n\rangle - a_n|a_n\rangle - b_n|u_{n-1}\rangle]. \tag{2.21}$$

The last relation is obtained by multiplying Eq. (2.20) by $\langle u_n|$ and accounting for the orthonormality of the basis. Multiplying by $\langle u_{n+1}|$ we find

$$b_{n+1} = \langle u_{n+1}|H|u_n\rangle. \tag{2.22}$$

The matrix H in the $\{|u_n\rangle\}$ vector representation is a tridiagonal matrix given as

$$H = \begin{vmatrix} a_0 & b_1 & 0 & \cdots & \\ b_1 & a_1 & b_2 & 0 & \cdots \\ 0 & b_1 & a_2 & b_3 & \cdots \end{vmatrix}. \tag{2.23}$$

From such a Hamiltonian we find

$$G(E) = \cfrac{b_0^2}{E - a_0 - \cfrac{b_1}{E - a_1 - \cdots}}. \tag{2.24}$$

Recasting Eq. (2.24) we get

$$G(E) = \cfrac{b_0^2}{E - a_0 - b_1 \cfrac{\ddots}{\cfrac{b_{n-2}^2}{E - a_{n-1} - b_{n-1}t(E)}}}, \tag{2.25}$$

where $t(E)$ is the cutoff function. For $t(E)=0$ we obtain the ordinary form of the density of states. With the Hamiltonian acting on the vector $|u_m\rangle$ we obtain a linear combination of vectors $|u_m\rangle$, $|u_{n-1}\rangle$, $|u_{n+1}\rangle$, where each is orthogonal to all preceding vectors, that is, the states $|u_n\rangle$ simulate a certain chain whose physics reflects the test system. It is therefore possible to investigate the eigenvectors and eigenvalues of the system using the chain method.

Therefore, in order to calculate the density of states $n(E)$ it is necessary to find the coefficients a_n and b_n by recurrent relation (2.20) and use the formula

$$n(E) = -\frac{1}{\pi}\,\mathrm{Im}\,\frac{1}{E - a_1 - \dfrac{b_1^2}{E - a^3 - b_2^2/(E - a_2\cdots)}}. \tag{2.26}$$

2.2. Methods of calculating the band structure of crystals

To date band theory has been successful for analyzing the electronic structure of crystals. It has become possible to use a variety of techniques to obtain quantitatively consistent results if, of course, the same effective potential is used in each case. This fact indicates that a certain closed structure consisting of several techniques has already been developed that can be called band theory.

These methods differ in the selection of basis functions and the form of the effective crystal potential. There are several powerful techniques in modern solid state band theory: as, for example, the orthogonal plane wave (OPW) method; the augmented plane wave (APW) method; Green's function (GF) method; the linear combination of atomic orbitals (LCAO) method; and the pseudopotential method.[20,146] These techniques can be classified by the type of effective potential used. The most commonly employed APW and GF techniques utilize the "muffin tin" (MT) form for the potential.[147] The OPW and LCAO do not use the MT approximation. In the pseudopotential scheme the effective potential is found by formulating a certain smooth function (pseudopotential) dependent on the parameters. The application of each method is limited not only by the approximation for the effective potential but also by purely technical details related to the selection of the basis functions. For example, although the OPW method also utilizes a common form for the potential, the use of the basis functions in the form of orthogonal plane waves substantially slows the convergence for the d and f states. It is therefore necessary to formulate a so-called combined scheme called the modified orthogonal plane wave (MOPW) method which can be treated as a combination of the orthogonal plane wave method and the linear combination of atomic orbitals method.[148–155] We will characterize each of the band theory methods. We begin the presentation with the Ritz variational method.

2.2.1. Ritz variational method

The problem of finding the eigenfunctions and eigenvalues of the Schrödinger equation with an effective crystal potential reduces to a variational problem of finding the minimum of the functional

$$D[\Psi] = \int_{\Omega_0} d^3r\,\Psi^*(H_{\mathrm{eff}} - \varepsilon)\Psi. \tag{2.27}$$

In a class of functions satisfying the Bloch boundary condition with the auxiliary normalization condition

$$\int d^3r|\Psi(r)|^2=1. \tag{2.28}$$

Let $\varepsilon_1,\varepsilon_2,\ldots,\varepsilon_n$ be the sequence of eigenvalues while Ψ_1, Ψ_2, Ψ_n are the corresponding eigenfunctions of the initial variational problem under the auxiliary condition

$$\int_{\Omega_0} d^3r\,\Psi_i^+\Psi=0, \quad i=1,2,\ldots,n-1. \tag{2.29}$$

We select for any integer valued N an arbitrary system of N linearly independent coordinate functions $\chi_\mu(\mathbf{r})$ and then find the minimum $D[\Psi]$, although not from all permitted functions Ψ, but rather only from the functions

$$\tilde{\Psi}=\sum_{\mu=1}^{N} c_\nu\chi_\mu(\mathbf{r}), \tag{2.30}$$

that is, the function $\tilde{\Psi}$ belonging to the Ritz subspace R, extended over the selected coordinate functions $\chi_1,\chi_2,\ldots,\chi_n$. Thus the initial variational problem is replaced by the following problem: Find the eigenvalue $\tilde{\varepsilon}_n$ and the corresponding eigenfunction $\tilde{\Psi}_n$ such that

$$\tilde{\varepsilon}_n=\min\left(\int_{\Omega_0} d^3r\,\tilde{\Psi}^+H_{\text{eff}}\tilde{\Psi}\Big/\int_{\Omega_0} d^3r\,\tilde{\Psi}^+\tilde{\Psi}\right) \tag{2.31}$$

under the conditions

$$\int_{\Omega_0} d^3r\,\tilde{\Psi}_i^+\tilde{\Psi}_n=0; \quad n=1,2,\ldots,N; \quad i=1,2,\ldots,n-1. \tag{2.32}$$

Then the coefficients C_μ can be determined from the following system of linear homogeneous equations:

$$\sum_{\nu=1}^{N} (H_{\mu\nu}-\tilde{\varepsilon}S_{\mu\nu})c_\nu=0 \quad (\mu,\nu=1,2,\ldots,N) \tag{2.33}$$

while the value $\tilde{\varepsilon}$ can be found from the equations

$$\det|H_{\mu\nu}-\tilde{\varepsilon}S_{\mu\nu}|=0, \tag{2.34}$$

$$H_{\mu\nu}=\int_{\Omega_0} d^3r\,\chi_\mu^+H_{\text{eff}}\chi_\nu, \tag{2.35}$$

$$S_{\mu\nu}=\int_{\Omega_0} d^3r\,\chi_\mu^+\chi_\nu. \tag{2.36}$$

As $N\to\infty$, $\tilde{\varepsilon}_n\to\varepsilon_n$. However, with a finite N, $\tilde{\varepsilon}_n\geqslant\varepsilon_n$.

2.2.2. *The strong coupling method*

Due to the regularity of the periodic structure of the metal the wave function $\Psi_k(\mathbf{r})$ must satisfy the Bloch theorem

$$\Psi_k(\mathbf{r}+\mathbf{R})=e^{i\mathbf{kR}}\Psi(\mathbf{r}). \tag{2.37}$$

The wave functions and energy levels of the inner electrons do not appear to differ significantly in an isolated metal atom or ion. Here the wave function of the internal electrons satisfies condition (2.37). We write the Schrödinger equation for an isolated atom:

$$H_{\mathrm{am}}\Psi_n^{\mathrm{am}}=\varepsilon_n\Psi_n^{\mathrm{am}}, \tag{2.38}$$

where Ψ_n^{am} is the wave function of the coupled level of the Hamiltonian of the atom.

If we construct from the atomic wave functions a combination of the type

$$\Psi_{nk}(\mathbf{r})=\frac{1}{\sqrt{N}}\sum_k e^{i\mathbf{kR}}\Psi_n^{\mathrm{am}}(\mathbf{r}-\mathbf{R}), \tag{2.39}$$

we can easily verify that $\Psi_{n\mathbf{k}}(\mathbf{r})$ satisfies condition (2.37) and conserves the properties of the atomic wave functions Ψ_n^{am} (N denotes the number of unit cells in the crystal).

The matrix elements of the Hamiltonian operator can be given as

$$\int d^3\mathbf{r}\,\Psi_{nk}^*(\mathbf{r})[-\nabla^2+V(\mathbf{r})]\Psi_{mk'}(\mathbf{r})=H_{nm}(\mathbf{k})\Delta(\mathbf{k}-\mathbf{k}'). \tag{2.40}$$

If $\mathbf{k}\neq\mathbf{k}'$ we can easily show that the matrix elements become equal to zero, that is,

$$\Delta(\mathbf{k}-\mathbf{k}')=\frac{1}{N}\sum_R \exp[i(\mathbf{k}-\mathbf{k}')\mathbf{R}]=\begin{cases}1, & \mathbf{k}=\mathbf{k}',\\ 0, & \mathbf{k}\neq\mathbf{k}'.\end{cases} \tag{2.41}$$

The overlap integral also vanishes:

$$\int d^3\mathbf{r}\,\Psi_{nk}^*(\mathbf{r})\Psi_{mk'}(r)=\Delta_{mn}(\mathbf{k})\Delta(\mathbf{k}-\mathbf{k}'). \tag{2.42}$$

Hence the electron wave function in this state $\mathbf{k}$ can be written only through the linear combination $\Psi_{n\mathbf{k}}(\mathbf{r})$ with the same $\mathbf{k}$:

$$\Psi_{\mathbf{k}}(\mathbf{r})=\sum_n c_n(\mathbf{k})\Psi_{n\mathbf{k}}(\mathbf{r}). \tag{2.43}$$

Using the Ritz variational method we can easily obtain[156]

$$\sum_n M_{nm}[E(\mathbf{k}),\mathbf{k}]c_n(\mathbf{k})=0, \tag{2.44}$$

where

$$H_{nm}(E,\mathbf{k})=H_{nm}(\mathbf{k})-E\Delta_{nm}(\mathbf{k}), \tag{2.45}$$

$$H_{nm}(\mathbf{k})=\sum_{\mathbf{R}}\exp(-i\mathbf{kR})h_{nm}(\mathbf{R}), \tag{2.46}$$

$$\Delta_{nm}(\mathbf{k})=\sum_{\mathbf{R}}\exp(-i\mathbf{kR})d_{nm}(\mathbf{R}), \tag{2.47}$$

$$h_{nm}(\mathbf{k}) = \int d^3\mathbf{r}\, \Psi^*_{nk}(\mathbf{r}-\mathbf{R})[-\nabla^2 + V(\mathbf{r})]\Psi_{mk}(\mathbf{r}), \tag{2.48}$$

$$d_{nm}(\mathbf{R}) = \int d^3\mathbf{r}\, \Psi^*_{nk}(\mathbf{r}-\mathbf{R})\Psi_{mk}(\mathbf{r}). \tag{2.49}$$

The roots of the secular equation

$$\det M(E,k) = 0 \tag{2.50}$$

determine the desired energy values. Here it is necessary to diagonalize both H and Δ. Since the atomic functions are localized, the overlap between these functions diminishes significantly with distance and ordinarily the calculations are limited to a certain number of neighbors. Slater and Koster[157] have demonstrated that two-center integrals can be expressed through a moderate number of independent parameters.

The strong coupling method yields a good description of bands below MT zero. The method is useful for describing the energy bands produced by the partially filled d shells of transition metal atoms.

The strong coupling method has nonorthogonality of the basis functions due to the nonorthogonality of the atomic orbitals localized at the different lattice sites. In the limit $V(\mathbf{r}) \to 0$, the method becomes meaningless from the physical viewpoint for testing an empty lattice.

The strong coupling method is best used as an interpolation scheme in which the following is the principal idea. $E(\mathbf{k})$ is calculated by an exact method for certain high symmetry points $\mathbf{k}$. The integrals $h^{nm}(\mathbf{k})$ and $d^{nm}(\mathbf{k})$ are treated as fitting parameters whose quantities are selected to match the energies $E(\mathbf{k})$ calculated by the strong coupling method to those found by other techniques.

Many attempts have been made to enhance the strong coupling method[158,159] proposed using so-called distorted orbitals in place of the atomic orbitals in expansion (2.43); these distorted orbitals are obtained from a solution of the Schrödinger equation with a potential that accounts for the environment of the given atom in the crystal.

A detailed survey of different modifications of the strong coupling method is given by Bullett *et al.*[160] In recent years the strong coupling approximation has been widely used in combination with other techniques. The LCAO coherent potential approximation has been applied by Richter *et al.*[161] to disordered Cu and Ni alloys, while the LCAO scheme was expanded for the relativistic case. It was possible to use nonrelativistic Slater–Koster integrals by Richter and Eschrig.[162]

2.2.3. The orthogonal plane wave (OPW) method and its modification (MOPW)

Functions of the following type are used as the coordinate functions of the Ritz routine in the OPW method:

$$\chi_{G\mu}(r) = \varphi_{G\mu}(r) - \sum_{c=1}^{n} a_c(\mathbf{G}_\mu)\Psi_c(\mathbf{r}), \tag{2.51}$$

where $\varphi_{G\mu}(r)$ are the wave functions of the core; Ψ_c is the plane wave:

$$\varphi_{G\mu}(r)=\frac{1}{\sqrt{\Omega_0}}\exp[i(\mathbf{k}+\mathbf{G}_\mu)\mathbf{r}]\equiv\frac{1}{\sqrt{\Omega_0}}\exp[i\mathbf{G}_\mu\mathbf{r}]. \tag{2.52}$$

The orthogonalization coefficients $a_c(G_\mu)$ are found from the orthogonality condition $\chi_{G\mu}$ on the core functions $\Psi_c(\mathbf{r})$:

$$a_c(\mathbf{G}_\mu)=\int_{\Omega_0} d^3\mathbf{r}\,\Psi_c^*(\mathbf{r})\varphi_{G\mu}(\mathbf{r}). \tag{2.53}$$

The matrix element of the OPW method which is based on symmetrized orthogonal plane waves takes the following form[163]:

$$\begin{aligned}H_{\mu\nu}-\varepsilon S_{\mu\nu}=&\sum_{(\alpha|\tau_\alpha)\in G_{\mathbf{k}_0}} C^j_{\alpha,Gv,l}[(G_\mu^2-\varepsilon)]\delta_{G\mu,\alpha Gv}+V(\mathbf{G}_\mu-\alpha\mathbf{G}_\nu)\\&+\frac{4\pi}{\Omega_0}\sum_s \exp\left[-i(\mathbf{G}_\mu-\alpha\mathbf{G}_\nu)\tau_s\sum_{nl}(2l+1)P_l(\cos\theta_{G\mu,\alpha Gv})\right.\\&\left.\times(\varepsilon-\varepsilon^S_{nl})A^S_{nl}(\mathbf{G}_\mu)A^S_{nl}(\mathbf{G}_\nu)\right],\end{aligned} \tag{2.54}$$

$$V(\mathbf{G}_\mu-\alpha\mathbf{G}_\nu)=\frac{1}{\Omega_0}\int d^3\mathbf{r}\,V(\mathbf{r})\exp[-i(\mathbf{G}_\mu-\alpha\mathbf{G}_\nu)\mathbf{r}]. \tag{2.55}$$

Here the first summation is carried out over all symmetry elements $\{\alpha|\tau_\alpha\}$ belonging to the point group of the wave vector G_{kc} and $C_{\alpha,Gv,1}$ are the symmetrization coefficients. The second summation is carried out over the unit cell atoms. The coefficients A^S_{ne} are determined by the formula

$$A^S_{nl}(\mathbf{G}_\mu)=\int d\mathbf{r}\,j_l(\mathbf{G}_\mu\mathbf{r})P^S_{nl}(\mathbf{r}), \tag{2.56}$$

$j_l(x)$is the spherical Bessel function, $P^S_{nl}(\mathbf{r})$ is the radial wave function of the core shell (nl) of the Sth atom, and $P_l\,(\cos\theta_{G\mu,\alpha Gv})$ is the Legendre polynomial.

In order to calculate the matrix elements it is necessary to know the core states with the following approximations used in the OPW method for calculating these states. First their wave functions are given as the Bloch sums

$$\Psi_c(\mathbf{r})=\sum_p \exp(i\mathbf{k}\mathbf{R}_p)\varphi^S_{nlm}(\mathbf{r}-\mathbf{R}_p-\tau_s), \tag{2.57}$$

where nlm are the atomic numbers, τ_s are the basis vectors indicating the atomic positions in the unit cell, and $\mathbf{R}_p$ is the lattice vector. Second, assuming a core bandwidth sufficiently small to neglect, that is,

$$\varepsilon_c(\mathbf{k})\simeq\varepsilon^S_{nl}, \tag{2.58}$$

$\varphi^S_{nlm}$and ε^S_{nl} are the pseudocore states calculated with the effective crystal potential

$$H^n_{\text{eff}}\Psi_c=\varepsilon_c\Psi_c. \tag{2.59}$$

Equation (2.59) is ordinarily not solved within the framework of an OPW calculation and different approximations are used for the core. In order to improve the convergence of the OPW method and obtain a more reliable band picture, the core states must be found by solving Eq. (2.59).[163]

A very significant drawback of the OPW method is its very weak convergence in calculating the band structure of *d* and *f* elements. This is because the wave functions of the *d* and *f* states are highly localized and, consequently, are very poorly represented by plane waves even if the latter are orthogonal to the core *d* or *f* state. Hence the ordinary OPW method can be used in practice with *s* and *p* elements only. The OPW basis is overcomplete.

The MOPW method was proposed by Herring[148] and further developed later.[149–155] Essentially, this method is a combination of two methods: the OPW and the LCAO. We write the wave function as

$$\Psi_{\mathbf{k}}(\mathbf{r})=\sum_{\mu=1}^{N} a_{\mu}^{(1)}\chi_{\mathbf{G}\mu}^{(1)}(\mathbf{r})+\sum_{s}\sum_{c} a_{sc}^{(2)}\chi_{sc}^{k(2)}(\mathbf{r}), \tag{2.60}$$

where $a_{\mu}^{(1)}$ and $a_{sc}^{(2)}$ are the variational parameters of the Ritz routine. $\chi_{\mathbf{G}\mu}^{(1)}(\mathbf{r})$ is the ordinary OPW, while

$$\chi_{sc}^{\mathbf{k}(2)}(\mathbf{r})=\chi_{sc}^{\mathbf{k}}(\mathbf{r})-\sum_{s''}^{\text{core}}\sum_{c''} b_{sc,s''c''}\Psi_{s''c''}(\mathbf{r}) \tag{2.61}$$

is the Bloch sum orthogonalized to the core. The function

$$\chi_{sc}^{k}(\mathbf{r})=\sum_{p}\exp(i\mathbf{k}\mathbf{R}_p)\Phi_{sc}(\mathbf{r}-\mathbf{R}_p-\tau_s) \tag{2.62}$$

is based on the wave functions $\phi_{sc}(\mathbf{r})$ that accurately describe the valence *d* and *f* states. The orthogonalization coefficients $b_{sc,s''c''}$ are calculated by the formula

$$b_{sc,s''c''}=\int_{\Omega_0} d^3r\,\Psi_{s''c''}^{*}(\mathbf{r}')\chi_{sc}^{\mathbf{k}}(\mathbf{r}). \tag{2.63}$$

Using the variational principle we obtain the determinant equation of the MOPW method

$$\det|\widetilde{H}-\varepsilon\widetilde{S}|=0, \tag{2.64}$$

$$\widetilde{H}=\begin{pmatrix}\widetilde{H}_{\mu\nu}^{\text{OPW}} & \widetilde{H}_{\mu,sc}\\ \widetilde{H}_{\mu,sc}^{+} & H_{s'c',sc}\end{pmatrix}, \tag{2.65}$$

$$\widetilde{S}=\begin{pmatrix}\widetilde{S}_{\mu\nu}^{\text{OPW}} & \widetilde{S}_{\mu,sc}\\ S_{\mu,sc}^{+} & S_{s'c',sc}\end{pmatrix}, \tag{2.66}$$

where $\widetilde{H}_{\mu\nu}^{\text{OPW}}$ and $\widetilde{S}_{\mu\nu}^{\text{OPW}}$ are the matrices of the ordinary OPW method:

$$H_{\mu,sc}=\int d^3r\,\chi_{\mathbf{G}\mu}^{(1)*}(\mathbf{r})H_{\text{eff}}\chi_{sc}^{\mathbf{k}(2)}(\mathbf{r}), \tag{2.67}$$

$$H_{s'c',sc}=\int_{\Omega_0} d^3r\,\chi_{s'c'}^{\mathbf{k}(2)*}(\mathbf{r})H_{\text{eff}}\chi_{sc}^{\mathbf{k}(2)}(\mathbf{r}), \tag{2.68}$$

$$S_{\mu,sc}=\int_{\Omega_0} d^3\mathbf{r}\,\chi_{G\mu}^{\mathbf{k}(1)*}(\mathbf{r})\chi_{sc}^{k(2)}(\mathbf{r}), \tag{2.69}$$

$$S_{s'c',sc}=\int_{\Omega_0} d^3\mathbf{r}\,\chi_{s'c'}^{k(2)*}(\mathbf{r})\chi_{sc}^{k(2)}(\mathbf{r}). \tag{2.70}$$

If no constraints are imposed on the selection of the functions $\phi_{sc}(\mathbf{r})$ then it is necessary to calculate the overlap integrals when calculating the matrix elements by formulas (2.67) and (2.68). Therefore, the MOPW method utilizes the following approximation: matrix elements are calculated by the so-called "cut-off" (co) functions which are selected to accurately describe the valence *d* or *f* states within an atomic sphere of radius $\mathbf{R}_s$ and at the same time are equal to zero outside this sphere. Since these functions are nonzero only in the spherical symmetry range of the potential

$$\Phi_{sc}^{\text{co}}(\mathbf{r})=\frac{1}{r}P_{nl}^{(S)\text{co}}(r)Y_{lm}(\theta,\varphi). \tag{2.71}$$

External integration of the radial Schrödinger equation with a fixed $P_{nl}^{(S)\text{co}}(\mathbf{r})$ is carried out in order to determine the functions $\varepsilon_{nl}^{(S)\text{co}}$. The resulting function $Q_{nl}^{S}(\mathbf{r})$ on the surface of the atomic sphere is nonzero. A certain point $\mathbf{R}_\mu < \mathbf{R}_S$ is then selected; this point is called the joining point and a co function that is nonorthogonal to the core is derived:

$$Q_{nl}^{(S)}(r)=\begin{cases} Q_{nl}^{S}(r), & 0\leqslant r\leqslant R_\mu, \\ b\{1-\cos[q(r-R_s)]\}, & R_\mu\leqslant r\leqslant R_s, \\ 0, & r>R_s. \end{cases} \tag{2.72}$$

The function on the atomic sphere surface is continuous with its first derivative. The parameters *b* and *q* are selected so that $Q_{nl}(\mathbf{r})$ is also continuous with its first derivative at the joining point. Then each co function will be orthogonal to the internal core functions:

$$P_{nl}^{(S)\text{co}}(r)=Q_{nl}^{(S)\text{co}}(r)-\sum_{n'}^{\text{core}} b_{nl,n'l'}^{S}P_{n'l}^{S}(r), \tag{2.73}$$

where

$$b_{nln'l'}^{S}=\delta_{ll'}\int_0^\infty dr\,P_{n'l'}^{S}(r)P_{nl}^{(S)\text{co}}(r). \tag{2.74}$$

The effective crystal Hamiltonian acts on the co function as

$$\hat{H}_{\text{eff}}\Phi_{sc}^{\text{co}}(r)=\frac{1}{r}\varepsilon_{nl}^{(S)\text{co}}G_{nl}^{(S)\text{co}}(r)Y_{lm}(\theta,\varphi). \tag{2.75}$$

Symmetry in the MOPW technique is accounted for by formulating symmetrized combinations of plane waves and symmetrized combinations of spherical harmonics.

2.2.4. The completely orthogonal plane wave (COPW) method

As noted above the OPW basis is overcomplete, which is responsible for the serious drawbacks of the OPW method. For example, the ambiguity of the OPW pseudopotentials is due specifically to the overcompleteness of the basis. Another serious drawback of the OPW is the nonorthogonality of the different orthogonal plane waves. This drawback is manifested in the band calculations. Reser and Dyakin[165] have demonstrated that the errors in calculations of the electronic spectrum of the conductivity and the lattice states are related by

$$\Delta E_{\mathbf{k}} \leqslant |\varepsilon_\alpha| \frac{1-\Delta_{\mathbf{k}}}{\Delta_{\mathbf{k}}}, \tag{2.76}$$

where ε_α is the error in determining E_α and $\Delta_{\mathbf{k}} = \det|\langle \mathrm{OPW}_{\mathbf{k}+\mathbf{G}}|\,|\mathrm{OPW}_{\mathbf{k}+\mathbf{G}}\rangle$ is the determinant of the nonorthogonality matrix. The determinant $\Delta_{\mathbf{k}}$ vanishes with increasing order due to the linear dependence of $\mathrm{OPW}_{\mathbf{k}+\mathbf{G}}$. In the final analysis this will cause significant errors in determining $E_{\mathbf{k}}$.

It is possible to significantly improve the OPW method if we use the ideas developed by Girardson.[166]

Let $\Psi_{nl}^{\mathrm{HF}}(\mathbf{r})$ be solutions of the Hartree–Fock problem for an isolated atom or ion. If we consider the nontransition metals, the functions $\Psi_{nl}^{\mathrm{HF}}(\mathbf{r}-\mathbf{R})$ are highly localized. We can then write

$$\int d^3\mathbf{r}\, \Psi_{nl}(\mathbf{r}-\mathbf{R})\Psi_{n'l'}(\mathbf{r}-\mathbf{R}') = \delta(\mathbf{R}-\mathbf{R}')\delta_{nln'l'}, \tag{2.77}$$

where Ψ_{nl} are the wave functions of the crystal bound states that are not substantially different from Ψ_{nl}^{HF}.

The states $\Psi_{nl}^{\mathrm{HF}}(\mathbf{r}-\mathbf{R})$ form the orthonormalized subspace $\{\alpha\}$ in accordance with relation (2.77).

We introduce the functions $|\mathrm{COPW}\rangle$ (Ref. 146):

$$|\mathrm{COPW}_{\mathbf{k}}\rangle = \mathbf{k} - \sum_{\alpha} c_{\alpha\mathbf{k}}(|\alpha\rangle - |k_\alpha\rangle), \quad \mathbf{k} \neq \mathbf{k}_\alpha. \tag{2.78}$$

Here $|\mathbf{k}_\alpha\rangle = (1/\sqrt{\Omega})\exp[i\mathbf{k}_\alpha \mathbf{r}]$ are plane waves. A certain plane wave $|k_\alpha\rangle$ is set to correspond to each $|\alpha\rangle$. We require orthogonality of the COPW wave functions of the core

$$\langle \alpha | \mathrm{COPW}_{\mathbf{k}} \rangle = 0, \tag{2.79}$$

or

$$\sum_{\alpha'} (\delta_{\alpha\alpha'} - \langle \alpha | \mathbf{k}_{\alpha'} \rangle) c_{\alpha\mathbf{k}} = \langle \alpha | \mathbf{k} \rangle. \tag{2.80}$$

We can easily obtain from Eq. (2.80):

$$\langle \mathrm{COPW}_{\mathbf{k}} | \mathrm{COPW}_{\mathbf{k}'} \rangle = \delta_{\mathbf{k}\mathbf{k}'}, \tag{2.81}$$

that is, unlike the OPW, the COPW are orthogonal.

The COPW basis can be treated as the direct sum of two orthogonal orthonormalized subspaces $\{\alpha\}$ and $\{\mathrm{COPW}\}$. The COPW basis is complete:

$$\sum_{\mathbf{k}}{}' \, |\mathrm{COPW}_{\mathbf{k}}\rangle\langle\mathrm{COPW}_{\mathbf{k}}| + \sum_{\alpha} |\alpha\rangle\langle\alpha| = 1. \tag{2.82}$$

The COPW can be represented as the result of the operation of a certain linear operator L on the plane wave

$$|\mathrm{COPW}(\mathbf{k})\rangle = L|k\rangle, \tag{2.83}$$

where L is the operator given in the plane wave space

$$L = 1 - S + SQ. \tag{2.84}$$

I is the unitary operator:

$$S = \sum_{\alpha} (|\alpha\rangle - |\mathbf{k}_{\alpha}\rangle) c_{\alpha}, \tag{2.85}$$

$$Q = \sum_{\alpha} |\mathbf{k}_{\alpha}\rangle\langle\mathbf{k}_{\alpha}|. \tag{2.86}$$

Operator L transforms the subspace $\{\mathbf{k}\}$ into the system of functions $|\mathrm{COPW}(\mathbf{k})\rangle$.

As a result the COPW basis is complete and orthonormal and the wave function of the valence electron is unambiguously summed over the COPW:

$$\Psi_{\mathbf{k}}(\mathbf{r}) = \sum c(\mathbf{k} + \mathbf{G}_i) L |\mathbf{k} + \mathbf{G}_i\rangle. \tag{2.87}$$

Z. A. Gurskiy generalized the COPW method for transition metals.[167,168] It was proposed that the electrons be divided into three groups: the core states, the states of the d shells, and the valence s and p states. The COPW basis in this case can be represented as the sum

$$\{\alpha\} + \{d\} + \{\mathrm{COPW}(\mathbf{k})\}, \tag{2.88}$$

where $\{\alpha\}$ is the subspace of the wave functions of the inner electrons and $\{d\}$ is the subspace of orthonormalized d states of the outer shells of the isolated ion. The wave function of the valence electron in the crystal in terms of this basis takes the form

$$\Psi_{\mathbf{k}}(\mathbf{r}) = \sum_{i} c(\mathbf{k} + \mathbf{G}_i) L |\mathbf{k} + \mathbf{G}_i\rangle + \sum_{d_j} a_{d_j} |i\tilde{d}_j\rangle, \tag{2.89}$$

where

$$|\tilde{d}_j\rangle = |d_j\rangle - \frac{1}{2} \sum_{i} \beta_{ij} |d_i\rangle. \tag{2.90}$$

β_{ij} is the overlap integral for d orbitals of the types σ, π, and δ. Ordinarily summation over i is limited to two coordinate shells. Then using the Ritz routine it is possible to obtain a secular equation that is suitable for both simple and transition metals.

2.2.5. The KKR method [Green's function (GF) method]

The formalism of the KKR method proposed by Korringa[170] and Kohn and Rostocker[169] is based on multiple scattering theory[176] which essentially first appeared as early as the studies by Rayleigh (1892). We will not outline the primary concepts of multiple scattering theory, assuming the reader is familiar with these principles. We represent the crystal potential as the sum of the MT potentials

$$V(\mathbf{r})=\sum_n \mathrm{v}(\mathbf{r}-\mathbf{R}_n). \tag{2.91}$$

Using a spherically symmetrical potential approximation

$$\mathrm{v}(\mathbf{r}-\mathbf{R}_n)=\mathrm{v}(|\mathbf{r}-\mathbf{R}_n|). \tag{2.92}$$

We assume that the wave at this point is a superposition of waves from all scatterers[156]:

$$\Psi(\mathbf{r})=\sum_n \Psi_n^0(\mathbf{r}). \tag{2.93}$$

Here a wave striking a scatterer labeled n is formed from all waves from the remaining centers:

$$\Psi_n^{(i)}=\sum_{p\neq n} \Psi_p^0(\mathbf{r}). \tag{2.94}$$

The wave reflected off the scatterer and the incident wave are related by the relation

$$\Psi_n^0(\mathbf{r})=\int\int d^3\mathbf{r}'\, d^3\mathbf{r}''\, G_0(E,\mathbf{r}',\mathbf{r}'')t_n(E,\mathbf{r}',\mathbf{r}'')\Psi_n^i(\mathbf{r}''), \tag{2.95}$$

where $G_0(E,\mathbf{r},\mathbf{r}')$ is the Green's function of the free particles:

$$G_0(E,\mathbf{r},\mathbf{r}')=-\frac{1}{4\pi}\frac{\exp(ik|\mathbf{r}-\mathbf{r}'|)}{|\mathbf{r}-\mathbf{r}'|}. \tag{2.96}$$

The function $t_n(E,\mathbf{r},\mathbf{r}')$ describes the scattering of the electron wave of potential $\mathrm{v}(\mathbf{r}-\mathbf{R}_n)$ and is called the t matrix.

In the case of a local potential $\mathrm{v}(\mathbf{r})$, $t_n(E,\mathbf{r},\mathbf{r}')$ satisfies the integral equation

$$t_n(E,\mathbf{r},\mathbf{r}')=\mathrm{v}(\mathbf{r}-\mathbf{R}_n)\delta(\mathbf{r}-\mathbf{r}')+\mathrm{v}(\mathbf{r}-\mathbf{R}_n)\int d^3r\, G_0(E,\mathbf{r},\mathbf{r}')t_n(E,\mathbf{r},\mathbf{r}'). \tag{2.97}$$

The functions $\Psi_n^i(\mathbf{r})$ and $\Psi_n^0(\mathbf{r})$ can be expanded in spherical harmonics:

$$\Psi_n^i(\mathbf{r})=\sum_{l,m} a_{lmj_l}^n(kr_n)Y_{lm}(\mathbf{r}_n), \tag{2.98}$$

$$\Psi_n^0(\mathbf{r})=\sum_{l,m} b_{lm}^n h_l^+(kr_n)Y_{lm}(\mathbf{r}_n), \tag{2.99}$$

where $j_l(\mathbf{kr}_n)$ is the spherical Bessel function, $h_l^+(\mathbf{kr}_n)$ is a spherical Hankel function corresponding to a diverging wave, and Y_{lm} are the spherical harmonics.

Substituting relations (2.98) and (2.99) into Eq. (2.95) it is possible to find the relation between the coefficients a^n_{lm} and b^n_{lm} (Ref. 156):

$$b^n_{lm} = -ikt_l a^n_{lm}, \tag{2.100}$$

where the elements of the t matrix t_l take the form

$$t_l = -\frac{1}{k}\exp(i\eta_l)\sin\eta_l. \tag{2.101}$$

$n_l(E)$ is the scattering phase shift that can be found from the formula

$$\operatorname{tg}\eta_l(E) = \frac{j_l'(\sqrt{E}r_{\text{MT}}) - j_l(\sqrt{E}r_{\text{MT}})\lambda_l(E)}{n_l'(\sqrt{E}r_{\text{MT}}) - n_l(\sqrt{E}r_{\text{MT}})\lambda_l(E)}, \tag{2.102}$$

where $\lambda_l(E)$ are the lognormal derivatives and r_{MT} is the radius of the MT sphere.

The relation between b^n_{lm} and a^n_{lm} will be different when Eqs. (2.98) and (2.99) are substituted into Eq. (2.94):

$$\sum_{lm} a^n_{lm} j_l(\mathbf{kr}_n) Y_{lm}(\mathbf{r}_n) = \sum_{p\neq n}\sum_{l'm'} b^p_{l'm'} h^+_{l'}(kr_p) Y_{l'm'}(\mathbf{r}_p). \tag{2.103}$$

The functions are singular for $\mathbf{r} = \mathbf{R}_p$. However, in the neighborhood of these points they can be expanded into Bessel functions:

$$-ikh^+_{l'}(kr_p) Y_{l'm'}(\mathbf{r}_p) = \sum_{lm} j_l(kr_n') Y_{lm}(\mathbf{r}_n) g^{np}_{lml'm'}, \tag{2.104}$$

where the coefficients $g^{np}_{lml'm'}$ are determined by the formula

$$g^{np}_{lml'm'} = -4\pi iki^{(l-l')}\sum_{LM} i^{-L} c^{LM}_{lml'm'} h^+_L(kR_{np}) Y_{lm}(\mathbf{R}_{np}). \tag{2.105}$$

$\mathbf{R}_{np} = \mathbf{R}_p - \mathbf{R}_n$, $c^{\text{LM}}_{lml'm'}$ are called the Gaunt factors:

$$c^{LM}_{lml'm'} = \int_0^{2\pi}\int_0^{\pi} d\theta\, d\psi\, Y_{LM}(\theta,\psi) Y^+_{lm}(\theta,\psi) Y_{l'm'}(\theta,\psi)\sin\theta. \tag{2.106}$$

From relation (2.96) we find

$$-ika^n_{lm} = \sum_{p\neq n} g^{np}_{lml'm'} b^p_{l'm'}. \tag{2.107}$$

Using the Bloch theorem

$$a^p_{lm} = \exp(i\mathbf{kR}_{np}) a^n_{lm}, \tag{2.108a}$$

$$b^p_{lm} = \exp(i\mathbf{kR}_{np}) b^n_l, \tag{2.108b}$$

we determine the energy-dependent coefficient $G_{lml'm'}(E,\mathbf{k})$:

$$G_{lml'm'}(E,\mathbf{k}) = \sum_{p\neq n}\exp(i\mathbf{kR}_{np}) g^{np}_{lml'm'}. \tag{2.109}$$

Then

$$-i\mathbf{k}a^n_{lm} = \sum_{l'm'} G_{lml'm'} b^n_{l'm'}. \tag{2.110}$$

Relations (2.107) and (2.110) are satisfied simultaneously if

$$\sum_{l'm'} [t_l^{-1}\delta_{ll'}\delta_{mm'} - G_{lml'm'}(E,\mathbf{k})] b^n_{l'm} = 0. \tag{2.111}$$

Then we derive the intermediate calculations for $n=0$ using the relation t_l^{-1} and ctg η_l (Ref. 156):

$$t_l^{-1} = -k \operatorname{ctg} \eta_l + ik \tag{2.112}$$

and we substitute Eq. (2.112) into Eq. (2.111) and introduce new coefficients:

$$A_{lmi'm'}(E,\mathbf{k}) = G_{lmi'm'}(E,\mathbf{k}) - ik\delta_{ll'}\delta_{mm'}, \tag{2.113}$$

which are called the structure constants. We obtain

$$\sum_{l'm'} [A_{lmi'm'}(E,\mathbf{k}) + k \operatorname{ctg} \eta_l \delta_{ll'}\delta_{mm'}] b_{l'm'} = 0. \tag{2.114}$$

The equation found here is also the primary equation in the KKR method. The primary field of applicability of the KKR method is to metals and alloys.

This completes our brief survey of several classical techniques in band theory. Our intent here was not to provide comprehensive coverage to all methods (e.g., the APW method with the same degree of accuracy as the KKR method, etc.). A rather complete analysis of all techniques can be found in the cited literature. We will now describe the most effective modern methods: linear methods.

2.3. Linear methods in band theory

The band theory of crystals can be divided into two stages based on historical development. The first stage is the *computational* implementation of techniques that largely were developed prior to the computer era. This period can also be called the era of nonself-consistent calculations. The second stage is the period of self-consistent calculations. Historically this existed from the late 1960s through the early 1970s.

The significant difficulties arising from self-consistent calculations using the APW and GF methods led to the development of linear schemes in 1976: the linear MT orbital (LMTO) method and the linear augmented plane wave (LAPW) method. The LMTO method is still called the linear analog of the GF. Linear methods have proven to be a powerful stimulus for development of self-consistent calculations.

An overall analysis of techniques from band theory suggests that if the nonlinear energy dependence of the matrix elements is eliminated from the APW and GF methods these techniques could be quite convenient in the sense of self-consistent calculations and could cover virtually all metals and a broad range of chemical compounds. The essence of the linear methods reduces to the idea voiced as early as 1967 by Marcus[172] where the energy derivatives of the radial wave functions are added to the basis. Then the problem of searching the energy eigenvectors and eigenvalues reduces to a single matrix diagonalization, and the problem of poles does not arise in the LAPW method.

2.3.1. The LMTO method

The LMTO method constructs energy-independent MT orbitals.[173] The basis radial wave function is represented as a Taylor series expansion near a certain fixed energy ε_i accurate to the quadratic terms, that is,

$$R_l(D,r) = R_l(\varepsilon_i,r) + \omega_l(D)\dot{R}_l(\varepsilon_i,r), \qquad (2.115)$$

where $D \equiv (dR_l/dr)R_l^{-1}$ is the logarithmic derivative $\dot{R}_l = \partial R_l/\partial\varepsilon$ and

$$\omega_l(D) = -\frac{R_i(\varepsilon_i,r)}{\dot{R}_l(\varepsilon_l,r)}\frac{D-D_i}{D-\dot{D}_i}, \qquad (2.116a)$$

$$D_i = R_s \cdot R_i'(\varepsilon_i,R_s)/R_l(\varepsilon_i,R_s), \qquad (2.116b)$$

$$\dot{D}_i = R_s \cdot \dot{R}_i'(\varepsilon_i,R_s)/\dot{R}_l(\varepsilon_i,R_s). \qquad (2.116c)$$

The MT orbital in the Andersen approximation takes the form

$$\chi_L^{\mathbf{k}}(\mathbf{r},D) \cong \frac{\omega_l(l) - \omega_l(D)}{\omega_l(l) - \omega_l(-l-1)}\chi_L^{\mathbf{k}}(r) \equiv \alpha_l(D)\chi_L^{\mathbf{k}}(\mathbf{r}), \qquad (2.117)$$

where the MT orbital $\chi_L^{\mathbf{k}}(\mathbf{r})$ is energy independent,

$$\chi_L^{\mathbf{k}}(\mathbf{r}) = R_l(\mathbf{r} - l - 1) - R_l(R_s, -l-1)\sum_{L'} S_{LL'}^{\mathbf{k}} \frac{R_{L'}(\mathbf{r},l')}{2(2l'+1)R_{l'}(R_s,l')}. \qquad (2.118)$$

Here $S_{LL'}^{\mathbf{k}}$ are the energy-independent structure constants.

After the MT orbitals are constructed we arrive at the secular equation of the LMTO method in the atomic sphere approximation (LMTO-ASA)[146] completely analogous to the LCAO method:

$$\det|H_{L'L}^{\mathbf{k}} + \varepsilon O_{L'L}^{\mathbf{k}}| = 0, \qquad (2.119)$$

where

$$H_{L'L}^{\mathbf{k}} = H_{l'}^{(1)}\delta_{L'L} + \left[-(H_{l'}^{(2)} + H_l^{(2)})S_{L'L}^{\mathbf{k}} + \sum_{L''} S_{L'L}^{\mathbf{k}} H_{L''}^{(3)} S_{L''L}^{\mathbf{k}}\right]$$
$$\times \frac{R_s}{2} R_{l'}(R_s, -l'-1)R_l(R_s, -l-1), \qquad (2.120a)$$

$$O_{L'L}^{\mathbf{k}} = O_{l'}^{(1)}S_{L'L} + \left[-(O_{l'}^{(2)} + O_l^{(2)})S_{L'L}^{\mathbf{k}} + \sum_{L''} S_{L'L}^{\mathbf{k}} O_{l''}^{(3)} S_{L''L}^{\mathbf{k}}\right]$$
$$\times \frac{R_s}{2} R_l(R_s - l' - 1)R_l(R_s, -l-1), \qquad (2.120b)$$

$$O_l^{(1)} = 1 + \langle \dot{R}_l^2(\varepsilon_v,R_s)\rangle \omega_l^2(-l-1), \qquad (2.120c)$$

$$O_l^{(2)}=\frac{1+\langle \dot{R}_l^2(\varepsilon_\nu,R_s)\rangle\omega_l(-l-1)\omega_l(l)}{\omega_l(-l-1)-\omega_l(l)}, \tag{2.120d}$$

$$O_l^{(3)}=\frac{1+\langle \dot{R}_l^2(\varepsilon_\nu,R_s)\rangle\omega_l^2(l)}{2R_s[(2l+1)R_l(\varepsilon_\nu,R_s)]^2}, \tag{2.120e}$$

$$H_l^{(1)}=\omega_l(-l-1)+\varepsilon_\nu O_l^{(1)}, \tag{2.121}$$

$$H_l^{(2)}=\frac{1}{2}+\frac{\omega_l(l)}{(\omega_l(-l-1)-\omega_l(l))}+\varepsilon_\nu O_l^{(2)}, \tag{2.122}$$

$$H_l^{(3)}=\frac{\omega_l(l)}{2R_s[(2l+1)R_l(\varepsilon_\nu,R_s)]^2}+\varepsilon_\nu O_l^{(3)}. \tag{2.123}$$

In the equations outlined above terms with $S^0_{L'L}$, $S^1_{L'L}$, and $S^2_{L'L}$ in Eq. (2.117) can be treated as one-, two-, and three-center integrals, respectively; such integrals are much easier to evaluate in the LMTO-ASA compared to the LCAO method due to the successful selection of basis functions (2.115).

If the atomic sphere approximation is neglected and we go over to the tangential MT spheres, we arrive at the LMTO method. Details on this method are described by Andersen.[173]

2.3.2. The LAPW method

A linear approach to calculating the band structure by the APW method has been proposed by Andersen[173] as well as by Koelling and Arbman.[174] In the LAPW method the problem of finding the electronic structure is reduced to single diagonalization of matrix (2.34) while the problem of poles does not arise in the Koelling and Arbman formulation.[174]

The LAPW method, like the LMTO method, introduces an energy derivative of the radial wave function $\dot{R}_l$ that satisfies the following equation:

$$\left[\frac{1}{r}\frac{d^2}{dr^2}r-\frac{l(l+1)}{r^2}-v_{\text{eff}}(r)+\varepsilon\right]\dot{R}_l(\varepsilon,r)=R_l(\varepsilon,r). \tag{2.124}$$

We require that

$$\int_0^{R_s} dr\, r^2R_l^2(\varepsilon,r)=1. \tag{2.125}$$

Then

$$\int_0^{R_s} dr\, r^2\dot{R}_l(\varepsilon,r)R_l(\varepsilon,r)=0. \tag{2.126}$$

The function $\dot{R}_l$ generally is nonnormalized:

$$\int_0^{R_s} dr\, r^2\dot{R}_l^2(\varepsilon,r)=N_l. \tag{2.127}$$

The normalization condition (2.127) in the LAPW method can be replaced by the following equation:

$$R_s^2[R_l'(\varepsilon_i,R_s)\dot{R}_l(\varepsilon_i,R_s)-R_l(\varepsilon_i,R_s)\dot{R}_l'(\varepsilon_i,R_s)]=1. \tag{2.128}$$

Equation (2.128) is exact and hence can serve as a good test for determining the numerically found functions R_l and $\dot{R}_l$.

We now fix a certain energy ε_i and formulate the energy-independent coordinate functions of the LAPW method:

$$\chi_{\mathbf{k}_\mu}(r)=\begin{cases}\sum\limits_{lm}[A_{lm}(\mathbf{k}_\mu)R_l(\varepsilon_i,\rho)+B_{lm}(\mathbf{k}_\mu)\dot{R}_l(\varepsilon_i,\rho)]Y_{lm}(\rho), |\rho|<R_s, \\ \exp(i\mathbf{k}_\mu r), \quad |\rho|>R_{s\cdot i}\end{cases}. \tag{2.129}$$

Unlike the APW method, the coefficients A_{lm} and B_{lm} are found based on the coincidence of two parts of the LAPW function and its first derivatives on the MT spherical surface. Therefore, the discontinuity of the normal derivatives is eliminated in the LAPW method. After joining we obtain the following final expression for the LAPW function:

$$\chi_{\mathbf{k}_\mu}(r)=\begin{cases}4\pi R_s^2\exp(i\mathbf{k}_\mu\tau_s)\sum\limits_{lm}i^l[a_l^{(S)}(k_\mu)R_l(\varepsilon_i,\rho) \\ \quad +b_l^{(S)}(k_\mu)\dot{R}_l(\varepsilon_i,\rho)]Y_{lm}^*(\mathbf{k}_\mu)Y_{lm}(\rho) \text{ where } |\rho|<R_s, \\ \exp[i\mathbf{k}_\mu(\rho+\tau_s)] \text{ where } |\rho|>R_s,\end{cases} \tag{2.130}$$

where

$$a_l^{(S)}(k_\mu)=j_l'(k_\mu R_s)\dot{R}_l(\varepsilon_i,R_s)-j_l(k_\mu R_s)\dot{R}'(\varepsilon_i,R_s) \tag{2.131}$$

and

$$b_l^{(S)}(k_\mu)=j_l(k_\mu R_s)R_l'(\varepsilon_i,R_s)-j_l'(k_\mu,R_s)R_l(\varepsilon_i,R_s). \tag{2.132}$$

Applying the variational principle for the functional (2.27) with the coordinate function (2.130) we find the determinant equation of the LAPW method:

$$\det|\Omega_0^{-1}(H-\varepsilon O)_{\mu\nu}^j|=0. \tag{2.133}$$

Here

$$\begin{aligned}\Omega_0^{-1}(H-\varepsilon O)_{\mu\nu}^j= &\sum_{\{\alpha|\tau_\alpha\}\in G_{\mathbf{k}_0}} c_{\alpha\mathbf{k}\nu,1}^j\Bigg\{(k_\nu^2-\varepsilon)\delta_{\mathbf{k}\mu,\alpha\mathbf{k}\nu}+\frac{4\pi}{\Omega_0}\sum_s R_s^2 \\ &\times\exp[-i(\mathbf{k}_\mu-\alpha\mathbf{k}_\nu)\tau_s]\Bigg[R_s^2\sum_{l=0}^{l_{\max}}(2l+1)P_l \\ &\times(\cos\theta_{\mathbf{k}_\mu,\alpha\mathbf{k}_\nu})(a_l^{(S)}(k_\mu)a_l^S(k_\nu)+b_l^{(S)}(k_\mu)b_l^{(S)}(k_\nu)N_l^{(S)} \\ &\times(\varepsilon_i-\varepsilon)+a_l^{(S)}(k_\mu)b_l^{(S)}(k_\nu)) \\ &-(k_\nu^2-\varepsilon)\frac{j_1(|\mathbf{k}_\mu-\alpha\mathbf{k}_\nu|R_s)}{|\mathbf{k}_\mu-\alpha\mathbf{k}_\nu|}\Bigg]\Bigg\}.\end{aligned} \tag{2.134}$$

As we see from Eqs. (2.133) and (2.134), the secular equation of the LAPW method is linear in ε and contains no poles. Moreover, l_{max} in Eq. (2.134) is less than in the APW method. Ordinarily it is sufficient to use $l_{max} = 7$–9 at the same time that in the APW method $l_{max} = 12$. The convergence of the LAPW method in the coordinate functions is approximately the same as in the APW method, that is, approximately 50 functions per atom. At first glance it may seem that difficulties will arise in the selection of ε_l when using the LAPW method. However, it turned out that the results obtained by the LAPW method as well as the LMTO method are weakly dependent on ε_l over a rather broad energy range.

The LAPW method can be generalized without any special difficulties to the relativistic case. Here it becomes much more complex in terms of calculations. It is sufficient to point out that in this case the LAPW matrices become complex and they double in size. A two-stage self-consistent relativistic approach (SRA) is used in this case to calculate the electronic structure. The SRA is used in the first stage of calculations to obtain the self-consistent effective crystal potential. Only the large component of the wave function dependent on the quantum numbers l and σ is used in the band calculations and the formalism of the LAPW method does not change in any way. The spin–orbit interaction is introduced at the end of the calculation when self-consistency has already been achieved. We briefly discuss the implementation of each stage.

Stage 1. We describe the SRA as applied to atomic structure calculations in the first chapter. In order to use the SRA in the LAPW method it is necessary to have a system of equations in addition to the equation system (1.94) to find the energy derivative of the large and small components. These equations can easily be obtained by derivation of Eq. (1.194) with respect to energy:

$$\frac{d\dot{P}_l}{dr} - \frac{\dot{P}_l}{r} = 2Mc\dot{Q}_l + \frac{1}{c}\,Q_l,$$

$$\frac{d\dot{Q}_l}{dr} + \frac{Q}{r} = \left[\frac{l(l+1)}{2Mcr^2} + \frac{1}{c}\,(\mathrm{v}_{\text{eff}} - \varepsilon)\right]\dot{P}_l - \left[\frac{l(l+1)}{(2Mcr)^2} + 1\right]\frac{1}{c}\,P_l. \quad (2.135)$$

For the valence states the normalization condition will be as follows:

$$\int_0^{R_s} dr(P_l^2 + Q_l^2) = 1. \quad (2.136)$$

In this case the orthogonality condition (2.126) is given as

$$\int_0^{R_s} dr(P_l\dot{P}_l + Q_l\dot{Q}_l) = 0. \quad (2.137)$$

The normalization constant N_l is calculated by the formula

$$N_l = \int_0^{R_s} dr(\dot{P}_l^2 + \dot{Q}_l^2). \quad (2.138)$$

The large component $g_l = P_l/r$ is used as the wave function in the SRA and the entire subsequent structure of the LAPW method remains unchanged. Since condition (2.128) will not hold in this case, coefficients (2.131) and (2.132) must be divided by $c = R_S^2(\dot{g}g_l' - g\dot{g}_l')$ and Eq. (2.134) is then used to calculate the matrix elements in the SRA. The eigenvalues and vectors are found by diagonalizing matrix (2.133). The algorithm for solving the generalized eigenvalue problem is described in Refs. 174 and 175.

It is necessary to calculate the electron density in each iteration in order to achieve self-consistency. We consider a crystal electron density averaged over the angles within the Sth MT sphere:

$$\rho_k^{(S)j}(r)=\frac{3\Omega_s R_s}{t_0(s)}\sum_{m=1}^{z}\sum_{n=1}^{z} v^{j*}(k_m)v^j(k_n)\sum_{q=1}^{q_0}\sum_{p=1}^{p_0} c_{mq}^{j*}c_{np}^{j}\sum_{t=1}^{t_0(s)}\exp[i(k_{np}$$
$$-k_{mq})\tau_{st}]\sum_{l=0}^{l_{\max}}(2l+1)P_l(\cos\theta_{k_{mq},k_{np}})[a_l^{(S)}(k_{m1})a_l^{(S)}(k_{n1})g_l^{(S)}(r)$$
$$+a_l^{(S)}(k_{m1})b_l^{(S)}(k_{n1})+a_l^{(S)}(k_{n1})b_l^{(S)}(k_{m1})g_l^{(S)}(r)\dot{g}_l^{(S)}(r)$$
$$+b_l^{(S)}(k_{m1})b_l^{(S)}(k_{n1})\dot{g}_l^{(S)2}(r)]. \tag{2.139}$$

We use the following designations. The symmetrized combinations of plane waves transformed in the jth irreducible representation are numbered in increasing order of their modulus (m, $n = 1,2,...,z$). Here p and q label the plane waves in the sets and $t_0(S)$ designates the number of S-type atoms. For clarity purposes a new designation, $v^{(j)}(k_n)$, is introduced in Eq. (2.139) for the eigenvectors; $\Omega_S = 4\pi R_S^3/3$ is the volume of the MT sphere.

The complete wave function will be normalized:

$$N_{\mathbf{k}}^j=\sum_{m=1}^{z}\sum_{n=1}^{z} v^{j*}(k_m)v^j(k_n)\int d^3r\,\chi_{\mathbf{k}_{m1}}^{j*}(\mathbf{r})\chi_{\mathbf{k}_{n1}}^{j}(\mathbf{r})$$
$$=\sum_{m=1}^{z}\sum_{n=1}^{z} v^{j*}(\mathbf{k}_m)v^j(\mathbf{k}_n)S_{mn}^j, \tag{2.140}$$

that is, is expressed through the overlap matrix 0.

We then introduce the following designations for convenience:

$$c_{mnl}^{(S)j}(\mathbf{k})=\sum_{q=1}^{q_0}\sum_{p=1}^{p_0} c_{mq}^{j*}c_{np}^{j}\left(\sum_{t=1}^{t_{0cs}}\exp[i(\mathbf{k}_{np}-\mathbf{k}_{mq})\tau_{st}]\right)P_l(\cos\theta_{\mathbf{k}_{mq},\mathbf{k}_{np}}), \tag{2.141}$$

$$H_{\mathbf{k},l}^{(S)j}=\sum_{m=1}^{z}\sum_{n=1}^{z} v^{j*}(k_m)v^j(\mathbf{k}_n)c_{mnl}^{(S)j}a_l^{(S)}(\mathbf{k}_{m1})a_l^{(S)}(k_n), \tag{2.142}$$

$$G_{\mathbf{k},l}^{(S)j}=\sum_{m=1}^{z}\sum_{n=1}^{z} v^{j*}(\mathbf{k}_m)v^j(\mathbf{k}_n)c_{mnl}^{(S)j}[a_l^{(S)}(k_{m1})b_l^{(S)}(k_{n1})+a_l^{(S)}(k_{n1})b_l^{(S)}(k_{m1}), \tag{2.143}$$

$$F_{\mathbf{k}}^{(S)j}=\sum_{m=1}^{z}\sum_{n=1}^{z} v^{j*}(\mathbf{k}_m)v^j(\mathbf{k}_n)c_{mnl}^{(S)j}b_l^{(S)}(k_{m1})b_l^{(S)}(k_{n1}). \tag{2.144}$$

Then the crystal electron density within the Sth MT sphere will take the form

$$\rho_{ar}^{S}(r)=\sum_{k}\sum_{j}^{occ} W^{j}\rho_{\mathbf{k}}^{(S)j}(r)$$

$$=\frac{3\Omega_s R_s}{t_0(s)}\sum_{i=0}^{i_{max}}(2l+1)\Big[g_l^{(S)2}(r)\sum_{\mathbf{k}}\sum_{j}^{occ} W_{\mathbf{k}}^{j}H_{\mathbf{k},l}^{(S)j}/N_{\mathbf{k}}^{j}+g_l^{(S)}(r)\dot{g}_l^{(S)}(r)$$

$$\times\sum_{k}\sum_{j}^{occ} WG_{\mathbf{k},l}^{(S)j}/N_{\mathbf{k}}^{j}+g_l^{(S,2)}(r)\sum_{\mathbf{k}}\sum_{j}^{occ} W_{\mathbf{k}}^{j}F_{\mathbf{k},l}^{(S)j}/N_{\mathbf{k}l}^{j}\Big], \quad (2.145)$$

where $W_{\mathbf{k}}^{j}$ are the weight multipliers. If we introduce the partial densities $\rho_l^S(r)$ into j the analysis, then

$$\rho_{ar}^{S}(r)=\sum_{l=0}^{l_{max}}\rho_l^S(r). \quad (2.146)$$

The electron charge within the Sth MT sphere is equal to

$$Q^S=\sum_{l_0}^{l_{max}}Q_l^S, \quad (2.147)$$

where

$$Q_l^S=\int_0^{R_s} dr\, r^2\rho_l^S(r). \quad (2.148)$$

The partial charge in the MT sphere for each vector **k** can be obtained if we integrate the equation for $\rho_{\mathbf{k},l}^{(S)j}$:

$$Q_{\mathbf{k},l}^{(S)j}=\frac{3\Omega_s R_s}{t_0(s)}(2l+1)(H_{\mathbf{k},l}^{(S)j}+N_l^S F_{\mathbf{k}l}^{(S)j}). \quad (2.149)$$

The total electron density within the Sth MT sphere takes the following form:

$$\rho^S(r)=\frac{1}{4\pi r^2}\sum_{nl}^{core}2(2l+1)\left\{Q_l^{(S)2}(r)+\left[\frac{l(l+1)}{(2M^{(S)}cr)^2}+1\right]P_l^{(S)2}(r)\right\}+\rho_{ar}^{(S)}(r). \quad (2.150)$$

After the electron density is found the crystal potential is formulated and the band structure as well as the electron density are again calculated in the new iteration. The core is recalculated with the new crystal potential in each iteration. Ordinarily the following approximation is used as the zeroth approximation for the electron density:

$$\rho^{(S)}(r)=\sum_{R_p}\rho_{ar}^{S}(|\mathbf{r}-\mathbf{R}_p-\tau|). \quad (2.151)$$

It is possible to calculate the atomic electron density $\rho_{am}(r)$ using, for example, the complex computer program set called "RATOM."[56]

Stage II. After the self-consistent potential is obtained a complete relativistic calculation of the band structure is carried out. The method[177] which will be given below is superior to standard techniques in terms of calculation speed and occupies very little computer memory. We will briefly describe its implementation.

The crystal wave function obtained from the SRA calculation in the first stage takes the form

$$\varnothing^{\mathbf{k}}_{n,\sigma}(\mathbf{r}) = \sum_{\mu=1}^{N} c^{\mathrm{SRA}}_{\mu}(\varepsilon_n)\chi_{\mathbf{k}\mu,\sigma}(\mathbf{r}). \tag{2.152}$$

We formulate the new wave function:

$$\Psi_{\mathbf{k},\sigma}(\mathbf{r}) = \sum_{n=1}^{M} \mathrm{v}_n(\mathbf{k})\varnothing^{k}_{n,\sigma}(\mathbf{r}). \tag{2.153}$$

Minimizing functional (2.27) in function (2.153) we obtain the following system of linear equations:

$$\sum_{n=1}^{M} (H_{nn',\sigma\sigma'} - \varepsilon' O_{nn',\sigma\sigma'})\mathrm{v}_n(\mathbf{k}) = 0. \tag{2.154}$$

The eigenvalues $\varepsilon_{\mathbf{k}}$ are found from the determinant equation:

$$\det|H_{nn',\sigma\sigma'} - \varepsilon' O_{nn',\sigma\sigma'}| = 0, \tag{2.155}$$

where

$$H_{nn',\sigma\sigma'} = \int_{\Omega_0} d^3\mathbf{r}\, \varnothing^{\mathbf{k}*}_{n,\sigma}(\mathbf{r}) H \varnothing^{\mathbf{k}}_{n',\sigma'}(\mathbf{r}) \tag{2.156}$$

and

$$S_{nn',\sigma\sigma'} = \int_{\Omega_0} d^3\mathbf{r}\, \varnothing^{\mathbf{k}*}_{n,\sigma}(\mathbf{r}) \varnothing^{\mathbf{k}}_{n',\sigma'}(\mathbf{r}). \tag{2.157}$$

The Hamiltonian $\hat{H}$ consists of two parts:

$$\hat{H} = \hat{H}^{\mathrm{SRA}} + \hat{H}^{\mathrm{co}}. \tag{2.158}$$

Then

$$H_{nn',\sigma\sigma'} = \sum_{\mu\nu} c^{*\mathrm{SRA}}_{\mu}(\varepsilon_n) c^{\mathrm{SRA}}_{\nu}(\varepsilon_n)[H^{\mathrm{SRA}}_{\mu\nu,\sigma\sigma'}\delta_{\sigma\sigma'} + H^{\mathrm{co}}_{\mu\nu,\sigma\sigma'}] \tag{2.159}$$

and

$$S_{nn',\sigma\sigma'} = \sum_{\mu\nu} c^{*\mathrm{SRA}}_{\mu}(\varepsilon_n) c^{\mathrm{SRA}}_{\nu}(\varepsilon_{n'}) S^{\mathrm{SRA}}_{\mu\nu}\delta_{\sigma\sigma'}. \tag{2.160}$$

The matrix elements $H^{\mathrm{SRA}}_{\mu\nu}$ and $S^{\mathrm{SRA}}_{\mu\nu}$ are calculated by Eq. (2.134). Since

$$\hat{H}^{\mathrm{co}} = \frac{1}{(2Mc)^2}\frac{1}{r}\frac{d\mathrm{v}_{\mathrm{eff}}}{dr}(\boldsymbol{\sigma}\cdot\mathbf{L}). \tag{2.161}$$

The spin–orbit coupling matrix element takes the form

$$H^{\mathrm{co}}_{\mu\nu,\sigma\sigma'} = \frac{16\pi^2}{\Omega_0}\sum_{s} R_s^4 \exp[-i(k_\mu - k_\nu)\tau_s]$$

$$\times \sum_{lm,m'} Y^*_{lm'}(k_\nu) Y_{lm}(k_\mu) T^S_{\mu\nu,l} I_{lmm',\sigma\sigma'}, \tag{2.162}$$

where

$$T^S_{\mu\nu,l} \equiv \{a_l^{(S)}(k_\mu) a_l^{(S)}(k_\nu) \xi_l^{(S)1} + [a_l^{(S)}(k_\mu) b_l^{(S)}(k_\nu) + a_l^{(S)}(k_\nu) b_l^{(S)}(k_\mu)] \xi_l^{(S)2} + b_l^{(S)}(k_\mu) b_l^{(S)}(k_\nu) \xi_l^{(S)3}\} \tag{2.163}$$

and

$$\xi_l^{(S)1} = \int_0^{R_s} dr\, P_l^2 \frac{1}{(2Mc)^2} \frac{1}{r} \frac{dv_{\text{eff}}}{dr},$$

$$\xi_l^{(S)2} = \int_0^{R_s} dr\, P_l \dot{P}_l \frac{1}{(2Mc)^2} \frac{1}{r^2} \frac{dv_{\text{eff}}}{dr},$$

$$\xi_l^{(S)3} = \int_0^{R_s} dr\, \dot{P}_l^2 \frac{1}{(2Mc)^2} \frac{1}{r^2} \frac{dv_{\text{eff}}}{dr}, \tag{2.164}$$

$$I_{lmm',\sigma\sigma'} \equiv \int d\Omega\, \chi^+(\sigma') Y^*_{lm'}(\boldsymbol{\sigma}\cdot\mathbf{L}) Y_{lm}\chi(\sigma) \tag{2.165}$$

[$\chi(\sigma)$ are the spin functions]. Operating on the product of the spherical and spin functions with $(\boldsymbol{\sigma}\cdot\mathrm{L})$, we obtain the following equation for matrix element (2.162):

$$H^{\text{co}}_{\mu\nu,\sigma\sigma'} = \frac{4\pi}{\Omega_0}\left[\delta_{\sigma\sigma'} D_{\mu\nu}\begin{bmatrix}\sigma\\ \sigma\end{bmatrix} + (1-\delta_{\sigma\sigma'}) D_{\mu\nu}\begin{bmatrix}\sigma\\ \sigma'\end{bmatrix}\right] \sum_s R_s^4 \exp[-i(\mathbf{k}_\mu - \mathbf{k}_\nu)\boldsymbol{\tau}_s]$$

$$\times \sum_{l=0}^{l_{\max}} (2l+1) P_l' (\cos\theta_{\mathbf{k}\mu,\mathbf{k}\nu}) T^S_{\mu\nu,l}. \tag{2.166}$$

The following designations are used in Eq. (2.166):

$$D_{\mu\nu}\begin{bmatrix}\sigma\\ \sigma\end{bmatrix} = -i\sigma[\mathbf{k}_\mu \times \mathbf{k}_\nu]_z/(k_\mu \cdot k_\nu),$$

$$D_{\mu\nu}\begin{bmatrix}+\\ -\end{bmatrix} = -i\{[\mathbf{k}_\mu \times \mathbf{k}_\nu]_x + i[\mathbf{k}_\mu \times \mathbf{k}_\nu]y\}/(\mathbf{k}_\mu \cdot \mathbf{k}_\nu),$$

$$D_{\mu\nu}\begin{bmatrix}-\\ -\end{bmatrix} = D^*_{\mu\nu}\begin{bmatrix}+\\ +\end{bmatrix},$$

$$D_{\mu\nu}\begin{bmatrix}+\\ -\end{bmatrix} = -D^*_{\mu\nu}\begin{bmatrix}-\\ +\end{bmatrix},$$

$$P_l'(x) = \frac{l}{1-x^2}[P_{l-1}(x) - xP_l(x)]. \tag{2.167}$$

Calculation of the relativistic band structure by the method outlined above will, in the final analysis, yield a complex $2M \times 2M$ matrix. For example, for a rare-earth compound $M=16[s(1)+p(3)+d(5)+f(7)]$, that is, the diagonalized matrix has an order of 32 while this figure would be 200 using the traditional method.

The LAPW method permits a simple generalization that makes it possible to avoid the MT approximation for the crystal potential.[178] The crystal electron density can be given in the form of the following expansions:

$$\rho(\mathbf{r})=\begin{cases}\sum\limits_{lm}\rho_{lm}(\mathbf{r})Y_{lm}(\hat{\mathbf{r}}), & r\leqslant R_s,\\ \sum\limits_{\mu}\rho(\mathbf{k}_\mu)\exp(i\mathbf{k}_\mu\mathbf{r}), & r>R_s.\end{cases} \tag{2.168}$$

Then the Coulomb part of the crystal potential can be expressed through Green's function

$$\mathrm{v}(\mathbf{r})=2\int_{\Omega_s}d\mathbf{r}'\,\rho(\mathbf{r}')G(\mathbf{r},\mathbf{r}')-\frac{R_s^2}{4\pi}\int_s ds\,\mathrm{v}_1(\mathbf{R})\frac{\partial G}{\partial n}, \tag{2.169}$$

where

$$G(\mathbf{r},\mathbf{r}')=4\pi\sum_{lm}\frac{Y_{lm}(\hat{\mathbf{r}})Y^*_{lm}(\hat{\mathbf{r}}')}{2l+1}\left(\frac{r^l_<}{r^{l+1}_>}-\frac{r^l_<r^l_>}{R_s^{2l+1}}\right), \tag{2.170}$$

$$\left.\frac{\partial G}{\partial n}\right|_{r=R_s}=-\frac{4\pi}{R_s^2}\sum_{lm}Y_{lm}(\hat{\mathbf{r}})Y^*_{lm}(\hat{\mathbf{r}})\left(\frac{r}{R_s}\right)^l, \tag{2.171}$$

$$\mathrm{v}_I(\mathbf{r})=\sum_{\mathbf{k}_\mu\neq0}\frac{8\pi\rho(\mathbf{k}_\mu)}{k_\mu^2}\exp(i\mathbf{k}_\mu\mathbf{r})+\mathrm{v}_I(0) \tag{2.172}$$

is the potential in the external domain between the MT spheres.

Substituting the expansion for $\rho(\mathbf{r})$ of Eq. (2.168) into Eq. (2.169) subject to Eqs. (2.169)–(2.172) and after some simple transformations, we have

$$\mathrm{v}(r)=\begin{cases}\sum\limits_{lm}\mathrm{v}_{lm}(\mathbf{r})Y_{lm}(\hat{\mathbf{r}}), & r\leqslant R_s,\\ \sum\limits_{\mu\neq0}\dfrac{8\pi\rho(\mathbf{k}_\mu)}{k_\mu^2}\exp(i\mathbf{k}_\mu\mathbf{r}), & r>R_s,\end{cases} \tag{2.173}$$

where

$$\begin{aligned}\mathrm{v}_{lm}(r)=\frac{8\pi}{2l+1}&\left\{\frac{1}{r^{l+1}}\int_0^r d\mathbf{r}'\,\rho_{lm}(r')r'^{l+2}+r^l\int_r^{R_s}dr'\,\rho_{lm}(r')r'^{1-l}\right.\\ &\left.-\frac{r^l}{R_s^{2l+1}}\int_0^{R_s}dr'\,\rho_{lm}(r')r'^{l+2}\right\}\\ &+32\pi^2i^l\left(\frac{r}{R_s}\right)^l\sum_{k_\mu\neq0}\frac{\rho(\mathbf{k}_\mu)}{\mathbf{k}_\mu^2}j_l(\mathbf{k}_\mu R_s)Y^*_{lm}(k_\mu)+\sqrt{4\pi}\mathrm{v}_I(0).\end{aligned} \tag{2.174}$$

Hence if we now use the LAPW functions as the wave functions and apply Eq.

(2.173) as the potential, two additions will appear in the matrix elements of the LAPW; these matrix elements can be calculated if we know the coefficients $v_{lm}(r)$ and $\rho(\mathbf{k}_\mu)$. The coefficients of the expansion of the Coulomb part are calculated by Eq. (2.174). A Taylor series expansion can be used for the exchange part of the potential within the MT spheres if we assume that the zero component of the electron density exceeds (to some degree) the subsequent components with $l\neq 0$. Ordinarily the least-squares method with random generation of points r in the unit cell is applied to find the Fourier coefficients of the exchange part of the potential in the intermediate region; in this case the Monte Carlo method is used for random generation of points $\mathbf{r}$.[179]

The primary problem of calculations using the general form of the crystal potential (2.173) arises in connection with the convergence of the series. Elyashar and Koelling[180] proposed adding and calculating an imaginary charge in the form of a Gaussian with the center in the MT sphere such that an electrically neutral pseudoatom would be analyzed within the MT sphere. This significantly accelerates the convergence within the MT sphere $[\rho_l \sim (1/R_S)^{l+1}]$, although it does not in practice solve the problems of the convergence of series (2.173) in $\mathbf{k}_\mu$. Hence Weinert[178] proposed a more effective method which can be explained as follows. In order to formulate the potential outside the MT spheres the true form of the charge inside the MT spheres is not significant but rather only its multipole moments are important:

$$v(\mathbf{r})=\sum_{lm}\frac{4\pi}{2l+1}q_{lm}Y_{lm}(\hat{\mathbf{r}})/r^{l+1}, \tag{2.175}$$

where

$$q_{lm}=\int_{\Omega_s}d\mathbf{r}\,\rho(\mathbf{r})\mathbf{r}^l Y^*_{lm}(\hat{\mathbf{r}}). \tag{2.176}$$

Hence the true charge density $\rho(\mathbf{r})$ within the MT sphere is replaced by the "pseudodensity" $\tilde{\rho}(\mathbf{r})$ with the same multipole moments as $\rho(\mathbf{r})$ but followed by a rapidly converging Fourier series. Moreover, this series is absolutely convergent, which makes it possible to carry out the termwise integration required to obtain the potential. Moreover, this method makes it possible to avoid the rather cumbersome calculations of the Ewald procedure.

We write the pseudodensity of the valence electrons as the sum of the densities in the intermediate range and the range within the MT spheres:

$$\tilde{\rho}(\mathbf{r})=\rho_I(\mathbf{r})\theta(\mathbf{r}\in I)+\sum_s \tilde{\rho}_s(\mathbf{r})\theta(\mathbf{r}\in\Omega_s), \tag{2.177}$$

where θ is a stepfunction and $\tilde{\rho}_S(\mathbf{r})$ has the same multipole moments as $\rho_S(\mathbf{r})$. We now apply $\rho_I(\mathbf{r})$ to the entire unit cell:

$$\rho(\mathbf{r})=\rho_I(\mathbf{r})+\sum_s [\rho_s(\mathbf{r})-\rho_I(\mathbf{r})]\theta(\mathbf{r}\in\Omega_s). \tag{2.178}$$

We then calculate the multipole moments of the true density q_{lm}^{S}, the density in brackets in Eq. (2.178) q_{lm}^{SI} and we formulate the multipole moments of the pseudodensity

$$\tilde{q}_{lm}^{S}=q_{lm}^{S}-q_{lm}^{SI}. \tag{2.179}$$

It is now possible to construct the pseudodensity Fourier transforms which have all the necessary properties [a rapidly divergent Fourier series and the same multipole moments as in $\rho(\mathbf{r})$]:

$$\begin{cases} \tilde{\rho}_p(k_\mu)=\dfrac{4\pi}{\Omega_0}\displaystyle\sum_{lm}\sum_{s}\dfrac{(-i)^l(2l+2n+3)!!!}{(2l+1)!!}\dfrac{j_{l+n+1}(k_\mu R_s)}{(k_\mu R_s)^{n+1}} \\ \qquad\qquad \times\tilde{q}_{lm}^{S}e^{-ik_\mu\tau_s}Y_{lm}(\hat{k}_\mu)/R_s^l \quad (k_\mu\neq 0), \\ \rho(0)=\dfrac{\sqrt{4\pi}}{\Omega_0}\displaystyle\sum_{s}\tilde{q}_{00}^{S}. \end{cases} \tag{2.180}$$

The parameter n in Eq. (2.180) is used to accelerate convergence. The desired pseudodensity Fourier transforms are calculated by the equation

$$\tilde{\rho}(k_\mu)=\rho_I(k_\mu)+\tilde{\rho}_p(k_\mu). \tag{2.181}$$

After the Fourier coefficients of the pseudodensity are calculated we can apply Eq. (2.174) to find the coefficients $v_{lm}(\mathbf{r})$ and to calculate the Coulomb potential in the intermediate range.

2.4. Electronic structure of crystal surfaces

The study of the electronic structure of surfaces by means of solid state computational physics began recently and at present is in the development stage. Appropriately modified techniques from band theory are used to calculate the electronic structure of surfaces: the LCAO, GF, LMTO, and LAPW methods. The modification of these methods is based on the elimination of the periodicity of the wave function perpendicular to the surface and ordinarily in place of a semi-infinite crystal a thin film consisting of several layers is examined.

One of the most common methods used to calculate the electronic structure of surfaces is the LCAO method. In the early studies utilizing the LCAO method,[181–185] no self-consistency was maintained and the Slater–Koster parameters were determined from band calculations of bulk crystals. The first self-consistent analysis was carried out by Gay *et al.*[186–189] In this case the atomic orbitals were approximated by linear combinations of Gaussian orbitals. The electronic structure of the (100) surface of nickel, copper, rhodium, palladium, and silver were calculated. The resulting theoretical work function was in moderate agreement with the experimental data. The electronic structure of the (111) surface of rhodium and the (001) surface of carbon-coated ruthenium was calculated by Kohn.[189] The theoretically derived quantities such as the work function, the surface shifts of the bonding energies of the core electrons, etc., are in good agreement with experiment. A very high chemical activity of the carbon monolayer on the test surfaces was identified which was confirmed in experiment.

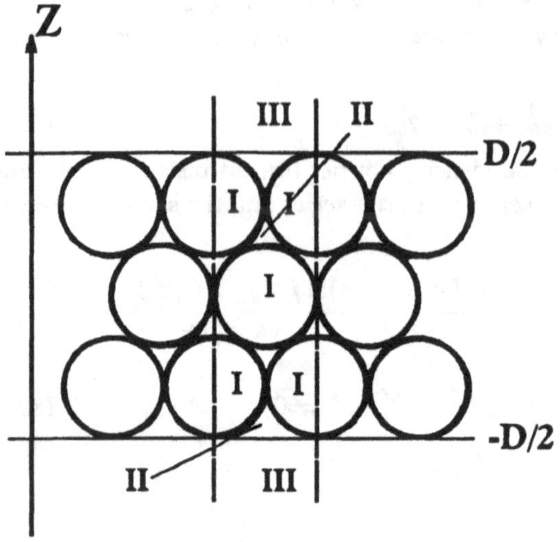

FIG. 2.1. Unit cell obtained in electron structure calculation of surfaces.

Overall the LCAO method has been rather successfully used to calculate the electronic structure of surfaces of both simple and transition light metals. At the same time the LCAO method is not used to calculate the electronic structure of heavy metal surfaces.

The GF and LMTO methods were also generalized to the two-dimensional case.[189–192] However, it should be noted that neither the GF nor the LMTO methods have been widely used for calculating the electronic structure of surfaces.

The most commonly used method of those specially developed in recent years for calculating the electronic structure of solid surfaces is the surface linear augmented plane wave method (SLAPW).[193–195] This method has a number of indisputable advantages over other methods. First it is comparatively simple from the algorithmic viewpoint which makes it possible to develop, using this method, an effective program set for self-consistent calculations of the electronic structure of surfaces. Second, the SLAPW method permits a comparatively simple generalization of a general potential based on the technique described in the preceding section. This fact is quite important for films where the MT effects are very weak.[196] Third, the SLAPW method is easily generalized to the relativistic case using the same technique as the LAPW method. Hence it can be successfully used to calculate the surfaces of heavy metals.

We consider the primary equations of the SLAPW method following Refs. 194, 197, and 198. The basis functions of the SLAPW method are constructed based on the form of the film MT potential (the MT-z potential). Figure 2.1 shows a unit cell and its primary regions; this cell is used in calculating the electronic structure of a surface by the SLAPW method. Since the film under analysis contains several atomic layers in the z direction and has translational symmetry in the (α,y) plane, the unit cell is a parallelepiped extended along the z axis from $-\infty$ to $+\infty$ and contains a certain quantity of atoms in the range $-D/2<z<D/2$. The cross-sectional area of the cell is equal to s and the film volume in the cell is equal to $\Omega_0=sD$.

In the MT-z approximation the entire cell volume is divided into three regions. In the vacuum region (III) the potential is averaged over (x,y) and is only a function of z. In regions I and II the potential takes the ordinary MT form. Therefore, the potential in the MT-z approximation takes the following form:

$$\mathrm{v}_{\mathrm{eff}}(\mathbf{r})=\begin{cases}\mathrm{v}(\rho), & \rho\in\Omega_s \quad (\rho\equiv\mathbf{r}-\mathbf{r}_s),\\ \mathrm{v}_0, & \mathbf{r}\in\Omega_0-\sum_s \Omega_s,\\ \mathrm{v}(z), & |z|>D/2.\end{cases} \tag{2.182}$$

As in the bulk LAPW method, the basis functions as plane waves in the intermediate range are added to functions constructed from exact solutions in spheres (region I) and vacuum regions (region III). These functions and their derivatives are joined with the basis functions as plane waves on the surface of MT spheres and along the boundary planes for $z=\pm D/2$. The resulting basis functions are continuous and differentiable everywhere.

In the intermediate domain the basis functions are defined as the product of the two-dimensional plane wave and a one-dimensional symmetrized plane wave:

$$\varphi^{\mathrm{II}}_{\mu n}(\mathbf{k},\mathbf{r})=\sqrt{\frac{2}{\Omega_0}}\exp[i\mathbf{k}_\mu\mathbf{r}]\varphi_n(z), \tag{2.183}$$

where

$$\varphi_n(z)=\cos(k_n z),\quad k_n=\frac{2\pi}{D}n \quad (n=0,1,2,...) \tag{2.184}$$

for the states that are symmetrical (even) relative to the $z\to -z$ reflection plane and

$$\varphi_n(z)=\sin(k_n z);\quad k_n=\frac{2\pi}{D}\left(n+\frac{1}{2}\right),\quad n=0,1,2,... \tag{2.185}$$

for the states antisymmetrical (odd) relative to the reflection. In Eq. (2.183) $\mathbf{k}_\mu = \mathbf{k} + \mathbf{k}_\mu$ ($\mathbf{k}$ is a two-dimensional wave vector and $\mathbf{k}_\mu$ is the two-dimensional vector of the reciprocal lattice).

Inside the Sth MT sphere the basis function takes the same form as in Eq. (2.129). In region II the basis function is defined as the product of a two-dimensional plane wave and a linear combination of the z-dependent function $R_\mu(\mathbf{k},\varepsilon,z)$ and its energy derivative $\dot{R}_\mu(\mathbf{k},\varepsilon,z)$:

$$\varphi^{\mathrm{III}}_{\mu n}(\mathbf{k},\mathbf{r})=\exp(i\mathbf{k}_\mu r)\,[V_{\mu n}R_\mu(\mathbf{k},\varepsilon_i,z)+W_{\mu n}\dot{R}_\mu(\mathbf{k},\varepsilon_i,z)], \tag{2.186}$$

where $V_{\mu n}$ and $W_{\mu n}$ are constant coefficients which are determined from the continuity condition of the functions $\varphi^{(\mathrm{II})}$ and $\varphi^{(\mathrm{III})}$ and their first derivatives along the film boundaries $z=\pm D/2$. The function $R_\mu(\mathbf{k},\varepsilon,z)$ is found by solving the Schrödinger equation in the vacuum region with energy ε:

$$\left[-\frac{d^2}{dz^2}+\mathrm{v}(z)+k_\mu^2-\varepsilon\right]R_\mu(\mathbf{k},\varepsilon,z)=0 \tag{2.187}$$

when

$$\int_{D/2}^{\infty} dz\, R_\mu^2(z) = 1. \tag{2.188}$$

The energy derivative of the function R_μ is sought by solving an inhomogeneous equation analogous to Eq. (2.124). Here the orthogonality condition is calculated:

$$\int_{D/2}^{\infty} dz\, R_\mu(z)\dot{R}_\mu(z) = 0. \tag{2.189}$$

Using the variational principle yields a secular equation of the SLAPW method:

$$\det|H_{\mu'n',\mu n} - \varepsilon O_{\mu'n',\mu n}| = 0, \tag{2.190}$$

where

$$H^{(\pm)}_{\mu'n',\mu n} = G^2_{\mu'n'}\delta_{\mu'n',\mu n}\varepsilon_n^{\pm} - \frac{4\pi}{\Omega_0}\sum_s \exp[-i(\mathbf{k}_\mu - k_{\mu'})\tau_s]R_s^2\{\cos[\mathbf{k}_n - \mathbf{k}_{n'})z_s]J_s(G^+_{\mu'n'},G^+_{\mu n})(G^+_{\mu'n'}G^+_{\mu n}) \pm \cos[(\mathbf{k}_n + \mathbf{k}_{n'})z_s]J_s(G^+_{\mu'a''},G^-_{\mu n}) \times (G^+_{\mu'n'}G^-_{\mu n})\} + \frac{4\pi}{\Omega_0}\sum_s \exp[-i(\mathbf{k}_\mu - \mathbf{k}_{\mu'})\tau]R_s^4 \sum_l (2l+1)[\varepsilon_l s_{sl} + \gamma_{sl}]P^{\pm}_{slj} + \varepsilon_0 S_{\mu'n',\mu n} + 2\delta_{\mu'\mu} S V_{\mu n} W_{\mu n'}; \tag{2.191}$$

$$O^{(\pm)}_{\mu'n',\mu n} = \delta_{\mu'n',\mu n}\varepsilon_n^{\pm} - \frac{4\pi}{\Omega_0}\sum_s \exp[-i(\mathbf{k}_\mu - \mathbf{k}_{\mu'})\tau_s]R_s^2\{\cos[(\mathbf{k}_n - \mathbf{k}_{n'})z_s]J_s(G^+_{\mu'n'},G^+_{\mu n})\} \pm \cos[(\mathbf{k}_n + \mathbf{k}_{n'})z_s]J_s(G^+_{\mu'n'},G^-_{\mu n}) + \frac{4\pi}{\Omega_0}\sum_s \exp[-i(\mathbf{k}_\mu - \mathbf{k}_{\mu'})\tau_s]R_s^4 \sum_l (2l+1)s_{sl}p_{sl} + S_{\mu'n',\mu n}, \tag{2.192}$$

$$S_{sl} = a_l^{(S)}(\mathbf{k}_{\mu'},\mathbf{k}_{n'})a_l^{(S)}(\mathbf{k}_\mu,\mathbf{k}_n) + b_l^{(S)}(\mathbf{k}_{\mu'},\mathbf{k}_{n'})b_l^{(S)}(\mathbf{k}_\mu,\mathbf{k}_n)N_l^{(S)}, \tag{2.193}$$

$$P^{\pm}_{sl} = \{\cos[(\mathbf{k}_n - \mathbf{k}_{n'})z_s]P_l(G^+_{\mu'n'}\hat{\cdot}G^+_{\mu n}) \pm \cos[(\mathbf{k}_n + \mathbf{k}_{n'})z_s]P_l(G^+_{\mu'n'}\hat{\cdot}G^-_{\mu n})\}, \tag{2.194}$$

$$S_{\mu'n',\mu n} = 2\delta_{\mu'\mu}S[V_{\mu n}V_{\mu n'} + W_{\mu n}W_{\mu n'}N_\mu], \tag{2.195}$$

$$\gamma_{sl} = \dot{R}_l^{(S)}R_l^{(S)\prime}[j_l'(G_{\mu'n'}R_s)j_l(G_{\mu n}R_s) + j_l(G_{\mu'n'}R_s)j_l'(G_{\mu n}R_s)] + [\dot{R}_l^{(S)\prime}R_l^{(S)\prime}j_l(G_{\mu'n'}R_s)j_l(G_{\mu n}R_s) + \dot{R}_l^{(S)}R_l^{(S)}j'(G_{\mu'n'}R_s)j'(G_{\mu n}R_s)], \tag{2.196}$$

$$J_s(x,y) = j_1(|x-y|R_s)/|x-y|, \tag{2.197}$$

$$\varepsilon_n^- = 1, \quad \varepsilon_n^+ = 1 + \delta_{n0}, \tag{2.198}$$

$$N_l^{(S)} = \int_0^{R_s} dr\, r^2 \dot{R}_l^{(S)^2}(r), \quad N_\mu = \int_{D/2}^{\infty} dz\, \dot{R}_\mu^2(k,\varepsilon,z), \quad G_{\mu n}^{\pm} \equiv \mathbf{k} + \mathbf{k}_\mu \pm \mathbf{k}_n. \tag{2.199}$$

$P_l(\mathbf{r},\mathbf{r}')$is the Legendre polynomial, $z_S - z$ is the coordinate of the vector $\mathbf{r}_s$ and $a_l^{(s)}$ and $b_l^{(s)}$ are the coefficients defined in Eqs. (2.131) and (2.132):

$$V_{\mu n}^{\pm} = \sqrt{\frac{2}{\Omega_0}}(-1)^n \dot{R}_\mu'(\mathbf{k},\varepsilon_\nu,D/2), \tag{2.200}$$

$$W_{\mu n}^{\pm} = -\sqrt{\frac{2}{\Omega_0}}(-1)^n R_\mu'(\mathbf{k},\varepsilon_\nu,D/2), \tag{2.201}$$

$$V_{\mu n}^{\pm}\Big|_{z=-D/2} = \pm V_{\mu n}^{\pm}\Big|_{z=D/2}; \quad W_{\mu n}^{\pm}\Big|_{z=-D/2} = W \pm {}^{\pm}_{\mu z}\Big|_{z=D/2}. \tag{2.202}$$

S is the cross-sectional area of the unit cell.

The formulation of the MT-z potential is described by Wolf and Korapanova.[199] As demonstrated by subsequent research,[198] the MT-z approximation for a film potential in the self-consistent calculations does not yield satisfactory results. The potentials of the spheres in both the surface and bulk layers are much deeper than the corresponding quantities for the potential in the bulk case. This means that the characteristics of the electronic structure calculated in such a potential will differ significantly from the observed values: therefore the work function exceeds the experimental value by a factor of several times.[198] Hence, in carrying out the self-consistent calculations it is necessary to use a potential of the general type. The MT-z potential is generalized in precisely the same manner as in the bulk case using the technique described above in the preceding section. Using the general potential in the SLAPW method yields a good agreement on virtually all experimental characteristics including the work function.[200]

2.5. Self-consistency in calculating the electronic structure of condensed matter

The most important (and at the same time cumbersome) element of electronic structure calculations of condensed matter is self-consistency, that is, deriving a self-consistent effective potential within the framework of the density-functional theory. There is always some degree of uncertainty in the formulation of the effective potential in nonself-consistent calculations. This uncertainty is particularly strongly manifested in systems with d and f elements. The problem is that in the DFT the process of deriving the potential in the final analysis reduces to calculating the electronic density which in turn is expressed through the atomic or ionic wave functions. It is therefore very important to represent the "proper" electronic configuration of the atom (the ion) at the outset, that is, the configuration which produces an electronic structure that is consistent with, for example, experiment. d and particularly f metals are extremely sensitive to changes in electron density (the electron configuration) (see the figures and explanations in the first chapter). This is due to the fact that such metals have a significant Coulomb electron–electron interaction level and hence their energy bands are very sensitive to the occupation numbers. The dependence of the electronic structure on the occupation numbers is

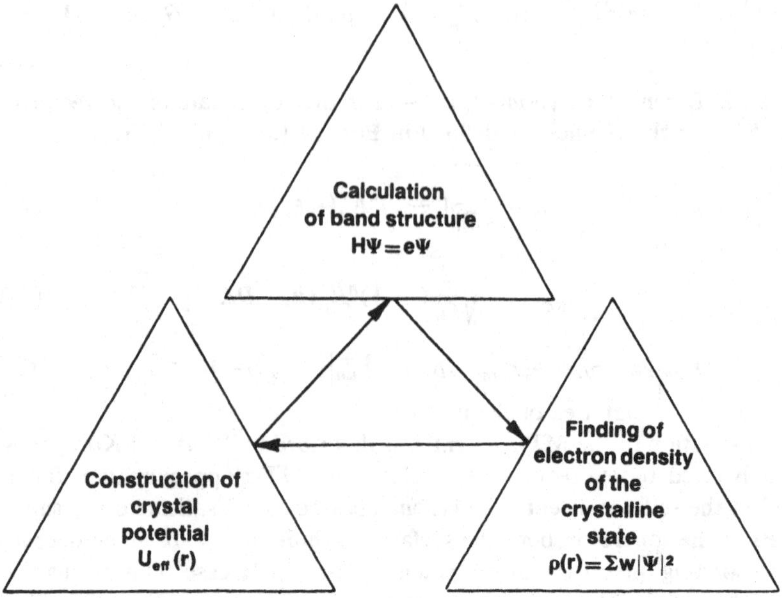

FIG. 2.2. Scheme of self-consistent routine.

weak for condensed matter with s and p elements and hence a nonself-consistent calculation for these objects yields a comparatively good representation of the energy bands. At the same time a self-consistent calculation must be carried out in order to obtain more exact wave functions.

The uncertainty in selecting the effective potential can be partially eliminated by means of the self-consistency. Schematically the self-consistency routine is shown in Figure 2.2. The majority of efforts in running self-consistent calculations are expended on deriving the electronic structure. Hence the theoretical method employed for this purpose must be efficient from the viewpoint of computer time while maintaining good accuracy since the wave function directly participates in the self-consistency process.

The primary problem of self-consistent calculations is the formulation of an iterative routine that would be stable and yield sufficiently rapid convergence. We consider the mth iteration. Let

$$\mathrm{v}^{mi}=\begin{pmatrix} \mathrm{v}(r_1) \\ \vdots \\ \mathrm{v}(r_n) \end{pmatrix}$$

be the column vector consisting of the potential values in a radial grid $(r_1,r_2,\ldots,r_n)$ at the beginning of the mth iteration, while v^{mf} is the same vector at the end of the mth iteration. If we use the final potential of one iteration as the initial potential for the next iteration, the convergence process will be unstable.[201] Hence the initial potential for the next iteration must be constructed using a convergence stabilization scheme. An arithmetic average scheme is most commonly used in self-consistent electron structure calculations; here the initial potential for the $(m+1)$th iteration $\mathrm{v}^{(m+1)i}$ is constructed as follows:

$$\mathrm{v}^{(m+1)i} = \mathrm{v}^{mf} - AF_m, \tag{2.203}$$

where $Fm = \mathrm{v}^{mf} - \mathrm{v}^{mi}$

$$A = \begin{pmatrix} a(r_1) & & 0 \\ & \ddots & \\ 0 & & a(r_n) \end{pmatrix}, \quad 0 < a(r_i) < 1. \tag{2.204}$$

With values of $a(r)$ close to one, the iterative process will be stable although there will be a low rate of convergence. If we reduce $a(r)$ the rate of convergence rises although beginning at a certain value of $a(r)$ the process begins to diverge. Therefore, by selecting $a(r)$ it is possible to achieve a balance between the stability of the iterative process and the convergence rate. In order to simplify the calculations we often assume $a(r) = \alpha(0.5 \leqslant \alpha \leqslant 1)$. The self-consistency criterion $\beta_m = \max|\mathrm{v}^{m_i} - \mathrm{v}^{m_f}|$ for $\alpha = 0.5$ drops by a factor of 2 for one iteration to another. In practice it is often necessary to select a rather large value of α. Ordinarily a value of $\alpha \sim 0.8$ is selected in band structure calculations of transition metals and compounds which on the average will result in 20–25 iterations. A value of $\alpha \sim 0.9$–0.95 is already required for f metals, compounds, and thin films. Such a selection of α will significantly slow the convergence and the attainment of self-consistency after only approximately 50 iterations.

Hence an important problem is improving the convergence in electronic structure calculations of condensed matter containing d and f elements. Current band calculation techniques utilize a so-called two-level scheme for this purpose.[202] An "intra-atomic" self-consistent calculation (neglecting the band structure calculation) is ordinarily carried out in the first stage in this routine, that is, the radial wave functions and their energy derivatives are recalculated from iteration to iteration within the framework of the linear method, while the multipliers in $\rho(\mathbf{r})$ dependent on $\mathbf{k}$ [see, e.g., formula (2.145) of the present section] remain unchanged. The relative density error is substantially reduced in this stage. A complete iteration including a band calculation at points $\mathbf{k}$ is already achieved here in the second stage. In this case the relative error in the first self-consistent calculation stage rises although it still remains much less than if we had carried out a complete iteration calculation prior to this stage. As we have seen from our experience with this scheme it works effectively in self-consistent calculations of metals and single-layered films. The scheme must be used with some care in compounds and multilayered films since its use will not yield the expected result in all cases. Moreover, this scheme can be used only in band self-consistent calculations which substantially narrows its field of application.

In 1965 Broyden[203] proposed one of the most effective methods of accelerating convergence which developed the well-known Newton–Raphson method.[204] We consider the Broyden scheme here as applied to self-consistent calculations following the presentations of Srivastava[205] and Singh *et al.*[206]

We represent the problem of a self-consistent solution to a certain equation in matrix form:

$$F(X)=0, \tag{2.205}$$

where

$$X^T=(x_1,x_2,\ldots,x_n), \tag{2.206}$$

$$F(X)=\begin{cases} f_1(x_1,x_2,\ldots,x_n), \\ \vdots \\ f_n(x_1,x_2,\ldots,x_n). \end{cases} \tag{2.207}$$

Expanding the function $F(X)$ near the exact solution X_0 we obtain accuracy in the quadratic terms

$$F(X)=F(X_0)+DF(X_0)(X-X_0), \tag{2.208}$$

where $DF(X_0)$ is the Jacobian matrix of the function $F(X)$ for $X = X_0$. Subject to Eq. (2.205) we have

$$F(X)+DF(X_0)(X-X_0)=0, \tag{2.209}$$

$$X=X_0-[DF(X_0)]^{-1}F(X_0). \tag{2.210}$$

The primary idea of the Broyden method is an initial approximation of the matrix $DF(X)$ with a certain matrix J^1 and formulation of recursion relations that relate the approximation of the Jacobian matrix J^m in the mth iteration with the initial iteration. These relations can be found from the variational principle which requires that two conditions be satisfied:

$$g=F_m-F_{m-1}-J^m(X_m-X_{m-1})=0, \tag{2.211}$$

$$Q=\min\|J^m-J^{m-1}\|^2. \tag{2.212}$$

We derive the functional

$$K=Q+\lambda^Tg, \tag{2.213}$$

where λ is the column of undefined Lagrangian multipliers. Varying with respect to J^m we obtain

$$J^m=J^{m-1}+\tfrac{1}{2}\lambda^T(X_m-X_{m-1}). \tag{2.214}$$

Using Eq. (2.211) it is possible to find λ:

$$\lambda=2\frac{F_m-F_{m-1}-J^{m-1}(X_m-X_{m-1})}{(X_m-X_{m-1})^T(X_m-X_{m-1})}. \tag{2.215}$$

Therefore,

$$J^m=J^{m-1}+\frac{[F_m-F_{m-1}-J^{m-1}(X_m-X_{m-1})](X_m-X_{m-1})^T}{(X_m-X_{m-1})^T(X_m-X_{m-1})}. \tag{2.216}$$

Consequently,

$$X_{m+1}=X_m-[J^m]^{-1}F_m. \tag{2.217}$$

The practical application of this scheme is complicated by two factors:
(1) it is necessary to store n^2 elements of the matrix J^m in memory;
(2) the high order matrix J^m must be inverted in each iteration.

Given these difficulties Singh *et al.*[206] proposed an alteration of the Broyden scheme. If Eq. (2.210) is written as

$$X_m=X_{m-1}-G^m(F_m-F_{m-1}), \tag{2.218}$$

where $G^m\equiv[J^m]^{-1}$, conditions (2.211) and (2.212) are transformed:

$$q=X_m-X_{m-1}-G^m(F_m-F_{m-1})=0, \tag{2.219}$$

$$Q=\min\|G^m-G^{m-1}\|^2. \tag{2.220}$$

In analogy to Eq. (2.213) we draft the functional

$$L=Q+\lambda^Tq \tag{2.221}$$

and allowing it to vary with respect to G^m we obtain

$$G^m=G^{m-1}+\frac{[X_m-X_{m-1}-G^{m-1}(F_m-F_{m-1})](F_m-F_{m-1})^T}{(F_m-F_{m-1})^T(F_m-F_{m-1})}. \tag{2.222}$$

Selecting the initial approximation G^1, Eq. (2.222) can be rewritten as

$$G^m=G^1+\sum_{i=2}^{m}U_iV_i^T, \tag{2.223}$$

where

$$V_i^T=(F_i-F_{i-1})^T[(F_i-F_{i-1})^T(F_i-F_{i-1})]^{-1}, \tag{2.224}$$

$$U_i=X_i-X_{i-1}-G^1(F_i-F_{i-1})-\sum_{j=2}^{i-1}U_jV_j^T(F_i-F_{i-1}). \tag{2.225}$$

It is possible to use the following diagonal matrix as the initial approximation of G^1:

$$G^1=-gI, \tag{2.226}$$

where $g=$ const. Finally we have

$$X_{m+1}=X_m-G^1F_m-\sum_{i=2}^{m}U_iV_i^TF_m. \tag{2.227}$$

This scheme requires storing a small quantity of information from the preceding iterations and does not require inversion of a high order matrix. Equations (2.223)–(2.227) can be reduced to a form that is more convenient for programming. For this purpose we recast Eq. (2.227) so that summation of the right-hand side of this equation begins with the index 1:

$$X_{m+1}=X_m - G^1F_m - \sum_{j=1}^{m-1} U_jV_j^TF_m, \tag{2.228}$$

and in precisely the same manner for Eq. (2.225):

$$U_{m-1}=X_m - X_{m-1} - G^1(F_m - F_{m-1}) - \sum_{j=1}^{m-2} U_jV_j^TF_m$$
$$+ \sum_{j=1}^{m-2} U_jV_j^TF_{m-1}. \tag{2.229}$$

The last sum in Eq. (2.228) appears only after the first iteration while the sum in Eq. (2.229) appears only after the second iteration. We introduce the new quantity

$$Z_m = \sum_{j=1}^{m-1} U_jV_j^TF_m = Z_m^1 + U_{m-1}V_{m-1}^TF_m, \tag{2.230}$$

where

$$Z_m^1 = \sum_{j=1}^{m-2} U_jV_j^TF_m. \tag{2.231}$$

Then

$$U_{m-1}=X_m - X_{m-1} - G^1(F_m - F_{m-1}) - Z_m - Z_{m-1} \tag{2.232}$$

and

$$X_{m+1}=X_m - G^1F_m - Z_m. \tag{2.233}$$

The Broyden scheme has been refined,[206] significantly increasing the rate of convergence and substantially reducing computer time, particularly in calculating thin films and bulk condensed matter with *d* and *f* metals.

An additional δ^2 Aitken transform procedure[207] can be used to accelerate convergence. This routine is particularly effective in a geometric progression convergence.[207] In the Aitken δ^2 transformation the new approximation is based on the results of the three preceding iterations:

$$v^{(m+1)i} = \frac{v^{(m-2)f}v^{mf} - [v^{(m-1)f}]^2}{v^{(m-2)f} + v^{mf} - 2v^{(m-1)f}}. \tag{2.234}$$

The Aitken scheme can also be applied in combination with the arithmetic average scheme.

2.6. The Green's function method in defect theory

The existence of a variety of defects in crystals will cause substantial changes in the electronic spectrum and, consequently, in the physical characteristics of the system. There are several theoretical approaches to calculating the spectrum of metals with different types of defects. Such approaches can arbitrarily be divided into two groups. The first group includes methods in which the defect and its environment are treated as a certain subsystem: a "defect molecule" system, and a cluster approach is used for its calculation. The methods classified under the second group consider the defect against an infinite crystal background. Such a system constructs Green's function and then determines all physical properties. Each of the approaches has its own advantages and drawbacks. For example, the advantage of the cluster method is that it is possible to investigate clusters of any composition or size (with the natural limits imposed by computer capabilities). We will deal with this issue in greater detail in the next section.

In many studies carried out in the 1970s and early 1980s,[208,218] the crystal was treated as a collection of periodically distributed MT potentials. An impurity was described by a perturbation potential selected in some specific form (a rectangular well, the Yukawa potential, and the MT potential). One general conclusion deriving from these studies is that the impurity potential must be constructed in a self-consistent manner and be based on the features of the electronic structure of the matrix. We note that it was possible to obtain a good agreement between the calculated characteristics (residual resistance, Knight shift, thermovoltage, etc.) with experiments in many of the cited studies. Evidently this was due to the use of Friedel sums in formulating the defect potential

$$\Delta z = \frac{2}{\pi} \Sigma(2l+1)[\eta_l^i(E_F) - \eta_i^H(E_F)], \tag{2.235}$$

where Δz is the difference in the valences of the impurity and the matrix and $\eta_l^I(E_F)$ and $\eta_l^H(E_F)$ are the phase shifts of the impurity and matrix, respectively, at the Fermi energy. This condition simulates the cumbersome self-consistency routine.[213]

In recent years a method was proposed by Dedericks and co-workers,[219,220] based on the KKR method (Green function method), and it has become popular. We will consider the primary features of this method. Assume a spherically symmetrical MT-type potential $v(\mathbf{r}-\mathbf{R})$ is formulated at each crystalline lattice site. We denote the lattice number of the crystal as m and record the atomic coordinates from the center of the cell. We define the Green function as $G(\mathbf{r} + \mathbf{R}_m, \mathbf{r}' + \mathbf{R}_m; E)$ and write the Schrödinger equation for the mth cell:

$$\left[-\frac{\partial^2}{\partial r^2} + v(\mathbf{r}) - E\right] G(\mathbf{r} + \mathbf{R}_m, \mathbf{r}' + \mathbf{R}'_m, E) = \delta(\mathbf{r}-\mathbf{r}')\delta_{mm'}. \tag{2.236}$$

We represent G as

$$G(\mathbf{r} + \mathbf{R}_m, \mathbf{r}' + \mathbf{R}_{m'}, E) = \sum_{LL'} R_l^m(\mathbf{r}, E) Y_L(r) G_{LL'}^{mm'}(E) R_{l'}^{m'}(r', E) Y_{L'}(\mathbf{r}')$$
$$+ \delta_{mm'} \frac{2m}{h^2} k \sum_L R_l^m(r_<, E) Y_L(r) H_l^m(r_>, E) Y_L(\mathbf{r}'), \tag{2.237}$$

where $\mathbf{R}_l^m$ is the regular solution of the Schrödinger equation for the potential $v_m(r)$; $L = \{l, m\}$; $r_< = \min(r, r')$; $r_> = \max(r, r')$; $k = (2mE/\hbar^2)^{1/2}$. Outside the MT sphere $\mathbf{R}_l^m$ is a combination of spherical Bessel and Neuman functions:

$\cos\eta_l^m j_l[(E)^{1/2}r] - \sin\eta_l^m n_l[(E)^{1/2}r]$; $H_l^m(r,E) = N_l^m(r,E) - iR_l^m(r,E)$; $N_l^m(r,E)$ is an irregular solution of the Schrödinger equation, which outside the MT sphere takes the form

$$N_l^{m'}(r,E)=\sin\eta_l^m(E)j_l(E^{1/2}r)+\cos\eta_l^m(E)n_l(E^{1/2}r). \tag{2.238}$$

The second term in Eq. (2.237) corresponds to Green's function for the MT potential in free space. The primary technical difficulty in this approach is the calculation of the coefficients $G_{LL'}^{mm'}$. We rewrite Eq. (2.237) as

$$G(\mathbf{r}+\mathbf{R}_m,\mathbf{r}'+\mathbf{R}_{m'})=\delta_{mm'}G_s^m(\mathbf{r}+\mathbf{R}_m,\mathbf{r}'+\mathbf{R}_{m'}) + \sum_{LL'} R_l^m(r,E)Y_L(r)G_{LL'}^{mm'}(E)R_{l'}^{m'}(r',E)Y_{L'}(\mathbf{r}'), \tag{2.239}$$

where

$$G_s^m(\mathbf{r}+\mathbf{R}_m,\mathbf{r}'+\mathbf{R}_{m'})=\frac{2m}{\hbar^2}k\sum_L R_l^m(r_<,E)Y_L(r)H_l^m(r_>,E)Y_{L'}(\mathbf{r}'). \tag{2.240}$$

Substituting Eq. (2.239) into the Dyson equation

$$G(E)=G_0(E)+G_0\Delta VG, \tag{2.241}$$

we obtain

$$G_s^m(\mathbf{r}+\mathbf{R}_m,\mathbf{r}'+\mathbf{R}_{m'})=G_s^0(\mathbf{r}+\mathbf{R}_m,\mathbf{r}'+\mathbf{R}_m)+\int dr''\, G_s^0(\mathbf{r}+\mathbf{R}_m,\mathbf{r}'' + \mathbf{R}_m)\Delta v_m(r'')G_s^m(\mathbf{r}''+\mathbf{R}_m,r'+R_m) \tag{2.242}$$

and

$$\begin{aligned} R_l^m(\mathbf{r})G_{LL'}^{mm'}R_{l'}^{m'}(\mathbf{r}')=&R_l^0(\mathbf{r})G_{LL'}^{0mm'}R_{l'}^0(\mathbf{r}')+\frac{2m}{\hbar^2}k\int_0^{2m_0} d\mathbf{r}''(\mathbf{r}'')^2R_l^0(\mathbf{r}_<'')\\ &\times H_l^0(r_>'')\Delta v_m(r'')R_l^m(\mathbf{r}'')G_{LL'}^{mm'}R_{l'}^{m'}(r')\\ &+\frac{2m}{\hbar^2}kR_l^0(r)G_{LL'}^{mm'}\int_0^{R_M} d\mathbf{r}''(\mathbf{r}'')^2R_{l'}^0(\mathbf{r}'')\Delta v_m(\mathbf{r}'')\\ &\times R_l^m(r_<'')H_l^m(r_>'')+R_l^0(r)\sum_{nL''}G_{LL''}^{0m''}\\ &\times\int_0^{R_{m_0}} d\mathbf{r}''(\mathbf{r}'')^2R_l^0(\mathbf{r}'')\Delta v_n(r'')R_{l''}^n(\mathbf{r}'')G_{L''L'}^{nm'}R_{l'}^{m'}(\mathbf{r}'). \end{aligned} \tag{2.243}$$

Then we can obtain

$$R_l^0(\mathbf{r})e^{-i\eta_l^m+i\eta_l^0}G_{LL'}^{mm'}R_{l'}^{m'}(\mathbf{r}')=R_l^0(\mathbf{r})G_{LL'}^{0mm'}e^{i\eta_l^0+i\eta_{l'}^{m'}}R_{l'}^{m'}(\mathbf{r}')$$

$$+R_l^0(r)\sum_{nL''}G_{LL''}^{0mn}\left\{\int_0^{R_{mn}}d\mathbf{r}''(\mathbf{r}'')^2\right.$$

$$\left.\times R_{l''}^0(\mathbf{r}'')\Delta v_n(r'')R_{l''}^n(\mathbf{r}'')\right\}$$

$$\times G_{L''L'}^{nm'}R_{l'}^{m'}(\mathbf{r}'). \tag{2.244}$$

Introducing a new designation:

$$\widetilde{G}_{LL'}^{mm'}=e^{-i\delta_l^m}G_{LL'}^{mm'}e^{-i\delta_{l'}^{m'}}, \tag{2.245}$$

$$\widetilde{G}_{LL'}^{0(m-m')}=e^{-i\delta_l^0}G_{LL'}^{0(m-m')}e^{-i\delta_{l'}^0}. \tag{2.246}$$

Then Eq. (2.242) will take the form

$$\widetilde{G}_{LL'}^{mm'}=\widetilde{G}_{LL'}^{0(m-m')}+\sum_{nL'}\widetilde{G}_{LL''}^{0(m-n)}\Delta t_{L''}^n\widetilde{G}_{L''L'}^{nm'},$$

where

$$\Delta t_l^m(E)=\int_0^\infty dr''(r'')^2\widetilde{R}_l^0(r'',E)\Delta v_m(r'')\widetilde{R}_{l''}^m(r'',E), \tag{2.247}$$

$$\widetilde{R}_l^0=e^{i\delta_l^0}R_l^0(r), \tag{2.248}$$

or introducing the t matrix $t_l^0(E)$ of an ideal crystal:

$$\Delta t_l^m(E)=t_l^m(E)-t_l^0(E)=-\frac{\exp[i(\delta_l^m+\delta_l^0)]}{k}\sin(\delta_l^m-\delta_l^0)\frac{\hbar^2}{2m}. \tag{2.249}$$

Green's function $\widetilde{G}_{LL'}^{mm'}$ can be written through Green's functions of free space:

$$\widetilde{G}_{LL'}^{mm'}=g_{LL'}^{(m-n)}(E)+\sum_{nL''}g_{LL'}^{(m-n)}t_{l''}^n\widetilde{G}_{L''L'}^{nm'}(E). \tag{2.250}$$

If we set $n=0$ in Eq. (2.250) and discard the remaining terms then we obtain the following for $l_{\max}<2$ and for crystals of cubic symmetry:

$$\widetilde{G}_{(L)}=\frac{1}{1-G_L^0\Delta T_l}\widetilde{G}_L^0(\widetilde{G}_{LL'}^0(E)=\widetilde{G}_{(L)}^0\delta_{LL'}). \tag{2.251}$$

Green's function of an ideal crystal is expressed through Green's function of free space:

$$G_{LL}^{0(m-m')}(E)=\frac{1}{v_{BZ}}\int_{v_{BZ}}d\mathbf{k}\sum_{L''}\left(\frac{1}{1-g(\mathbf{k},E)t(E)}\right)_{LL}g(\mathbf{k},E)e^{i\mathbf{k}(\mathbf{R}^m-\mathbf{R}^{m'})}. \tag{2.252}$$

After determining Green's function it is possible to find the density of electronic states

$$N(E) = -\frac{2}{\pi} \int d\mathbf{r}\, \mathrm{Im}\, G(\mathbf{r},\mathbf{r},E) \tag{2.253}$$

as well as the electron density

$$\rho(\mathbf{r}) = -\frac{2}{\pi} \int_0^{E_F} dE\, \mathrm{Im}\, G(\mathbf{r},\mathbf{r},E). \tag{2.254}$$

The calculation scheme is as follows: Green's function of an ideal crystal is determined and the scattering phases for the point defects are found; Eq. (2.251) is then used to determine Green's function and Eq. (2.254) is applied to determine the electron density. The self-consistency here lies in the fact that the derived density is used to find the new potential, while the new scattering phases and the new $\widetilde{G}_L$ as well as $\rho(\mathbf{r})$ are again found. The cycle is repeated until the results of two successive iterations differ by less than a given small quantity.

The primary drawback of this calculation scheme is the assumption of strong defect localization.

This was done in an additional step in Ref. 221: The perturbation of the potentials was accounted for in the neighborhood of the defect. The primary difficulty here lies in inverting the matrix

$$\delta_{nn'}\delta_{LL'} - \widetilde{G}_{LL'}^{nn'}(E)\Delta t_{l'}^{n'}(E), \tag{2.255}$$

whose rank is determined by the number of "perturbed" potentials and the number of nonzero elements $\Delta t_l^n(E)$. If we operate within the framework of a strongly localized defect model it is necessary to invert a 9×9 matrix for $l \leqslant 2$. When accounting for perturbations of the nearest neighbor atoms to the defect the matrix size will jump significantly. For example, a 117×117 matrix is required for a fcc lattice. However, quasidiagonalization of the matrix using group theory makes it possible to examine a total of ten 9×9 submatrices (the number of different irreducible representations).

The next serious problem is calculating the charge density $\rho(\mathbf{r})$. Here approximately 400 energy values are required in the impurity cell while in the neighborhood approximately 1000 energy values are required. Since these calculations are carried out in each iteration step, it is clear that the difficulties are too severe. One successful escape from this situation is the solution offered by Zeller *et al.*[222] to integrate over a complex domain:

$$\rho(\mathbf{r}) = \frac{i}{\pi} \oint dz\, G(\mathbf{r},\mathbf{r},z). \tag{2.256}$$

The contour is the range $E_F - i\varepsilon$ and $E_F + i\varepsilon (\varepsilon \to 0)$ which makes it possible to limit the analysis to 20–40 complex energy values.

A recent study[223] postulated the problem of accounting for lattice relaxation in a dilute alloy within the framework of Green's function formalism. We will briefly examine a few of its more significant details. A shift of an atom at site $\mathbf{R}_n$ makes it necessary to formulate Green's function in a "shifted" coordinate system. If the new radius vector is denoted by $\widetilde{\mathbf{R}}_n$ then

$$\widetilde{\mathbf{R}}_n = \mathbf{R}_n + \mathbf{S}_n, \tag{2.257}$$

where S_n is the atomic displacement vector at site n. In the new variables $\tilde{\mathbf{r}} = \mathbf{r} - S_n$ the functions $\mathbf{R}_L^n(\tilde{\mathbf{r}},E)$ and j_l take the form

$$\tilde{R}_L^n(\tilde{r},E) = i_l(r\sqrt{E})\,Y_L(\tilde{\mathbf{r}}) + \int d\tilde{r}' \int dr''\, g(\tilde{r},\tilde{r}',E)\, z^n(\tilde{r}',\tilde{r}'',E)\, j_l(\tilde{\mathbf{r}}''\sqrt{E})\,Y_L(\tilde{\mathbf{r}}''), \tag{2.258}$$

$$j_l(|\tilde{\mathbf{r}}+s|\sqrt{E})\,Y_L(\tilde{\mathbf{r}}+\mathbf{s})$$

$$= \sum_{L'L''} 4\pi i^{l'+l''-l} C_{LL'L''}\, j_{l''}(\mathbf{s}\sqrt{E})\,Y_{L''}(s)\, j_{l'}(\tilde{\mathbf{r}}\sqrt{E})\,Y_{L'}(\mathbf{r}'), \tag{2.259}$$

where

$$G_{LL'L''} = \int d\omega_r\, Y_L(\mathbf{r})\,Y_{L'}(\mathbf{r})\,Y_{L''}(\mathbf{r}),$$

$$R_L^n(\mathbf{r},E) = R_l^n(\mathbf{r},E)\,Y_L(\mathbf{r}) \tag{2.260}$$

for a spherically symmetrical scatterer. Substituting Eq. (2.259) into Eq. (2.258) we obtain

$$R_L^n(\mathbf{r},E) = \sum_{L'} \tilde{R}_{L'}^n(\tilde{r},E)\,U_{L'L}(s^n,E), \tag{2.261}$$

where

$$U_{L'L}(s,E) = 4\pi \sum_{L''} i^{l'+l''-l} c_{LL'L''}\, j_{l''}(\mathbf{s}\sqrt{E})\,Y_{L''}(\hat{s}). \tag{2.262}$$

The fundamental properties of the matrix U are

$$U_{LL'}(\mathbf{s},E) = (-1)^{l+l'}\,U_{L'L}(\mathbf{s},E), \tag{2.263}$$

$$[U^{-1}(s,E)]_{LL'} = U_{L'L}(-s,E) = U_{L'L}(s,E). \tag{2.264}$$

If the atomic shifts are small it is convenient to utilize an approximate expression in these calculations for the matrix U:

$$U_{L'L}(s,E) \cong \delta_{LL'} + i^{l'-l+1}\frac{4\pi}{3} s\sqrt{E} \sum_{L''} c_{LL'L''}\,Y_{L''}(s)\,\delta_{l''l}. \tag{2.265}$$

Green's function $G(\mathbf{r} - \mathbf{R}^n, \mathbf{r}' - \mathbf{R}^{n'}, E)$ (accounting for the functions outlined above) can be given as

$$G(\mathbf{r}+\mathbf{R}^n,\mathbf{r}'+\mathbf{R}^{n'},E) = G(\tilde{\mathbf{r}}+\tilde{\mathbf{R}}^n,\tilde{\mathbf{r}}'+\tilde{\mathbf{R}}^{n'},E)$$

$$= \delta_{nn'}G^{S_n}(\tilde{\mathbf{r}},\tilde{\mathbf{r}}',E) + \sum_{LL'} \tilde{R}_L^n(\mathbf{r},E)\,G_{LL'}^{nn'}(E)\,\tilde{R}_{L'}^{n'}(\tilde{\mathbf{r}}',E), \tag{2.266}$$

where

$$\tilde{G}_{LL'}^{nn'}(E) = \sum_{L''L'''} U_{LL''}(s^n,E)\,G_{L''L'''}^{nn'}(E)\,U_{z'z'''}(s^{n'},E). \tag{2.267}$$

In the next step it is necessary to formulate Green's function G for the system of MT potentials which undergo a rigid shift from the $\mathbf{R}^n$ position to the $\tilde{\mathbf{R}}^n = \mathbf{R}^n + S^n$ position.

Similar to Eq. (2.237) we give the function G as

$$\overline{G}(\tilde{\mathbf{r}} + \tilde{R}^n, \tilde{\mathbf{r}}' + \tilde{\mathbf{R}}^{n'}, E) = \delta_{nn'} \overline{G}^{S_n}(\tilde{\mathbf{r}}, \tilde{\mathbf{r}}', E) + \sum_{LL'} R_L^n(\tilde{\mathbf{r}}, E) \overline{G}_{LL'}^{nn'}(E) R_{L'}^{n'}(\tilde{\mathbf{r}}', E) \tag{2.268}$$

or represent this function through the function $\tilde{G}_{LL'}^{nn'}$:

$$\overline{G}_{LL'}^{nn'}(E) = \tilde{G}_{LL'}^{nn'} + \sum_{n'L''L''} \tilde{G}_{LL''}^{nn'}(E) [t_{l''}^{n''}(E)] \overline{G}_{L''L'}^{n''n'}(E). \tag{2.269}$$

The matrices $t_l^n(E)$ and $\tilde{t}_{LL'}^n(E)$ are determined from the equation

$$t_l^n(E) = \int d\mathbf{r} \int d\mathbf{r}' \, j_l(r\sqrt{E}) Y_L(r) t^n(r, r', E) j_{l'}(r'\sqrt{E}) Y_{L'}(\mathbf{r}'),$$
$$\tilde{t}_{LL'}^n(E) = \int d\tilde{\mathbf{r}} \int d\tilde{\mathbf{r}}' \, j_l(r\sqrt{E}) Y_L(\mathbf{r}) t^n(\mathbf{r}, \mathbf{r}', E) j_{l'}(\tilde{\mathbf{r}}'\sqrt{E}) Y_L'(\tilde{\mathbf{r}}). \tag{2.270}$$

Or in terms of the matrix $U_{LL''}$:

$$\tilde{t}_{LL'}^n(E) = \sum_{L''} U_{LL''}(s^n, E) t_{l''}^n(E) U_{L'L''}(s^n, E). \tag{2.271}$$

However, this formalism does not yet permit a self-consistent representation of the relaxation of a lattice containing an impurity metal, but rather only makes it possible to account for the effect of distortions on the physical properties of dilute alloys. We can hope that any further development of this approach will be quite useful in substantiating and predicting the properties of alloys. We believe that the most reasonable approach is to incorporate experimental information on lattice distortions (e.g., from EXAFS spectroscopy, x-ray diffuse scattering, etc.).

2.7. The cluster method for calculating the electronic structure of condensed matter

The majority of problems associated in the condensed matter theory are from the violation of ideal three-dimensional translational symmetry of the crystalline lattice. A cluster method is used to analyze the electronic structure in such cases. A cluster ordinarily will refer to a system consisting of one or many atoms, ions, or molecules, together with the corresponding boundary conditions. The boundary conditions are selected to account for the properties of the entire system simulated by the cluster. The range of application of the cluster model for describing physical phenomena is always limited by its size. Hence this method is best applied to phenomena whose characteristic lengths are less than the geometric dimensions of the cluster (e.g., localized excitations in solids, formation of self-localized quasi-particles, point defects, chemisorption processes, etc.).

The central issue in the cluster approach to calculating the electronic structure of condensed matter is the cluster size. On the one hand, it is intuitively clear that the larger the fragment of condensed matter, the more adequate the description of the electronic structure. However, on the other hand, the problem of calculating the electronic structure becomes insoluble with an unlimited increase in cluster size. Here it is important to remember that an additional problem of the so-called surface states arises in the cluster method; these states are localized along the peripheral atoms of the cluster and are related to bond breaking along the boundary of the crystal. The potential cluster effect is related to the limit on cluster size. The essence of this effect is that the equivalent atoms in a crystal may have a different potential in a cluster due to their different positions relative to the cluster boundary. Hence the potentials of atoms in the interior of a crystal may have deeper "wells" than the atoms along the external cluster boundary. As was demonstrated,[224] such a distortion of the potential in a cluster may play the dominant role in the formation of the electronic structure of the cluster.

A fragment approximation[224] is commonly used to eliminate the potential cluster effect. Using the MT approximation the potential within the jth MT sphere can be given as

$$v(r) = v^j(r) + \Delta^j_{cr}, \tag{2.272}$$

where $v^j(r)$ is the potential created by the charge distribution within the jth sphere and Δ^j_{cr} is the addition that correctly accounts for the contribution of the long-range part of the Coulomb interaction to the potential from the infinite number of atoms in the crystal. In order to calculate Δ^j_{cr} it is necessary to know the geometry and parameters of the crystal unit cell, the Madelung potential, and the electron densities within the MT spheres.

The problem of the surface states is often solved as follows: pseudoatoms (such as hydrogen atoms) are placed on an external coordination shell of a cluster, that is, on unsaturated bonds; these pseudoatoms form a chemical bond with the internal atoms of the cluster and thereby stabilize the electrons localized at the broken bonds. There is also a second method of suppressing edge effects.[225] Here, instead of the hydrogen atoms the same atoms as in the crystal are placed on the external coordination shell although the valence electrons of the edge atoms that participate in the chemical bond with the nearest neighbors and are not included in the cluster, and are excluded from the analysis.

Occasionally it is convenient to use the quasimolecular expanded unit cell (QEUC) method[226] to calculate the electronic structure of condensed matter, that is, to consider a quasimolecule whose atoms are distributed in the same manner as in the corresponding range of the model crystal, which in turn is represented by means of an expanded unit cell. In the QEUC model unlike the band transition the wave vector does not explicitly figure into the calculation. At the same time the quasimolecule analyzed in the QEUC model is (strictly speaking) a molecular system since the introduction of cyclic boundary conditions ensures a Hamiltonian symmetry that makes it possible to use the quasimolecule to successfully calculate certain states that belong to the space group of the initial crystal.[226] A significant element in the QEUC method is the electronic structure calculated using states found in a "narrowed" Brillouin zone.

Unlike the QEUC method, in the periodic cluster model,[227] the problem is solved in a direct lattice. Here a cyclic system containing a comparatively small number of cells of minimum size is considered instead of the primary crystal region. The remaining crystal is roughly accounted for as an effective field surrounding the periodic cluster.

The theoretical techniques for calculating the electronic structure of clusters can be divided into two major groups: empirical and nonempirical methods. The first group includes a rather large number of techniques. Here we will only briefly review the most common methods. Nonempirical methods can in turn be divided into two classes of techniques: those based on Hartree–Fock theory and methods utilizing local potentials dependent on electron density.

The wave function or molecular orbital describing the electron state in the molecular cluster is represented as an expansion in the basis functions (atomic orbitals):

$$\varphi_i(\mathbf{r}) = \sum_{A} \sum_{\mu} c_{i\mu}^{A} \chi_{\mu}^{A}(\mathbf{r}). \tag{2.273}$$

Here i labels the molecular orbitals while A_μ corresponds to the μth orbital of atom A designated χ_μ^A. The coefficients $C_{i\mu}^A$ are found by a variational technique from secular Hartree–Fock–Rutan equations corresponding to Hartree–Fock equation (1.19) in the LCAO approximation:

$$\det|F_{\mu\nu} - \varepsilon S_{\mu\nu}| = 0, \tag{2.274}$$

where

$$F_{\mu\nu} = h_{\mu\nu} + G_{\mu\nu}, \tag{2.275}$$

$$h_{\mu\nu} = \left\langle \chi_\mu \left| -\nabla^2 - 2 \sum_{A} \frac{z_A}{|r - R_A|} \right| \chi_\nu \right\rangle, \tag{2.276}$$

$$G_{\mu\nu} = \sum_{\varkappa\lambda} P_{\varkappa\lambda}\{2\langle \mu\nu|g|\varkappa\lambda\rangle - \langle \mu\lambda|g|\nu\varkappa\rangle\}, \tag{2.277}$$

$$g = 1/|\mathbf{r} - \mathbf{r}'|; \quad P_{\varkappa\lambda} = 2 \sum_{i=1} n_i c_{i\varkappa}^{*} c_{i\lambda}. \tag{2.278}$$

The approach based on self-consistent solution (2.274) is called the MO LCAO technique. The primary difficulty of this method lies in the evaluation of the integrals in Eq. (2.276). Hence analytic atomic functions as Slater orbitals or Gaussian orbitals are used in practical implementations. In order to obtain an identical level of accuracy using the Gaussian function bases and the Slater function bases, it is necessary to use a much larger number of iterations in the first case. However, the mathematical simplifications associated with using Gaussian functions compensate this inconvenience.

One method of simplifying the "first principles" calculation of the electronic structure of molecular clusters is to use the X_α method (see Chapter 1). Jones used the X_α method[228] to develop a routine for calculating the electronic structure of molecules based on multiple scattering theory (the X_α-RS method). The X_α-RS method is an analog of Green's function method used in band theory. The one-electron wave functions in the neighborhood of each atom decompose into partial waves and they are joined on the basis of Green's function formalism. The MT approximation is used in its simplest version. In this case the secular equation is obtained simply by joining the equations at the MT sphere boundary:

$$\det|t^{-1}(\varepsilon) - G(\varepsilon)| = 0, \tag{2.279}$$

where t is a diagonal matrix with matrix elements

$$t^{ij}_{lm,l'm'} = \delta_{ij}\delta_{ll'}\delta_{mm'}\frac{W^j[R_l, i_l]}{W^{-j}[R_l, K_l]}. \tag{2.280}$$

$W^j[R_l, i_l]$ is the Wronskyan matrix calculated at the boundary of the jth MT sphere with radial wave functions R_l and modified Bessel i_l and Hankel K_l functions. The matrix

$$G = \{G^{ij}_{lm,l'm'}\}$$

in Eq. (2.279) is the structure factor dependent solely on the molecular cluster energy and geometry:

$$G^{ij}_{lm,l'm'}(\mathbf{R},\varepsilon) = (1-\delta_{ij})(-1)^{l+l'}\sum_L I_L(lm,l'm')K_L(\sqrt{-\varepsilon}R)Y^*_{l,m-m'}(\hat{R}), \tag{2.281}$$

where I_L are the Gaunt factors, $\mathbf{R} = \mathbf{r}_i - \mathbf{r}_j$, and K_L is the modified Hankel function.

The primary calculation difficulty of the X_α-RS method is the nonlinear energy dependence of the matrix elements of secular matrix (2.279). After the linear techniques from band theory appeared, efforts were made to apply analogous techniques in cluster methods of calculating the electronic structure of condensed matter. A linearized version of the X_α-RS method is described in detail by Gunnarson *et al.*[229]

The primary drawback of the X_α-RS method is the use of the MT approximation. The multiple scattering formalism to the case of a random potential has been generalized[230] while conserving the form of secular equation (2.279). The approximation of overlapping MT spheres is also applied.[231]

In order to overcome the difficulties that arise from calculating the electronic structure of a cluster with a general potential, the discrete variation (DV) method was proposed.[232,233] In this method the Schrödinger equation is not solved in all space but rather only within a certain set of discrete points $\mathbf{r}_k$. The principle element underlying the DV method is the use of molecular orbitals (2.273) as the wave functions of the cluster electrons. The matrix elements of secular matrix (2.274) include the weight multipliers $\omega(\mathbf{r}_k)$. The primary problem of the DV method is the selection of $\mathbf{r}_k$ and $\omega(\mathbf{r}_k)$. In the majority of studies the selection of points $\mathbf{r}_k$ is determined by a stochastically distributed set in the selected cluster volume in accordance with the given distribution density function $D(r)$. The weight multiplier $\omega(\mathbf{r})$ is defined here as

$$\omega(\mathbf{r}_k) = [ND(\mathbf{r})]^{-1}, \tag{2.282}$$

where N is the total number of points $\mathbf{r}_k$. In ordinary calculations $N \sim 1500$–3000.[233]

The exchange potential in the DV method is as a rule selected as a local X_α approximation.

We now consider the empirical methods of calculating the electronic structure of clusters. All these reduce to a simplification of solving the problem described by Eq. (2.274) using two approximations: the Malliken approximation and the zero-differential overlap (ZDO) approximation.

The essence of the Malliken approximation lies in the substitution of two-electron and three- and four-center integrals with one- and two-center integrals using the following equations:

$$\langle\mu\nu|g|\lambda\nu\rangle = -S_{\mu\nu}[\langle\mu\nu|g|\mu\nu + \lambda\nu|g|\lambda\nu\rangle], \tag{2.283}$$

$$\langle\mu\nu|g|\varkappa\mu\rangle = -S_{\mu\varkappa}S_{\mu\nu}[\langle\mu\mu|g|\mu\mu\rangle + \langle\mu\varkappa|g|\mu\varkappa\rangle + \langle\nu\mu|g|\nu\mu\rangle + \langle\nu\varkappa|g|\nu\varkappa\rangle], \tag{2.284}$$

$$\langle\mu\nu|g|\lambda\varkappa\rangle = \tfrac{1}{4}S_{\mu\lambda}S_{\nu\varkappa}[\langle\mu\nu|g|\mu\nu\rangle + \langle\mu\varkappa|g|\mu\varkappa\rangle + \langle\lambda\nu|g|\lambda\nu\rangle + \langle\lambda\varkappa|g|\lambda\varkappa\rangle]. \tag{2.285}$$

Using these equations it is possible to write Eq. (2.277) as

$$G_{\mu\nu} = \sum_{\lambda} Q_\lambda[2\langle\mu\lambda|g|\nu\lambda\rangle - \langle\mu\lambda|g|\lambda\nu\rangle], \tag{2.286}$$

where $Q_\lambda = \Sigma_i n_i q_i(\lambda)$ is the sum population of the orbital χ_λ.

A semiempirical calculation method was proposed[234] in which

$$H_{\mu\mu} = -I_\mu, \tag{2.287}$$

$$H_{\mu\nu} = \tfrac{1}{2}kS_{\mu\nu}(H_{\mu\mu} + H_{\nu\nu}) \quad (\mu \neq \nu), \tag{2.288}$$

where k is the fitting parameter and I_μ is the ionization potential of orbital μ. This method has been called the extended Hückel method (EHM). The EHM method for $k = 2$ and 0 atomic charges becomes the method described by Eq. (2.274) with matrix elements (2.286). Ordinarily k is taken over the range 1.5–2.5 in practice. The variation of k from 2 in the EHM calculations can be attributed to the attempt to account for kinetic energy integrals in the nondiagonal elements. When the charge $q(\lambda)$ is calculated in the Malliken approximation the population of the atomic orbital overlap region $2\,\Sigma_\lambda P_{\varkappa\lambda}S_{\varkappa\lambda}$ [see Eq. (2.278)] is divided in half between the two atoms. Such an approximation works well in the case of a symmetrical electron configuration in the interatomic region such as covalent molecules (clusters). Otherwise this scheme will assign excess charge to the atom in which the orbitals have a higher degree of diffusion. Therein lies the primary drawback of the Malliken scheme.

The ZDO approximation is also widely used in electronic structure calculations of molecules and clusters. The differential overlap of two atomic orbitals, χ_μ and χ_ν, refers to the probability of the ith electron being in a volumetric element common to χ_μ and χ_ν which is given by the equation

$$\delta_{\mu\nu} = \chi_\mu(i)\chi_\nu(i). \tag{2.289}$$

The ZDO approximation assumes that $\delta_{\mu\nu} \neq 0$ if $\mu = \nu$. The result of this approximation is a neglect of the overlap integrals between different atomic orbitals, that is, $S_{\mu\nu} = \delta_{\mu\nu}$. Moreover, the matrix elements of the secular equation do not account for all integrals containing the orbital product $\chi_\mu \chi_\nu$ for $\mu \neq \nu$. The matrix elements of the Hamiltonian in the ZDO approximation take the form

$$F_{\mu\mu} = U^{\mathrm{A}}_{\mu\mu} + \sum_{\mathrm{B}\neq\mathrm{A}} V^{\mathrm{B}}_{\mu\mu} + \sum_{\lambda} P_{\lambda\lambda}\Gamma_{\mu\lambda} - \tfrac{1}{2}P_{\mu\mu}\Gamma_{\mu\mu}, \tag{2.290}$$

$$F_{\mu\nu} = H_{\mu\nu} - \tfrac{1}{2}P_{\mu\nu}\Gamma_{\mu\nu}, \tag{2.291}$$

where

$$U^{\mathrm{A}}_{\mu\mu} = \left\langle \mu \left| -\nabla^2 - \frac{2z_A}{r_A} \right| \mu \right\rangle, \quad V^{\mathrm{B}}_{\mu\mu} = \left\langle \mu \left| \frac{2z_B}{r_B} \right| \mu \right\rangle, \tag{2.292}$$

$$\Gamma_\nu = \langle +gm\nu | g | \mu\nu \rangle.$$

A number of different versions of methods utilizing the ZDO have been proposed. In some of these the requirements of the ZDO are completely satisfied as is the case in the CNDO method (complete neglect of differential overlap).[235] In other methods, such requirements are applied only to certain types of integrals as is the case in the PNDO method (partial neglect of differential overlap).[235]

Variable parameters are used to estimate the integrals that remain within the framework of the ZDO approximation. Here it is assumed that the two-electron integrals $\Gamma_{\mu\nu}$ depend only on the type of atoms, that is, $\Gamma_{\mu\nu} = \Gamma_{\mathrm{AB}}$ in the CNDO method. An analogous approximation is also used for the $V^{\mathrm{B}}_{\mu\mu} = V^{\mathrm{B}}_{A}$ integrals. The $U_{\mu\mu}$ integrals are evaluated through the ionization potentials I_μ (CNDO scheme/1) or the ionization potentials as well as the electron affinity A_μ (CNDO scheme/2).[235]

2.8. Approaches to calculating the electronic structure of alloys

The development of techniques for calculating electronic structure (OPW, APW, LAPW, LMTO, the pseudopotential, and other methods), as well as increasing computer capabilities, has given rise to a favorable situation in the theory of metals. Indeed it is now possible to obtain metal lattice parameters with good accuracy and to determine (as well as predict) existing crystalline structures; the existence and temperature of polymorphic transformations, phonon spectra, etc. Here either calculations from first principles are used or calculations are carried out with a minimum number of fitting parameters (e.g., the pseudopotential parameters can be determined from a single measured characteristic of the metal while the derived potential is used in calculating other properties). Such extraordinary success has not yet been achieved for alloys. This is due primarily to the lack of translational symmetry in disordered alloys (in the complete sense of this term) which makes it impossible to use standard band techniques. We will briefly review the primary theoretical schemes used to calculate the electronic structure of alloys.[236–238]

The Hamiltonian of a disordered alloy $A_{1-x}B_x$ can be given as

$$H = \frac{p^2}{2m} + \sum_i V^\alpha(\mathbf{r} - \mathbf{R}_i), \quad \alpha = (\mathrm{A},\mathrm{B}). \tag{2.293}$$

The second term represents the sum of the potentials of atoms A and B randomly distributed in a regular lattice. Ordinarily it is assumed that these potentials overlap and have a MT form. After determining the Hamiltonian it is possible to unambiguously determine Green's function. Its pole determines the energy of the alloy bands while the matrix elements in r or P space make it possible to find many properties of the alloys. However, it is necessary to carry out averaging over all possible configurations of atoms A and B in disordered alloys. Ordinarily the scheme for formulating the averaged Green's function $\langle G\rangle$ is as follows: we represent the T matrix through the single-site t matrices:

$$T=\sum_i\left(t_i+\sum_{j\neq i}t_iG_0t_j+\sum_{j\neq i}\sum_{k\neq j}t_iG_0t_jG_0t_k+\cdots\right),\quad t_j=V_i(1-G_0V_i). \tag{2.294}$$

We average T over all possible configurations:

$$\langle T\rangle=\sum_i\left(\langle t_i\rangle+\sum_{j\neq i}\langle t_iG_0t_j\rangle+\sum_{j\neq i}\sum_{k\neq j}\langle t_iG_0t_jG_0t_k\rangle+\cdots\right). \tag{2.295}$$

This equation still does not contain any approximations. Within the framework of the single-site approximation $\langle T\rangle$ is substituted by T_{eff}:

$$T_{\text{eff}}=\sum_i\left[t_i^{\text{eff}}+\sum_{j\neq i}t_i^{\text{eff}}G_0t_j^{\text{eff}}+\sum_{j\neq i}\sum_{k\neq j}t_i^{\text{eff}}G_0t_j^{\text{eff}}G_0t_k^{\text{eff}}+\cdots\right]. \tag{2.296}$$

The meaning of such a substitution lies in the fact that a disordered system is replaced by a medium where each site in the medium contains an effective scatterer, T_{eff}, that is selected so that scattering by the effective medium will be the same as the averaged scattering of the actual system. By substituting $\langle T\rangle$ into the expression averaged over the configurations for Green's function

$$\langle G\rangle=\langle G_0\rangle+\langle G_0TG_0\rangle, \tag{2.297}$$

we can obtain[239]

$$\langle G(\mathbf{p},\mathbf{p}',E)\rangle=\frac{\delta_{pp'}}{E-p^2}+\sum_k\frac{4\pi^2N\delta_{p-p'}k}{(E-p)^2(E-p'^2)}\times\sum_{LL'}Y_L(p)\left[\left\{t_l(\mathbf{p},\mathbf{p}')-\frac{t_l(\mathbf{p},\varkappa)t_l(\varkappa,\mathbf{p}')}{t_l(\varkappa,\varkappa)}\right\}\delta_{LL'}+\frac{t_l(\mathbf{p},\varkappa)}{t_l(\varkappa,\varkappa)}[t^{-1}(\varkappa,\varkappa)-B(\mathbf{p},E)]_{LL'}^{-1}\frac{t_{l'}(\varkappa,\mathbf{p}')}{t_{l'}(\varkappa,\varkappa)}\right]Y_{L'}(\mathbf{p}'), \tag{2.298}$$

where $\mathbf{p}=\mathbf{k}+\mathbf{G}$; $\mathbf{k}$ is the wave vector in the first Brillouin zone, $\varkappa=\sqrt{E}$, N is the number of lattice sites, and $A(\mathbf{p},E)=A(\mathbf{k},E)$ are the matrices of the structure constants. The pole of Green's function and, consequently, the energy bands can be found from a solution of the secular equation

$$\det|\tau^{-1}(E)-B(\mathbf{k},E)|=0. \tag{2.299}$$

The solution of this equation in the case of alloys yields complex bands. The imaginary part of the energy characterizes the attenuation of the quasiparticle states. If the weight of each state is equal to unity for an ordered alloy system, the area under the peak of its integral corresponds to the Bloch spectral density.

There are also other simpler approaches to calculating the electronic structure of alloys.

The simplest approach that is suitable for analyzing the electronic structure of alloys is the strong coupling model. The difference in the electronic structure of the alloy components is neglected within the framework of this model. Information on the components of the alloy and their concentration is contained solely in the quantity $\bar{n}$: concentration of electrons per atom

$$\bar{n}=cN_{\mathrm{A}}+(1-c)N_{\mathrm{B}}, \tag{2.300}$$

where N_{A} and N_{B} is the number of electrons per atom in pure metals A and B and c is the concentration of A atoms in the alloy.

Knowing $\bar{n}$ it is possible to determine the Fermi energy of the alloy. In this approach it is not possible to adequately convey the features of the electronic structure of the alloys in the overwhelming majority of the test systems. Moreover, in this case when the components have a similar band structure there is hope for a qualitatively accurate description of the alloying effects.

An approximation that is conceptually close to this model is the virtual crystal approximation which was used most commonly in the first studies on pseudopotential alloy theory. In this method the potential is represented at a random point as a sum of the component potentials for the disordered alloy with the corresponding weight multipliers c and $(1-c)$:

$$V(r)=cV_{\mathrm{A}}(r)+(1-c)V_{\mathrm{B}}. \tag{2.301}$$

The alloy is therefore substituted by a "single component" crystal with an averaged component potential. In this case the difference between the band structures of the components is not ignored (at least partially). Deriving potential (2.301) makes it possible to then use any band method to determine $E(k)$ and other characteristics of the alloy. This approach assumes that each crystal site contains an identical scatterer characterized by the t matrix

$$t_l^{\mathrm{cont}}(E)=-\frac{1}{k}\exp(i\eta_l^{\mathrm{AB}})\sin\eta_l^{\mathrm{AB}}, \tag{2.302}$$

where η^{AB} are the scattering phase shifts found for the potential $V(r)$.

An analysis of the research results from the virtual crystal approximation have demonstrated that meaningful results are obtained only when $V_{\mathrm{A}}(r)$ and $V_{\mathrm{B}}(r)$ do not differ substantially. Moreover, this approximation generally becomes meaningless in alloys where short-range order (SRO) effects are noticeable.

The next approximation in terms of complexity is the average t matrix approximation (ATA) in which t matrices are defined for each potential A and B while the t matrix of the alloy is found as the weighted average of t_A and t_B, that is,

$$t^{\mathrm{ATA}}(E) = -\frac{1}{k}\,[c\exp(i\eta_l^{\mathrm{A}})\sin\eta_l^{\mathrm{A}} - (1-c)\exp(i\eta_l^{\mathrm{B}})\sin\eta_l^{\mathrm{B}}]\,. \tag{2.303}$$

Essentially this is a single-site approximation like the virtual crystal model.

If we formulate a potential that yields an average t matrix this potential will be complex and energy-dependent:

$$U(E,\mathbf{r}) = \sum_n V(E,|\mathbf{r}-\mathbf{R}_n|). \tag{2.304}$$

The matrix elements of Green's function in the site number orbital moment representation are given as

$$G_{LL'}^{nn'} = g_{LL'}^{nn'}(E) + \sum_{\substack{L_1L_2\\ n_1n_2}} g_{LL'}^{nn'}(E)\,T_{L_1L_2}^{n_1n_2}(E)\,g_{L_2L_1}^{n_2n_1}(E), \tag{2.305}$$

where $T_{LL'}^{nn'}$ is the scattering matrix of the system satisfying the equation

$$T_{LL'}^{nn'}(E) = t_l^n(E)\,\delta_{LL'}\delta_{nn'} + \sum_{L_1n_1} t_l^n g_{LL'}^{nn_1}(E)\,T_{L_1L'}^{n_1n'}(E). \tag{2.306}$$

If we introduce the averaged scattering matrix at the site then by carrying out the Fourier transform (2.306) we obtain

$$\langle T(E)\rangle_{LL'} = \frac{1}{\tau}\int d\mathbf{k}\left[\frac{1}{\langle t(E)\rangle^{-1} - g(E,\mathbf{k})}\right]_{LL'}. \tag{2.307}$$

If the averaged t matrix $t^{(\mathrm{ATA})}$ is substituted into this equation in place of $\langle t(E)\rangle$ it is possible to find $\langle T(E)\rangle_{LL'}$. This approach is not self-consistent. It can, however, be used to successfully solve many problems for alloys. A self-consistent definition of the scattering matrix of the system can easily be found by means of a coherent potential approximation (CPA) (Refs. 236,and 238) in which $\langle T(E)\rangle_{LL'}$ is implicitly expressed through

$$\begin{aligned}\langle T(E)\rangle_{LL'} = c&\left[\frac{\langle T(E)\rangle}{1-(\langle t(E)\rangle^{-1} - t^{\mathrm{A}}(E)^{-1})\langle T(E)\rangle}\right]_{LL'}\\ &+ (1-c)\left[\frac{\langle T(E)\rangle}{1-(\langle T(E)\rangle^{-1} - t^{\mathrm{B}}(E)^{-1})\langle T(E)\rangle}\right]_{LL'}.\end{aligned} \tag{2.308}$$

Such an analysis of the coherent potential method and all its possible approximations can be found in Ehrenreich and Schwartz's comprehensive review.[236]

2.9. Calculation of the electronic structure of liquid metals and melts

Investigating the electronic structure of liquid metals and melts as representatives of a disordered system has its own special feature. In this case finding the function

$n(E)$ is a rather cumbersome problem that is solved using the apparatus of Green's function. All existing methods of calculating $n(E)$ can be divided into self-consistent and nonself-consistent methods. The nonself-consistent methods are much more rapidly implemented numerically. These include Lax's quasicrystal approximation[240]; the modified Schwartz, Peterson, and Bansil approximation[241]; the Cyrot–Lackman chain approximation (recursion) method[242]; the Anderson–McMillan averaged t matrix method[243]; and the Ballentine nonself-consistent variant of calculating $n(E)$ for nontransition metals.[244] The self-consistent methods include the Gyorffy quasicrystal approximation[245]; the Ehrenreich and Schwartz self-consistent approximation[246]; and the Roth effective medium approximation.[247] A self-consistent method of cluster the self-energy part in the Ballentine scheme is used for transition metals. There are also intermediate Lloyd[248] and Ishida and Yonesawa[249] techniques. For nontransition liquid metals and their alloys it has been possible to properly describe the electronic structure within the framework of nearly free electron (NFE) theory. The NFE approximation is not suitable in the case of strong scattering systems (transition metals, noble metals).

The discussion below will be as follows: we consider a method of calculating $n(E)$ of nontransition liquid metals based on Edward's theory after which the Anderson–McMillan method will be given, which is suitable for transition and noble metals.

2.9.1. *Edward's method*[250]

We define Green's function in the standard manner:

$$G(E)=(E-H)^{-1}=\sum_n \frac{|\varphi_n\rangle\langle\Psi_n|}{E-\varepsilon_n} \tag{2.309}$$

together with the spectral operator

$$\rho(E)=\lim_{\eta\to 0}\left(-\frac{1}{2\pi i}\right)[G(E-i\eta)-G(E-i\eta)]. \tag{2.310}$$

We will use the momentum representation. We will require the following quantities:

$$G(\mathbf{k},E)=\langle \mathbf{k}G|(E)|\mathbf{k}\rangle, \tag{2.311}$$

$$\mathscr{G}(\mathbf{k},E)=\langle G(\mathbf{k},E)\rangle, \tag{2.312}$$

where $\mathscr{G}(\mathbf{k},E)$ is Green's function averaged over the ensemble. The spectral function yields the momentum representation of electrons of energy E and takes the form

$$\rho(\mathbf{k},E)=\langle\mathbf{k}|\rho(E)|\mathbf{k}\rangle=-\frac{1}{\pi}\,\mathrm{Im}\,G(\mathbf{k},E+i0)=\sum_n |\langle\mathbf{k}|\Psi_n\rangle|^2\delta(E-\varepsilon_n). \tag{2.313}$$

The following sum rule derives from the completeness of the eigenfunctions $\{\Psi_n\}$ for $\rho(\mathbf{k},E)$:

$$\int_{-\infty}^{\infty}\rho(\mathbf{k},E)dE=1. \tag{2.314}$$

The density of the electron states per unit of energy and volume (for the one-spin orientation) is given by the equation

$$n(E)=\Omega^{-1}\,\mathrm{Sp}\,\rho(E)=(2\pi)^{-3}\int d^3\mathbf{k}\,\rho(\mathbf{k},E). \qquad (2.315)$$

Ω is the volume of the system. Edwards proposed partial summing of the series of Green's function. Here it is proven that all subsequent odd degrees of the perturbation are quantities of the same order in the density of states as the even terms that precede them.

We assume that the electron travels in a field generated by the potentials of all atoms. We write in the momentum representation for the matrix element of the potential

$$\langle\mathbf{k}|\mathrm{v}|\mathbf{k}+\mathbf{q}\rangle=\langle\mathbf{k}|\mathrm{v}|\mathbf{k}+\mathbf{q}\rangle\sum_j e^{i\mathbf{q}\mathbf{R}_j}, \qquad (2.316)$$

where $\mathrm{v}(\mathbf{r})$ is the potential centered at a single site and $\mathbf{R}_j$ is the radius vector of the j site.

We treat the liquid metal as an ensemble of identical neutral pseudoatoms.[251] Matrix element (2.316) fluctuates about the mean zero value. Hence in considering averaging over the ensemble it is more correct to utilize the squared matrix element

$$\langle|\langle\mathbf{k}|\mathrm{v}|\mathbf{k}+\mathbf{q}\rangle|^2\rangle=\frac{1}{N}\,|N\langle\mathbf{k}|\mathrm{v}|k+q\rangle|^2A(q), \qquad (2.317)$$

where

$$A(\mathbf{q})=N^{-1}\left\langle\left|\sum_j e^{i\mathbf{q}\mathbf{R}_j}\right|^2\right\rangle. \qquad (2.318)$$

The primary calculation problem in finding the density of the electron states is averaging Eq. (2.311) over the ensemble. Within the framework of Green's function formalism we have

$$n(E)=-\frac{1}{\pi}\,\mathrm{Im}\,\mathrm{Sp}\,\langle G(\mathbf{k},E)\rangle, \qquad (2.319)$$

$$G(\mathbf{k},E)=\left[E-k^2-\sum(\mathbf{k},E)\right]^{-1}. \qquad (2.320)$$

In the second order of perturbation theory the self-energy part of Σ takes the form[250]

$$\Sigma_2(\mathbf{k},E)=-\frac{n}{8\pi^2k}\int_0^\infty d\mathbf{q}\,\mathbf{q}\,|u(\mathbf{q})|^2A(\mathbf{q})\,\ln\left|\frac{(|\mathbf{q}+\mathbf{k}|^2-E)}{(|\mathbf{q}-\mathbf{k}|^2-E)}\right|, \qquad (2.321)$$

where n is the atomic density ($n=N/\Omega$) and q is the scattering vector

$$u(\mathbf{q})=\Omega\langle\mathbf{k}|\mathrm{v}|\mathbf{k}+\mathbf{q}\rangle=\Omega\mathrm{v}(\mathbf{q}). \qquad (2.322)$$

The spectral density function is given by the expression[244]

$$\rho(\mathbf{k},E)=\frac{2\ \mathrm{Im}\ \Sigma_2\ (\mathbf{k},E)}{[E-k^2-\mathrm{Re}\ \Sigma_2\ (\mathbf{k},E)]^2+[\mathrm{Im}\ \Sigma_2\ (\mathbf{k},E)]^2}. \tag{2.323}$$

The imaginary part of $\Sigma_2(\mathbf{k},E)$ represents the magnitude of smearing associated with the average path length of the free electrons

$$\mathrm{Im}\ \Sigma_2\ (\mathbf{k},E)=-\frac{n\pi}{8\pi^2 k}\int_{|k-\sqrt{\varepsilon}|}^{k+\sqrt{\varepsilon}} d\mathbf{q}|u(\mathbf{q})|^2 A(\mathbf{q})\mathbf{q}. \tag{2.324}$$

The real part of Σ_2 is given as

$$\mathrm{Re}\ \Sigma_2\ (\mathbf{k},E)=-\frac{n}{8\pi^2 k}\int_0^\infty d\mathbf{q}\ \mathbf{q}|\mathrm{v}(\mathbf{q},\mathbf{k})|^2 \ln\left|\frac{|\mathbf{q}+\mathbf{k}|^2-E}{|\mathbf{q}-\mathbf{k}|^2-E}\right| A(\mathbf{q}). \tag{2.325}$$

Henceforth $\mathrm{v}(\mathbf{q},\mathbf{k})$ will denote the pseudopotential that in the general case is nonlocal. In Chap. 3 we carry out a detailed analysis of modern pseudopotential theory.

Therefore, the scheme for calculating the density of electron states assumes knowledge of only the structure factor and the pseudopotential. The structure factor can be taken either from experiment or can be obtained theoretically in the hard sphere approximation.

2.9.2. The Anderson–McMillan method

Assume the medium contains randomly distributed scatterers with potentials $V(\mathbf{r}-\mathbf{R}_n)$. The potential of the medium at point $\mathbf{r}$ takes the form

$$V(\mathbf{r})=\sum_n \mathrm{v}(\mathbf{r}-\mathbf{R}_n)=\sum_n \mathrm{v}_n(\mathbf{r}). \tag{2.326}$$

Green's function of the system can be given as

$$G=\sum_\alpha G_0 t_\alpha G_0+\sum_{\alpha\neq\beta} G_0 t_\alpha G_0 t_\beta G_0+\sum_{\alpha\neq\beta\neq\gamma} G_0 t_\alpha G_0 t_\beta G_0 t_\gamma G_0+\cdots$$
$$=\left[\frac{1}{1-\sum_\alpha G_0 t_\alpha}\right] G_0, \tag{2.327}$$

where

$$G_0=(E-H_0)^{-1}. \tag{2.328}$$

α and β are the types of scattered particles. Equation (2.327) is recast as

$$G=\frac{1}{E-H_0-\sum_\alpha t_\alpha}. \tag{2.329}$$

We often assume for a system of arbitrary scatterers that

$$\left(\sum_\alpha t_\alpha\right)_{kk'}=nt_{kk}\delta(\mathbf{k}-\mathbf{k}'), \tag{2.330}$$

that is, only the diagonal matrix elements are conserved although here α-type scattering may be accounted for twice. Hence Anderson and McMillan proposed that in order to avoid accounting for α-type scattering twice the α portion of the description of the medium should be replaced by true scattering, that is,

$$(t)_{\text{scattering}} = (t)_{\text{medium}}$$

or (2.331)

$$(t)_{\text{scattering outside the medium}} = 0.$$

Then Green's function can be given as

$$G(\mathbf{k},\varepsilon) = \left\{ \varepsilon - k^2 - 2\pi N \sum_l [(2l+1)\exp(i\delta_l)\sin\delta_l] H^{-1} \right\}^{-1}. \tag{2.332}$$

The matrix elements t_{kk} are given as

$$t_{kk} \sim \frac{1}{k} \sum_l (2l+1)\exp[i\delta_l(\mathbf{k},\varepsilon)]\sin\delta_l(\mathbf{k},E) = 0. \tag{2.333}$$

This relation is a dispersion relation for random energy.

We find the density of the electron states $n(E)$:

$$n(E) = \frac{1}{\pi}\,\mathrm{Sp}\,\mathrm{Im}\,G(\varepsilon) = \frac{1}{\pi}\int dr\, G(\mathbf{r},\mathbf{r}',\varepsilon) = \frac{1}{\pi}\sum_l (2l+1)\int_0^{r_c} d\mathbf{r}\, G_l(\mathbf{r},\mathbf{r}')r^2. \tag{2.334}$$

where $\mathbf{r}_c$ is the radius of the atomic cluster within the Wigner–Seitz sphere and $G_l(\mathbf{r},\mathbf{r}')$ is the radial Green's function satisfying the equation

$$\frac{1}{r^2}\frac{d}{dr}\left[r^2\frac{dG(\mathbf{r},\mathbf{r}')}{dr}\right] + \left[E - \frac{l(l+1)}{r^2} - V(r)\right]G(r,r') = \frac{\delta(r-r')}{r^2}. \tag{2.335}$$

Here $V(\mathbf{r})$ is a potential of the type

$$V(r,E) = \begin{cases} V_{\mathrm{MT}}, & r < r_c, \\ V_c, & r > r_c, \end{cases} \tag{2.336}$$

where V_{MT} is the crystal MT potential and V_c is a complex quantity representing the form of the medium selected. In defining this form, Anderson and McMillan[243] required plane waves propagating in the effective medium to undergo no forward scattering. For the potential this means

$$V(r,E) = V(r,E) - V_c(E), \tag{2.337}$$

where V_c is found from the dispersion relation [see Eq. (2.333)]

$$k_c^2(E) = E - V_c(E). \tag{2.338}$$

Here k_c is the complex wave vector in the effective medium. Since we are interested in the natural electron waves, that is, solutions for $r > r_{\mathrm{MT}}$ (the MT radius), the following function will satisfy the radial Schrödinger equation:

$$\mathscr{T}_l(kr)=\cos\eta_l\cdot j_l(kr)-\sin\eta_l n_l(kr). \tag{2.339}$$

We seek Green's function in the form

$$G_l(r,r')=kF_l(k,r_>)\mathscr{T}_l(k,r_<), \tag{2.340}$$

where $r_>$, $r_<$ is the largest and smallest from $\{r,r'\}$. In order to conserve the normalization of $G_l(r,r')$ the following condition must hold:

$$W(\mathscr{T}_l,F_l)=W(j_l,n_l)=\text{const} \tag{2.341}$$

or

$$W(\mathscr{T}_l,F_l)=\mathscr{T}_l(kr)F_l'(kr)-F_l(kr)\mathscr{T}_l'(kr). \tag{2.342}$$

This gives the discontinuity of the function G_l for $r=r'$ which is such that it satisfies the equation

$$\lim_{\varepsilon\to 0}\int_{r'-\varepsilon}^{r'+\varepsilon}dr\frac{\delta(r-r^2)}{r^2}=\lim_{\varepsilon\to 0}\int_{r'-\varepsilon}^{r'+\varepsilon}dr\frac{1}{r^2}\frac{d}{dr}\left[r^2\frac{dG_l(r,r')}{dr}\right]+o(\varepsilon)$$

$$=k^2[\mathscr{T}(kr')+F'(kr')-F(kr')\mathscr{T}'(kr')]$$

$$=\frac{1}{(k^2r'^2)}. \tag{2.343}$$

The integral on the left-hand side is equal to $1/r'^2$ and Eq. (2.342) can be satisfied by setting

$$F_l=N_l+a_l\mathscr{T}_l, \tag{2.344}$$

where

$$N_l(kr)=-[(\cos\eta_l)n_l(kr)-(\sin\eta_l)j_l(kr)],\quad r>r_{\rm MT}. \tag{2.345}$$

where $a_l=1$ we obtain

$$F_l=N_l+i\mathscr{T}_l=e^{i\eta_l}h_l^{(+)}(kr), \tag{2.346}$$

where $h^{(+)}(kr)$ is the Hankel function. Using Eqs. (2.334) and (2.340) we obtain the density of states

$$n(E)=\frac{k}{\pi}\sum_l(2l+1)\left[\int_0^{r_c}dr\,r^2\mathscr{T}_l^2(kr)\right]\text{Im}\,a_l. \tag{2.347}$$

We can show that

$$\int_0^{r_c}dr\,r^2\mathscr{T}_l^2(kr)=r_c^2\mathscr{T}_l^2(kr_c)\frac{d\gamma_l(F_{rl})}{dE}, \tag{2.348}$$

where $d\gamma_l/dE$ is the logarithmic derivative at the boundary of the atomic sphere r_c. Finally we obtain

$$n(E)=\frac{k}{\pi}r_c^2\sum_l(2l+1)\mathscr{T}_l^2(kr_c)\frac{d\gamma_l(E_{rl})}{dE}\text{Im}\,a_l. \tag{2.349}$$

After some simple transforms[243] we have

$$n(E) = \frac{1}{\pi} \sum_l (2l+1) \left[\frac{d\eta_l}{dE} - \frac{1}{2} r_c^2 \mathscr{T}_l^2(kr_c) \frac{\partial \gamma_l(k\eta_l)}{\partial k} \right]. \tag{2.350}$$

This equation can be simplified by using the relation*

$$\operatorname{Im} ka_l = [r_c^2 \mathscr{T}_l^2(kr_c)]^{-1} \operatorname{Im}[(\gamma_l - \gamma_l^c)^{-1}]^{-1}, \tag{2.351}$$

where γ_l and γ_l^c are the real and complex variables, respectively. Finally we find

$$n(E) = -\frac{\operatorname{Im}}{\pi} \sum_l (2l+1) \frac{d\gamma_l/dE}{\gamma_l - \gamma_l^c}. \tag{2.352}$$

The quantity a_l through which the dispersion law of the medium enters into the densities of states is determined by the equality of the logarithmic derivatives at the boundary of a cluster of radius:

$$k[N_l(kr_c) + a_l \mathscr{T}_l(kr_c)]/(N_l + a_l \mathscr{T}_l) = k_c h_l^{(+)}(k_c r_c)/h_l^{(+)}(k_c r_c). \tag{2.353}$$

The scheme for calculating $n(E)$ is rather transparent: we formulate the atomic potentials followed by the MT potentials, obtain the real logarithmic derivatives, and use the equations given above to find $n(E)$. Below we illustrate the application of the Anderson–McMillan method to metal melts.

2.10. Application of density functional theory to calculations by means of molecular dynamics

Electronic structure calculations based on the DFT and computer modeling of complex systems at finite temperatures by means of molecular dynamics (MD) (Ref. 252) have facilitated the rapid progress in our understanding of the physical properties of condensed matter. However, the MD calculations provide no information on the electronic properties of these systems. At the same time the MD method can be used to rather easily calculate the equations of state and the dynamic properties of both disordered and ordered systems. On the other hand, although the DFT provides us with direct information on the electronic structure of condensed systems, it cannot be used to calculate dynamical properties or thermodynamic properties in the majority of interesting cases. Therefore, Yastrebov and Katsnelson have proposed[255] a MD method based on the DFT. This method can be used to obtain the properties of the ground state of both ordered and disordered systems and to implement "first principles" MD modeling without relying on empirical Lennard-Jones potentials.

Unlike the self-consistent field method we write the total energy functional which now includes the nuclear coordinates $\{R_I\}$ and certain parameters $\{\alpha_j\}$ related to, for example, the volumetric properties, voltages, etc., as

$$E(\{\varphi_i\},\{R_I\},\{\alpha_\nu\}) = \sum_i \int_\Omega dr\, \varphi_i^*(\mathbf{r})(-\nabla^2)\varphi_i(\mathbf{r}) + U[\rho(r),\{R_I\},\{\alpha_\nu\}]. \tag{2.354}$$

*See, for example, J. J. Olson, Phys. Rev. B. 12, 2908 (1975).

The functional U contains the nuclear Coulomb repulsion, and the effective interaction between electrons, including interaction with nuclei, the Hartree term, exchange, and correlation.

Hence minimization of Eq. (2.354) is treated as a complex optimization problem with efficiency function E and variational parameters $\{\varphi_i\}$, $\{R_I\}$, and $\{\alpha_\nu\}$. Assuming these parameters are time-dependent we introduce the Lagrangian[253]

$$\mathscr{L}=\frac{1}{2}\sum_i \mu \int_\Omega d\mathbf{r}|\dot{\varphi}_i|^2+\frac{1}{2}\sum_I M_I\dot{R}_I^2+\frac{1}{2}\sum_\nu \mu_\nu\dot{\alpha}_\nu^2-E[\{\varphi_i\},\{R_I\},\{\alpha_\nu\}], \tag{2.355}$$

$$\int_\Omega dr\,\varphi_i^*(\mathbf{r},t)\varphi_j(\mathbf{r},t)=\delta_{ij}. \tag{2.356}$$

In Eq. (2.355) the dot designates the time derivative, M_I are the ion masses, and μ and μ_ν are arbitrary parameters. We obtain from Lagrangian (2.355) an "equation of motion" for the parameters $\{\varphi_i\}$, $\{R_i\}$, and $\{\alpha_\nu\}$[253]:

$$\mu\ddot{\varphi}_i(\mathbf{r},t)=-\delta E/\delta\varphi_i^*(\mathbf{r},t)+\sum_k \Lambda_{ik}\varphi_k(\mathbf{r},t), \tag{2.357}$$

$$M_I\ddot{R}_I=-\nabla_{R_I}E, \tag{2.358}$$

$$\mu_\nu\ddot{\alpha}_\nu=-(\partial E/\partial\alpha_\nu), \tag{2.359}$$

where Λ_{ik} are the Lagrange multipliers introduced to satisfy condition (2.356). Of the last three equations only Eq. (2.358) has a real physical meaning to describe ion motion. The dynamics associated with $\{\varphi_i\}$ and $\{\alpha_\nu\}$ [Eqs. (2.357) and (2.358)] are imaginary and are treated only as a means of relaxing the computational scheme.

After the parameters $\{\varphi_i\}$, $\{R_I\}$, and $\{\alpha_\nu\}$ are found by solving Eqs. (2.357)–(2.359) it is possible to calculate the classical kinetic energy

$$k=\frac{1}{2}\sum_i \mu \int_\Omega dr|\dot{\varphi}_i|^2+\frac{1}{2}\sum_I M_I\dot{R}_I^2-\frac{1}{2}\sum_\nu \mu_\nu\dot{\alpha}_\nu^2. \tag{2.360}$$

Then the equilibrium value $\langle k\rangle$ is calculated as the time average along the trajectories obtained by solving Eqs. (2.357)–(2.359). The equilibrium value $\langle k\rangle$ is related to the temperature of the system under analysis by appropriate normalization. Varying the rates $\{\varphi_i\}$, $\{R_I\}$, and $\{\alpha_\nu\}$ it is possible to continuously diminish the temperature of the system and reach an equilibrium state $\{\ddot{\varphi}_i\}=0$ as $t\to\infty$ and then Eq. (2.357) coincides up to the unitary transform with the Kohn–Sham DFT equation, while the Lagrange multipliers coincide with the Kohn–Sham energy eigenvalues ε_i for the occupied states (see Chap. 1).

The primary advantage that makes this a universal method is that the self-consistency, ion relaxation, and volumetric and stress relaxation are achieved simultaneously within the framework of a unified scheme. Here the kinetic energy (2.360) is a measure of the deviation of the system from the self-consistent total energy minimum.

This method was applied by Car and Parrinello[253] to electronic structure calculations of the ground state of silicon and to investigating the effect of ion motion on this state. In all likelihood we can expect more extensive research within the framework of this scheme in the future.

Chapter 3

The pseudopotential method

3.1. Introduction, certain modern problems of the pseudopotential method

An analysis of extensive experiments in the field of solid state physics has demonstrated that solid state theory requires using both rigorous exact quantitative methods of solving various problems as well as substantially simpler methods that permit identifying the fundamental qualitative mechanisms of various phenomena, yet do not claim to be quantitative in nature. Such methods include the pseudopotential method although more precisely it is an "intermediate"-type method: in many cases it can be used to solve qualitative problems and to identify certain mechanisms, although many researchers have attempted to go beyond this limit and focus on quantitative problems. They have been somewhat successful in many cases.

In these fields of scientific investigation a special role is played by model pseudopotentials that will be "fit" to specific characteristics of atoms or solids and are widely used to analyze or calculate other physical characteristics or properties. An attractive feature of the pseudopotential method is that it can be used with a diffraction model to analyze the characteristics of alloys including multicomponent alloys for which even a consistent experimental analysis of the interrelationship and physical properties present significant difficulties.

A traditional treatment of the pseudopotential method begins with a discussion of its axiomatic principles and occasionally with an historical analysis. We believe that today, when this method has already been described in a number of monographs and textbooks, it would be more convenient to first consider certain possible applications of the method as well as certain new ideas in the field of pseudopotential theory and to then systematically present the primary principles of this theory.

The parameters of model pseudopotentials such as the Ashcroft, Shaw, and Krasko-Gurskiy pseudopotentials are determined by different properties of solids, and if these parameters are found based on properties similar to those under test, the possibility for using the pseudopotential to calculate the test characteristics is beyond doubt (particularly for nontransition metals) and has been successfully confirmed in several cases.[20,254–258] The use of model pseudopotentials that are fit based on ion characteristics rather than solid state characteristics have stimulated extensive interest; these include above all the Heine–Abarenkov pseudopotentials for nontransition metals and the Heine–Abarenkov–Animalu pseudopotentials for b transition metals.[263] In spite of the large volume of research, the range of applicability of these pseudopotentials has not yet been established and here we consider the results of recent research in order to illustrate the capabilities of the pseudopotential method and to indicate the current direction of development. Here we focus solely on the possibility of using these pseudopotentials to calculate the characteristics of transition metals and their alloys. Here we consider such issues based on a sample calculation of the stability and polymorphism of the crystalline structures of pure metals and the short-range order problems in binary metallic alloys.

Katsnelson *et al.*[259] carried out an analysis of the applicability of Heine–Abarenkov–Animalu (HAA) pseudopotentials[263] for calculating the stable crystalline structures of 24 transition metals. The calculations were carried out in the second order of perturbation theory using the familiar equation

$$U_{tot} = U_{es} + U_0 + U_1 + U_2 + U_3, \qquad (3.1)$$

where U_{es} is the electrostatic energy, U_0 is the free electron gas energy (accounting for exchange-correlated interactions), and U_2 is the band structure energy. In order to determine U_1, Katsnelson *et al.*[259] utilized a routine used by Ashcroft and Langreth[260] in place of a consistent calculation of U as the long-wavelength limit of the function $\langle \mathbf{k} + \mathbf{q}|\omega|\mathbf{k}\rangle_1$; according to the former scheme $U_1 = Bz/\Omega$ and the parameter "*b*" were found from the condition

$$\left.\frac{\partial U_{tot}}{\partial \Omega}\right|_{\Omega_{peq}} = -P=0, \qquad (3.2)$$

which satisfies the equilibrium condition.

Various authors[261] have demonstrated that a consistent calculation of $U_1 = \langle \mathbf{k}|\omega|\mathbf{k}\rangle$ for many pseudopotentials will not satisfy equilibrium condition (3.2) without which the total energy calculation becomes meaningless.[261] Hence, in order to satisfy this condition the majority of authors will either perform an additional fit of the pseudopotential parameters or, leaving the parameters unchanged, will fit U_1. In both routines the fitting is carried out to satisfy condition (3.2). The calculation results for U_{tot} are shown in Table 3.1.

A correct prediction of the stable crystalline structure was achieved for 10 metals of the 24 elements (*3d, 4d, 5d*) as a rule along the ends of the isoelectronic series.[259] The phase observed at comparatively high temperatures was more stable for four elements, and for two elements the difference in energies between phases of minimum energy in calculation and experiment was less than 10^{-3} Ry/at. At such a level of discrepancy in the phase energy observed experimentally and the minimum calculation energy such differences are less than the contribution of third order perturbation theory to the phase energy and the study accounts for the fact that incorporating higher orders may result in a fit between calculation and experiment. A similar result had already been previously obtained by the authors of this same study[262] for Co. It was also demonstrated for Co that an additional (moderate) variation in the pseudopotential parameters will cause the results obtained with a standard form factor in third order perturbation theory to be obtained as early as the second order of perturbation theory. Overall either a limited agreement with experiments is observed for two-thirds of the elements or data are obtained

TABLE 3.1. Phase transition heats (ΔE,Ry/at.) and atomic volumes (Ω, atomic units) calculated by the Animalu method.

Lattice	Parameter	Sc	Ti	V	Cr	Mn	Fe	Co	Ni
FCC	ΔE	0.00087	0.01633	0	0	0	0.00005	0/0.00035	0/0
	Ω	168.8	122.0	94.2	77.9	72.8/97.3	79.6	75.2/75.2	73.6/73.6
HCP	ΔE	0.00011/0	0.00419/0	0.00295	0.00093	0.00020	0	0.00024/0	0.00002/0.00029
	Ω	168.7/168.7	119.0/119.0	94.6	78.5	72.5	79.7	74.9/74.9	73.3
	c/a	1.590/1.593	1.306/1.593	(1.633)	1.637	1.642	1.560	1.630/1.632	6(1.633)
BCC	ΔE	0/0.00306	0/0.00255	0.0003/0	0.00498/0	0.00484/0.00217	0.00127/0	0.00343	0.00339
	Ω	169.5-[15]	118.1/120.0	93.9/93.9	80.6/80.6	72.5/98.8	79.8/79.8	75.4	73.6
		Y	Zr	Nb	Mo	Tc	Ru	Rh	Pd
FCC	ΔE	0.00017	0.01945	0.02231	0	0	0.03312	0/0	0/0
	Ω	214.5	157.5	121.0	102.4	96.7	92.2	92.6/92.6	99.3/99.3
HCP	ΔE	0/0	0/0	0.01905	0.07774	0.00024/0	0.00659/0	0.0109	0.00289
	Ω	223.1/223.1	157.0/157.0	121.0	103.9	96.5/96.5	91.9/91.9	92.6	101.0
	c/a	2.060/1.571	1.450/1.593	(1.633)	(1.633)	1.633/1.603	1.580/1.583	1.700	1.660
BCC	ΔE	0.01451/0.00379	0.00303/0.00287	0/0	0.12666/0	0.02150	0	0.0313	0.00546
	Ω	222.7/229.4	150.4/158.8	121.3/121.3	105.5/105.5	97.2	91.6	93.0	101.9
		La	Hf	Ta	W	Re	Os	Ir	Pt
FCC	ΔE	0.0064/0.00030	0.02637	0.02277	0	0.05678	0.05609	0.05641/0	0/0
	Ω	244.2/250.8	151.5	121.1	104.2	99.8	95.8	95.5/95.5	101.6/101.6
HCP	ΔE	0/0	0.00221/0	0.02091	0.06046	0/0	0.02026/0	0.02845	0.00996
	Ω	252.2/252.2	150.2/150.2	121.2	105.4	99.3/99.3	94.8/94.8	94.8	108.6
	c/a	2.120/1.610	1.388/1.581	(1.633)	1.641	1.760/1.615	1.410/1.579	1.306	(1.633)
BCC	ΔE	0.01310/0.0024	0/0.00514	0.0	0.09904/0	0.03185	0	0	0.01302
	Ω	252.8/261.1	148.5/159.6	121.3/121.3	106.5/106.5	99.2	93.1	92.8	133.6

Note: The experimental results are indicated by a sign / if only one value is shown that is calculated.

suggesting that such agreement can be achieved after carrying out some simple procedures. These include the elements along the ends of the isoelectronic series and the first half of these series. It is interesting that according to Animalu his parameter z is treated as the effective chemical valence only for elements with a fully filled d shell or elements with less than a half-filled d shell. In large measure this correlates with the observation that the convergence or proximity of phases that are stable in calculation and experiment has been observed for the first half and the end of each of the isoelectronic series. However, the calculation results explicitly diverge from experiment for 8 (i.e., 1/3) of the elements and this is clearly related to the excess crudity of the HAA pseudopotential. In polyvalent transition metals in which the d band is far from filled, $U_3 \sim U_2$, and therefore it is necessary to go beyond second order and possibly third order perturbation theory. It is also possible that a significant redefinition of z is required.

However, regardless of the results, from any future analysis of this problem it is clear that using the HAA pseudopotential (as it is ordinarily defined) to calculate phase energy will not always yield results that are in agreement with experiment. On the other hand, the data reported above demonstrate that there is not yet sufficient evidence to assume the model pseudopotential method is unfit even for describing the atomic properties of transition metals. On the other hand, these data suggest that there is not yet enough evidence to classify the model pseudopotential method as unfit for describing the atomic properties of transition metals.

We now consider the possibility of using these same pseudopotentials to describe one of the most common phenomenon in crystalline alloys: the formation of atomic short-range order. The most comprehensive relations of short-range order pseudopotential theory are given by Katsnelson *et al.*,[267,268] where they are provided in second and third order perturbation theory; these expressions are provided in second order perturbation theory accounting for atomic shift effects.[266–268] The configurational energy of alloys with short-range order (SRO) in second order perturbation energy neglecting static atomic shifts takes the form

$$U_{\text{conf. SRO}}=\frac{1}{2}\left\{\overline{Z^{*2}}\left[\frac{4\pi}{\Omega}\sum_{n,f}{}'\frac{1}{g_{n,f}^2}\exp(-g_{n,f}^2/4\eta)\,|S(\mathbf{g}_{n,f})|^2-\frac{\pi}{\eta\Omega}-2\sqrt{\eta/\pi}\right]\right.$$
$$\left.+\frac{\overline{\Omega}}{4\pi}\sum_{n,f}{}'g_{n,f}^2\frac{1-\varepsilon(\mathbf{g}_{n,f})}{\varepsilon^*(\mathbf{g}_{n,f})}\,|S(\mathbf{g}_{n,f})|^2\,\overline{|W^b(\mathbf{g}_{n,f})|^2}\right\}$$
$$+\frac{1}{2}\left\{\Delta Z^{*2}\left[\frac{4\pi}{\Omega}\sum_{\mathbf{q}}{}''\frac{1}{q^2}\exp(-q^2/4\eta)\,|c(\mathbf{q})|^2-2c(1\right.\right.$$
$$\left.\left.-c)\sqrt{\eta/\pi}\right]+\frac{\overline{\Omega}}{4\pi}\sum_{\mathbf{q}}{}''\mathbf{q}^2\frac{1-\varepsilon(\mathbf{q})}{\varepsilon^*(\mathbf{q})}\,|c(\mathbf{q})|^2|\Delta W^b(\mathbf{q})|^2\right\}. \tag{3.3}$$

The first part of this equation represents the average crystal energy while the second part represents the energy related to the onset of short-range order. Strictly speaking, the average crystal energy must also to one degree or another be related to short-range order since the onset or change in short-range order will have an effect on interatomic distances. However, since this effect is small it has not yet been accounted for in specific calculations. In this cycle of research the equations that account for the contribution of atomic shift effects to crystal energy due to the differences in the atomic radii of the components were obtained for the first time.[266–268] The specific calculations of this contribution have only begun and their application to a wide range of material is of indisputable interest.

If the ordering energy and short-range order parameters are independent of the orientation, the part of the configurational energy attributable to short-range order can be given as (calculated per single atom)

$$U_{\text{conf. SRO}}=c(1-c)\sum_i c_i\alpha_i V(\rho_i), \tag{3.4}$$

where C_i and α_i is the coordination number and short-range order parameter for the i coordination shell, $V(\rho_i)$ is the ordering energy given as

$$V(\rho_i)=\frac{\bar{\Omega}}{\pi^2}\int dq\, q^2 F_p(q)\sin q\rho_i/q\rho_i, \tag{3.5}$$

where

$$F_p(q)=\frac{\bar{\Omega}}{8\pi}|\Delta W^b(\mathbf{q})|^2 q^2\frac{1-\varepsilon(q)}{\varepsilon^*(q)}+\frac{2\pi}{\Omega}(\Delta Z^*)^2\frac{1}{q^2}\exp(-q^2/4\eta)$$

$$=F_{pbs}(\mathbf{q})+F_{pes}(\mathbf{q}). \tag{3.6}$$

The equilibrium values of the short-range order parameters correspond to minimum free energy, that is, the condition

$$F_{\text{conf}}=U_{\text{conf}}-TS_{\text{conf}}, \tag{3.7}$$

where S is the entropy. Assuming independent short-range order parameters we can obtain

$$\frac{\alpha_1}{(1-\alpha_1)^2}=c(1-c)\{\exp[-V(\rho_i)/k_BT]-1\}. \tag{3.8}$$

In a rigorous analysis we can write the following for the short-range order parameter of coordination shell (i) (Ref. 269):

$$\alpha(\rho_i)=\int\frac{dq\, e^{-iq\rho_i}}{1+[c(1-c)/k_BT]\Sigma_{\rho_i\neq 0}V(\rho_i)e^{iq\rho_i}}. \tag{3.9}$$

It is convenient to carry out the following estimates to determine the contribution of short-range order to the energy of the solid solution. We have already accounted for the fact that the total energy of a single-component crystal takes the form

$$U_{\text{tot}}=U_{\text{es}}+U_0+U_1+U_2+U_3+\cdots. \tag{3.10}$$

Obviously the "average" crystal energy will be approximately the same as the energy of a single-component metal. Rigorous calculations were carried out for such pure metals as Al and Na. For example, for Al (Ref. 270) we have in Ry/at.: $U_{\text{es}}=3.417$, $U_0=0.054$, $U_1=1.473$, $U_2=0.121$, $U_3=0.016$, so $U_{\text{tot}}=4.103$ Ry/at. The experiment for this quantity yields 4.142. An error of only 1% provides indisputable evidence of the correctness of this method.

These data make possible certain additional estimates of the contribution of different factors to the energy of the crystallization phases. Based on the contributions analyzed here U_0 and U_1 will formally depend solely on volume and will be independent of structure. Although this situation is generally well known, a stipulation will be made here. The issue is that structural changes (a transition from one phase to another and ordering in alloys) ordinarily are accompanied by changes in interatomic distances (i.e., the effective atomic radius which will depend on the quantity and type of neighboring atoms). In actual calculations such changes must be rigorously accounted for; U_2 and U_3 are largely determined by structure. The electrostatic Ewald energy also explicitly contains a volume-dependent part and a part determined by the structure itself. This is because the Madelung constant can be treated as an entity consisting of a fraction determined by the interaction of a point charge with its surrounding sphere $\alpha_{sp} = -1.8$ and an addition α_{str} to this quantity that depends on the character of sphere packing in the periodic structure. Only this part can be considered to be structurally independent. Hence the contributions for Al can be relabeled as follows:

$$U_{tot}=U_{es\ vol}+U_{es\ str}+U_0+U_1+U_2+U_3, \tag{3.11}$$

where

$$U_{es\ vol}=1.8(-Ze^2/r);\quad U_{es\ str}=\alpha_{str}(-Ze^2/r). \tag{3.12}$$

Here $U=-0.054$, $U_{es\ vol}=-5.4419$, $U_1=1.473$, $U_{es\ str}=0.249$, $U_2=0.121$, and $U_3=0.016$, which yields $U_{vol}=4.0229$ and $U_{str}=-0.0801$.

Therefore, the structurally dependent part of the crystal phase energy amounts to only 2.5% (for this case) of the total energy. If we then proceed to estimate the accuracy of the calculation of U_{tot} from the viewpoint of reproducing the structural part of the energy, the error of 1% of the total energy cannot be treated as extremely small. Obviously this is why the volume-dependent part of energy must be primarily responsible for the discrepancy in the results of total energy derived by calculation and by experiment and why it must be precisely treated. This makes it necessary, for example, when using certain pseudopotentials, to find U_1 from the condition $\partial U_{tot}/\partial\Omega = 0$ and not the long-wavelength limit $U_1 = \langle \mathbf{k} - \mathbf{q}|\omega|\mathbf{q}\rangle$. In analyzing polymorphism or other structurally sensitive phenomena it is necessary, above all, to use such qualitative factors as the stable phase sequence, the energy difference of the various structural states, etc. Quantitative differences should be given secondary consideration.

It is from such a position that we should proceed with analyzing the energy states of structural changes in alloys. Here we will only focus on an analysis of energy conditions associated with the establishment of short-range order. In order to estimate the contribution of short-range order we assume, in accordance with standard calculations and measurements, that $\alpha_1 \simeq -0.1$, $V(\rho_1) \simeq 10\times10^{-3}$ atomic units. This yields for an alloy with 15 at. % of one of the components

$$U_{n.o.}\sim0.15\times0.85\times12(-0.1)\times10\times10^{-3}\sim1.5\times10^{-3}\ \text{Ry/at.}$$

This quantity seems very small particularly compared to the total crystal energy. However, if we recall on the average the energy difference between the stable phase and the nearest phase in terms of energy in terms of pure metals (see above) is

$\sim 10^{-2}$ Ry/at., it becomes clear that the short-range order energy is only an order of magnitude (or even a lower quantity) less than the energy responsible for polymorphism of the phases. The ordering energy (long-range ordering) of the phase is several times less than the energy of the band structure.[271] Therefore, a calculation of the energy associated with the onset of short-range or long-range order by means of the pseudopotential method demonstrates that this method makes it possible to obtain reasonable values of the ordering energy (in spite of its obvious small value).

It is therefore interesting to determine the capabilities of pseudopotential theory for predicting a variety of short-range order characteristics. Extensive numerical calculations based on the Madelung HAA potential and generalized by Katsnelson *et al.*[255,256,265–268,272,273] make it possible to explain the following behavioral mechanisms of short-range order.

(a) The quasioscillating dependence of the ordering energy on the interatomic distance caused by the dielectric screening effect is explained. For nontransition metals this dependence is due to Friedel oscillations.

(b) The dependence of the ordering energy and the reduced short-range order parameters $[\alpha_1/c(1 - c)]$ on composition is explained. The dependence of $V(\rho_i)$ on the interatomic distance underlies this explanation.

(c) It is demonstrated that the static atomic shifts have an effect on ordering energy and that they may both elevate and diminish this quantity. This means that the idea that the dimensional effect is a factor that directly enhances short-range order has, in general, been refuted. The calculated and measured values and signs of the short-range order parameters on the first coordination shell have also been compared. It has been demonstrated that the signs of both the calculated and measured values of these parameters are in agreement in approximately 70%–75% of the cases. The percentage of coincidence rises for alloys whose components have identical valency. Accounting for third order perturbation theory, atomic shifts, nonlinear screening, and rigorous selection of exchange-correlated corrections, although yielding very close values between calculation and experiment did not result in a radical growth of the fraction of signs of α_1 coinciding with experiment. Groups of alloys have been identified where there has been virtually no coincidence: these are noble metal alloys as well as simple alloys with noble or transition elements. The calculated and measured numerical values of α_1 were very rarely quite close.

One area of significant interest involves efforts to attempt agreement between calculation and experimental results either by radically changing the Animalu parameters or by going to other potentials or variations of the pseudopotential method. In order to calculate the short-range order parameters, the phase shift method was utilized[274] which in principle is suitable either for dilute solutions or for alloys of equiatomic composition. It turned out that for the majority of alloys for which the signs of α_1 that were measured and calculated by the HAA model potential did not coincide well in this case. An analogous result could also be obtained in the nonrigid ion model,[276] which applied specific rules to vary the valency of the ions and the ionic (model) radii. A good agreement between the values of $V(\rho_i)$ can also be obtained in this model for a number of alloys under certain conditions. A good agreement was also obtained, yet only in sign for this group of alloys using the resonance potential.[276]

An important conclusion from these studies is that the pseudopotential method is suitable for describing even very subtle effects in metals and alloys although a systematic routine has not yet been found to make it possible to precisely calculate the various physical properties without preliminary studies of its suitability in this case. The search for such a routine (including a set of calculation formulas, optimum pseudopotentials, different corrections, etc.) is indisputably an important problem which remains unsolved in general form. It is only clear that the model pseudopotential method in principle may serve as the basis for such a procedure. We therefore consider several pseudopotentials developed in recent years.

These pseudopotentials can be divided, albeit it somewhat arbitrarily, into three primary groups: norm-conserving pseudopotentials, pseudopotentials constructed within the framework of concepts of completely orthogonal plane waves,[279] and resonance-type pseudopotentials.[280,281] In this section we will focus solely on the first of these groups.

We consider the pseudopotentials of the first group based on Chulkov's studies.[278] It was demonstrated[282] that in order to formulate a pseudopotential it is possible to use inversion of the pseudowave problem for a free atom (ion):

$$\left[-\frac{1}{2}\frac{d^2}{dr^2}+\frac{l(l+1)}{2r^2}+V^l_{\rm ps}(r)\right]P_{{\rm ps},l}(r)=\varepsilon_l P_{{\rm ps},l}(r), \tag{3.13}$$

This yields

$$V^l_{\rm ps}(r)=\varepsilon-\frac{l(l+1)}{2r^2}+\frac{1}{2}\frac{P''_{ps,l}(r)}{P_{{\rm ps},l}(r)}, \tag{3.14}$$

where ε_l is the energy of the valence state with orbital quantum number l, $V^l_{\rm ps}(r)$ is the screened pseudopotential component with a given l, and $P_{{\rm ps},l}(r)$ is the radial pseudowave function.

We know from general pseudopotential theory[254,255] that the problem of pseudopotential formulation reduces to finding (selecting) the appropriate pseudowave function that is identical to the true function with large r without sites within the core. The following equation was used[283] as such a function:

$$P_{{\rm ps},l}(r)=\begin{cases} r^{l+t}f(r), & r<r_c,\\ P_l(r), & r>r_c,\end{cases} \tag{3.15}$$

where $P_l(r) = rR_l(r)$, $R_l(r)$ is the radial part of the true wave function of the valence state with orbital quantum number l; the function $f(r)$ is selected as

$$f(r)=\exp\left(\sum_{i=0}^{N} a_i r^i\right). \tag{3.16}$$

The parameter t will adopt two values: 1 or 2. For $t = 2$ a singular potential with the asymptotics $\sim B_l/r^2 (B_l > 0)$ as $r\to 0$; if $t = 1$ the singularity of this type vanishes although a type c/r singularity is conserved. Setting $a_1 = 0$ it is possible to obtain a nonsingular potential (soft core). The coefficients a_i are determined from the coincidence of the true function and the pseudowave function at point r_c as well as the coincidence of their first two derivatives. An auxiliary condition here is norm conservation: the pseudowave density integral from zero to $r = r_c$ must be equal to the value of this integral of the true density. Formally this condition is an auxiliary condition and the pseudopotential can also be constructed neglecting the condition. However, its physical meaning is clear from the relation

$$\int_0^{r_c} dr\, P_l^2(\varepsilon,r) = -P_l^2(\varepsilon,r_c)\frac{d}{d\varepsilon}\left[\frac{d}{dr}\ln\left(\frac{P_l(\varepsilon,r)}{r}\right)\right]_{r_c}, \tag{3.17}$$

which shows that under conservative conditions, the norm and the logarithmic derivatives of the wave and pseudowave functions on a sphere of radius r_c, which determine the scattering properties of the ionic core and will vary identically as a function of energy in the neighborhood of ε.

It is then necessary to formulate the ionic pseudopotential by subtracting from the screened potential the Coulomb–Hartree potential and the exchange-correlation potential based on the pseudocharge density

$$W^l_{\text{ps}}(\mathbf{r}) = V^l_{\text{ps}}(\mathbf{r}) - V_H(\rho_v) - V_{\text{exc}}(\rho_v), \tag{3.18}$$

where

$$V_H(\rho_v) = \int dr' \frac{\rho_v(r)}{|\mathbf{r}-\mathbf{r}'|}. \tag{3.19}$$

The exchange-correlation pseudopotential is divided into the sum of the exchange and correlation potentials:

$$V_{\text{exc}}(\rho_v) = V_x(\rho_v) + V_c(\rho_v). \tag{3.20}$$

The Gaspari–Kohn–Sham[284] approximations are used for the first of these potentials while the approximation introduced by Hedin and Lundquist[285] is used for the correlation potential. If the core and valence densities strongly overlap it is necessary to account for the nonlinearity of the exchange-correlation potential

$$W^l_{\text{ps}}(\mathbf{r}) = V^l_{\text{ps}}(\mathbf{r}) - V_H(\rho_v) - V_{\text{exc}}(\rho_c + \rho_v). \tag{3.21}$$

The equations are somewhat more complex for the pseudopotentials of heavy atoms; they were also derived and generalized.[278]

We provide an example[278] to illustrate the results given above. Figures 3.1 and 3.2 show the wave and pseudowave functions together with the components of the ionic pseudopotential of Al. It is clear that the joining point (r_c) is taken between the last zero and the maximum for the wave function. For singular potentials r_c virtually corresponds to the maximum while for nonsingular potentials r_c may vary over a significant range. The selection of r_c has a significant effect (Fig. 3.2) on the form of the ionic pseudopotential components (which is particularly strong for the *S* state). This makes it possible in the pseudopotential development stage to select the pseudopotentials so that the depths of the potential wells are minimized with minimum nonlinearity determined by the difference of the ionic pseudopotential components. We can easily see that the potential shown in the first of the figures has optimum properties. Here the selected values of r_c for the different components differed by a factor of approximately 3: $r_{cs} = 1.702$, $r_{cp} = 2.097$, and $r_{cd} = 4.690$.

The Chulkov group research[278] formulated the ionic pseudopotentials for a wide range of elements including transition elements and demonstrated the similarity of

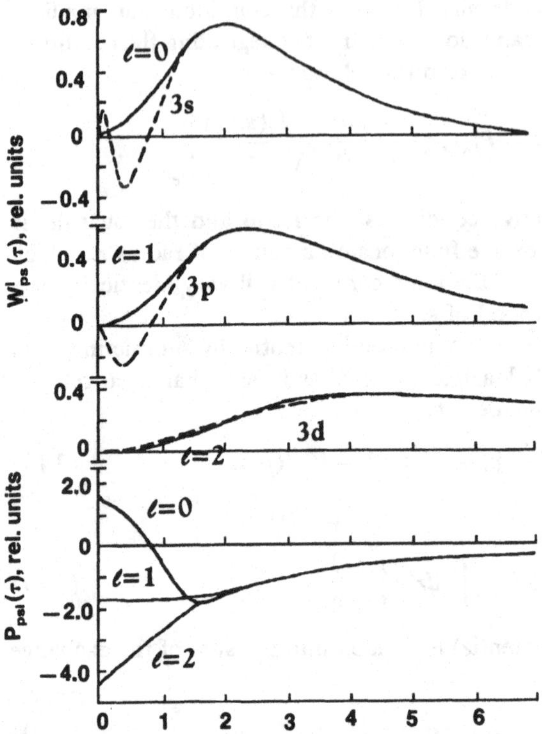

FIG. 3.1. Wave (---) and pseudowave (—) functions of the electron valence states in an aluminum atom. The lower section of the figure provides the ion pseudopotential component of Al (Ref. 278).

the derived one-electron energy values to those obtained from more exact multi-electron calculations, which indicates the promise of using norm-conserving pseudopotentials to solve various problems in solid state physics. Such characteristics as the equilibrium lattice parameter, elastic properties, frequencies of the

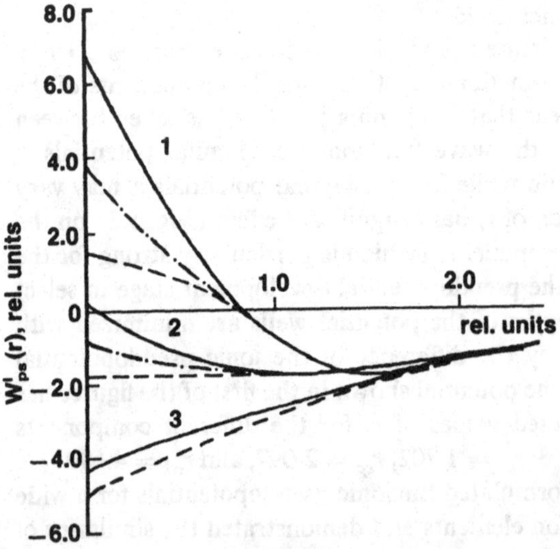

FIG. 3.2. Ion pseudopotential components of aluminum for different r_c: 1—s component (--- 1.975; ----- 1.753 ---- 1.702); 2—p component (---- 2.434; 2.160; ---- 2.097); 3—d component ×----- 5.44, 4.960) (Ref. 278).

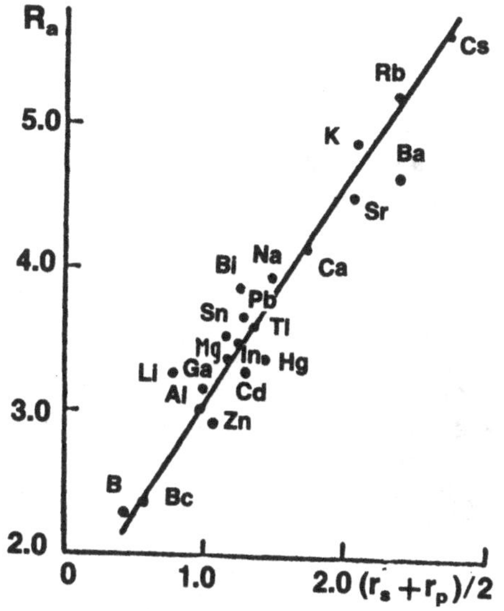

FIG. 3.3. Relationship between the half-sum of orbital radii $1/2(r_s + r_p)$ and the radius of the Wigner–Seitz sphere for nontransition metals (Ref. 278).

phonon spectra, the Grüneisen constant, the basic dependence of the lattice parameters, and the structure types have been successfully described using such pseudopotentials to analyze the atomic properties of single-component crystals (metals, semiconductors). A number of self-consistent calculations of the electronic structure of simple and transition metals, elementary semiconductors, intermetallides, alloys, and hydrides were carried out and found to agree with experiment. The electronic structure of the surface of a number of metals was calculated and the pseudocharge density distribution at various distances from the surface was identified. However, in order to calculate the atomic properties of nonsingle-component systems, and moreover, disordered systems, the authors propose, instead of direct energy calculations and the associated characteristics, a phenomenological approach based on the semiempirical estimates and the results from the pseudopotential method.

A correlation between the half-sum of the orbital radii $(r_s + r_p)/2$ and the radius of a Wigner–Seitz sphere for many nontransitions has been established[278] (Fig. 3.3) and, more interestingly, transition elements (Fig. 3.4). This makes it possible to provide a nonempirical scale of the dimensional effect and to analyze the solubility of the elements (within a range of 0.01 at. %, etc.). The situation for solid solutions is the most unclear. Significant difficulties may result from charge transfer effects. As demonstrated in the same study[278] the calculation result may be very sensitive to changes in the ionic charge state. On the one hand, this makes it possible to analyze these states for systems for which all states, aside from the given state, are known yet creates additional difficulties in analyzing systems where the charge state is not the only unknown. Clearly such issues are worthy of this kind of analysis. Certain difficulties may also arise due to the nonlocality of this group of pseudopotentials. Further research is therefore required in order to successfully utilize the pseudopotentials in the theory of alloys (and compounds).

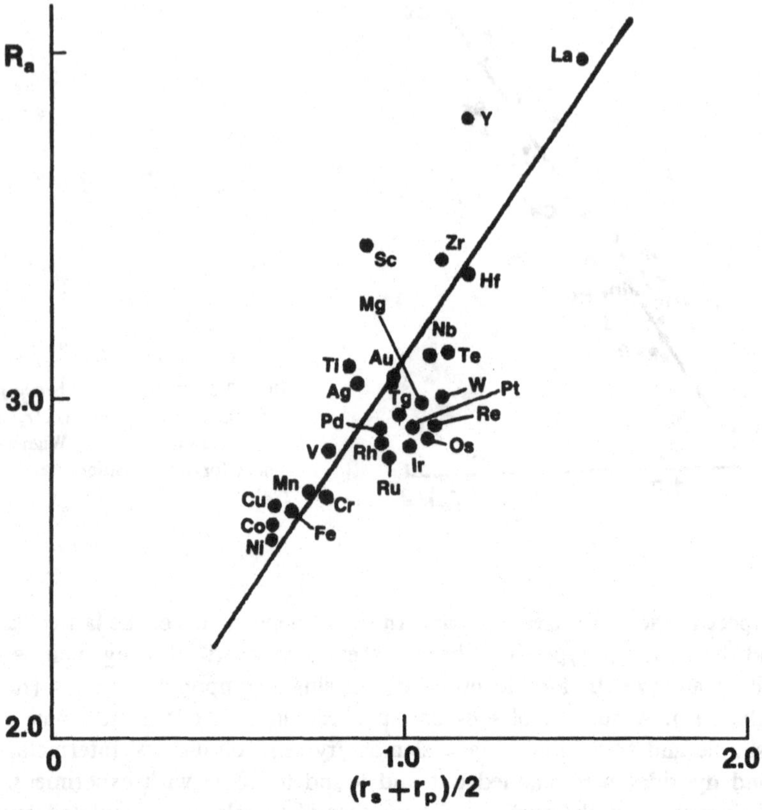

FIG. 3.4. Interrelationship between the half-sum of the orbital radii $1/2(r_s + r_p)$ and the radius of the Wigner–Seitz sphere for transition metals (Ref. 278).

In the following sections we examine the primary concepts of the pseudopotential method. An understanding of the principles of the pseudopotential concept has become important for all researchers actively involved in the field of solid state computational physics.

3.2. The OPW: Formulation of the pseudopotential approximation

We consider the Schrödinger equation for an electron in state **k** in the periodic field $V(\mathbf{r})$:

$$-\nabla^2\psi_{\mathbf{k}}(\mathbf{r}) + V(\mathbf{r})\Psi_{\mathbf{k}}(\mathbf{r}) = \varepsilon\Psi_{\mathbf{k}}(\mathbf{r}). \tag{3.22}$$

If we represent the wave function of the conduction electrons Ψ_k as an expansion in plane waves:

$$\Psi = \sum_n B_n|\mathbf{k}+\mathbf{g}_n\rangle, \tag{3.23}$$

it will not be possible in practice to achieve convergence in Eq. (3.23) (10^6 terms are required). We therefore consider a different method of expansion.[286] The wave functions of the inner electrons vary little in the transition from an isolated atom to a metal. The Hamiltonian of the metal describes the core and valence states and hence the wave functions of both groups of electrons will be orthogonal. It can be shown that the functions $\chi_{\mathbf{k}+\mathbf{g}_n}$ [see Eq. (2.51)]:

$$\chi_{\mathbf{k}+\mathbf{g}_n}=|\mathbf{k}+\mathbf{g}_n\rangle - \sum_{\alpha} \langle\alpha|\mathbf{k}+\mathbf{g}_n\rangle|\alpha\rangle \tag{3.24}$$

are orthogonal to the core functions, that is,

$$\begin{aligned}\langle\chi_{\mathbf{k}+\mathbf{g}_n}|\alpha\rangle&=\langle\mathbf{k}+\mathbf{g}_n|\alpha\rangle - \sum_{\alpha'} (\langle\alpha|\mathbf{k}+\mathbf{g}_n\rangle)^*\langle\alpha|\alpha'\rangle\\ &=\langle\mathbf{k}+\mathbf{g}_n|\alpha\rangle - \langle\mathbf{k}+\mathbf{g}_n|\alpha\rangle=0.\end{aligned} \tag{3.25}$$

The function $\chi_{\mathbf{k}+g_n}$ is called an orthogonalized plane wave. It is these functions that will be used to expand Ψ_k:

$$\Psi_k= \sum_n B_n|\mathbf{k}+\mathbf{g}_n\rangle - \sum_{\alpha,n} B_n\langle\alpha|\mathbf{k}+\mathbf{g}_n\rangle|\alpha\rangle. \tag{3.26}$$

We introduce the projection operator $\widehat{P}$:

$$\widehat{P}= \sum_n |\alpha\rangle\langle\alpha| = \sum_l \widehat{P}_l, \tag{3.27}$$

$$\widehat{P}_l= \sum_n^z Y_{lm}(\theta,\varphi) \int d\theta'\, d\varphi'\, Y_{lm}(\theta',\varphi'), \tag{3.28}$$

where $Y_{lm}(\theta,\varphi)$ are spherical harmonics, and P_l selects from any wave function the component with orbital number l.

By definition of the projection operator for any function,

$$\widehat{P}f(r)= \sum_{\alpha} |\alpha\rangle\langle\alpha|\, f\rangle. \tag{3.29}$$

The first term of function (3.26):

$$\varphi= \sum_n B_n|\mathbf{k}+\mathbf{g}_n\rangle \tag{3.30}$$

is called the pseudowave function. Accounting for the definition of Eq. (3.29) we can easily find

$$\Psi_{\mathbf{k}}(1-\widehat{P})\varphi_{\mathbf{k}}=\varphi_{\mathbf{k}} - \sum_{\alpha} |\alpha\rangle\langle\alpha|\varphi_{\mathbf{k}}\rangle. \tag{3.31}$$

Substituting Eq. (3.20) into Eq. (3.21) we find

$$(-\nabla^2 + V(\mathbf{r}) - E)\varphi_{\mathbf{k}} - \sum_{\alpha} (E_{\alpha} - E)|\alpha\rangle\langle\alpha|\varphi_{\mathbf{k}}\rangle=0, \tag{3.32}$$

where E_{α} are the eigenvalues of the state $|\alpha\rangle$:

$$(-\nabla^2 + V(\mathbf{r}))|\alpha\rangle = E_\alpha|\alpha\rangle. \tag{3.33}$$

By comparing Eq. (3.31) to Eq. (3.32) we see that the same eigenvalue corresponds to different Hamiltonians H_1 and H_2:

$$H_1 = (-\nabla^2 + V(\mathbf{r})), \tag{3.34}$$

$$H_2 = \left(-\nabla^2 + V(\mathbf{r}) - \sum_\alpha (E_\alpha - E)|\alpha\rangle\langle\alpha|\varphi\rangle\right), \tag{3.35}$$

with different eigenfunctions Ψ_k and φ_k.

We recast Eq. (3.32) as

$$(-\nabla^2 + W)\varphi_\mathbf{k} = E\varphi_\mathbf{k}, \tag{3.36}$$

where we introduce the designation

$$Wf(\mathbf{r}) = V(\mathbf{r})f(\mathbf{r}) - \sum_\alpha (E_\alpha - E)|\alpha\rangle\langle\alpha|f\rangle. \tag{3.37}$$

In Eq. (3.37) $f(\mathbf{r})$ is any function.

The operator W is called the pseudopotential. In the general case, W is a nonlocal integral operator dependent on energy E. Here and henceforth the operator V will be treated as a local operator if its action of $\Psi(\mathbf{r})$ reduces to multiplication of the function $\Psi(\mathbf{r})$ by a certain given function of the same argument $\mathbf{r}$ [say, $V(\mathbf{r})$]:

$$V\Psi(\mathbf{r}) = V(\mathbf{r})\Psi(\mathbf{r}). \tag{3.38}$$

All other operators are nonlocal. The nonlocality of the potential significantly complicates the calculations although it was possible to establish that in describing many phenomena the effect of nonlocality can be neglected.

The crystal potential $V(\mathbf{r})$ is attractive, that is, negative. The quantity $E_\alpha - E$ is also negative. This means that when formulating the pseudopotential, the potential of electron interaction with the core electrons is subtracted from the electron interaction potential with the nucleus and the remaining electrons, that is, partial compensation of these potentials is used. Therefore, the orthogonalization routine is used to make the pseudopotential more weak than the true potential. An important feature here is the invariance of the eigenvalues to the introduction of the pseudopotential.

Figure 3.5 provides a schematic representation of the wave function $\Psi(\mathbf{r})$ and the pseudowave function $\varphi(\mathbf{r})$: the $\Psi(\mathbf{r})$ function in the core region is characterized by oscillations which are determined by the presence of bound states; the pseudowave function $\varphi(\mathbf{r})$ outside the core largely coincides with $\Psi(\mathbf{r})$ and has no oscillations in the vicinity of the core.

It was demonstrated that using a pseudopotential that is weaker compared to the true potential makes it possible to calculate the dispersion law. Here in many cases this calculation is possible in second order perturbation theory:

$$E(\mathbf{k}) = \mathbf{k}^2 + \langle\mathbf{k}|V|\mathbf{k}\rangle - \sum_{q\neq 0} \frac{|\langle\mathbf{k}|V|\mathbf{k}+\mathbf{q}\rangle|^2}{|\mathbf{k}+\mathbf{q}|^2 - \mathbf{k}^2}. \tag{3.39}$$

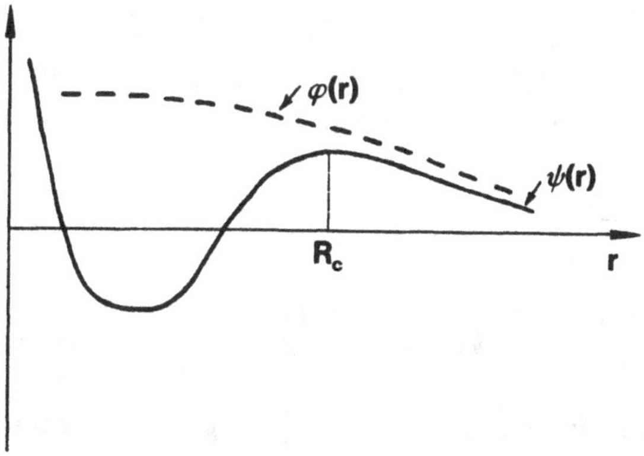

FIG. 3.5. Wave and pseudowave functions.

We note that strong localization of the wave functions of the core electrons was assumed in formulating pseudopotential (3.37). This made it possible to divide the atomic electrons into core and valence electrons and to initially treat the wave functions as fixed when going from the atom to the crystal. Such an approximation is not suitable for the wave functions of *d* electrons which are poorly localized even in a free atom.

It is possible to approach the problem of introducing the pseudopotential from the viewpoint of scattering theory as well[254,255]; in this respect the essence of the pseudopotential lies in the fact that it is possible to find a sufficiently weak pseudopotential W in which the electrons will scatter in the same manner as in the true potential V. The scattering amplitude of a plane wave through angle θ is equal to

$$f(\theta)=(2i\sqrt{E})^{-1}\sum_{l=0}(2l+1)\,[\exp(2i\eta(E))-1]P_l(\cos\theta), \tag{3.40}$$

where $\eta_l(E)$ are the scattering phase shifts represented as

$$\eta_i(E)=n_l\pi+\delta_l(E). \tag{3.41}$$

Here n_l is equal to the number of nodes of wave function Ψ within the core or, which is the same thing, the number of bound states with a given l. If the pseudopotential W is defined so that all resulting phase shifts are equal to $\delta_l(E)$ $[|\delta_l(E)<\pi/2]$, such a potential does not generate bound states, that is, it is weak. Moreover, when going from phase shifts $\eta_l(E)$ to $\delta_l(E)$ the scattering amplitude $f(\theta)$ remains unchanged.

We note that if $(n_l-1,)\pi$, $(n_l-2)\pi$ is subtracted from $\eta_l(E)$ rather than $n_l\pi$, etc., the scattering properties of the ion will remain unchanged although the pseudowave function will oscillate in the vicinity of the core. This is responsible for the ambiguity of the pseudopotential.

An important feature of introducing the pseudopotential is the appearance of the orthogonalization hole.

When the true potential is substituted by the pseudopotential and the wave function is replaced by the pseudowave function, the eigenvalues in the Schrödinger equation remain unchanged although the electron density found by functions χ and φ will be different since, as noted above, the functions are substantially different in the vicinity of the core. We find the charge attributable to these differences. We differentiate Eqs. (3.22) and (3.32) with respect to energy:

$$[-\nabla^2+V]\frac{\partial\Psi_{\mathbf{k}}}{\partial\varepsilon_{\mathbf{k}}}+\frac{\partial V}{\partial\varepsilon_{\mathbf{k}}}\Psi_{\mathbf{k}}=\Psi_{\mathbf{k}}+\varepsilon_{\mathbf{k}}\frac{\partial\Psi_{\mathbf{k}}}{\partial\varepsilon_{\mathbf{k}}}, \tag{3.42}$$

$$[-\nabla^2+W]\frac{\partial\varphi_{\mathbf{k}}}{\partial\varepsilon_{\mathbf{k}}}+\frac{\partial W}{\partial\varepsilon_{\mathbf{k}}}\varphi_{\mathbf{k}}=\varphi_{\mathbf{k}}+\varepsilon_{\mathbf{k}}\frac{\partial\psi_{\mathbf{k}}}{\partial\varepsilon_{\mathbf{k}}}. \tag{3.43}$$

Multiplying each of these equations by Ψ_k^* and φ_k^* and integrating over the core region,

$$\int_{V_{\text{res}}}|\Psi_{\mathbf{k}}|^2\,d^3r=-\oint_{V_{\text{res}}}dS\left[\Psi_k^*\frac{\partial}{\partial r}\frac{\partial\Psi_{\mathbf{k}}}{\partial\varepsilon_{\mathbf{k}}}-\frac{\partial\Psi_{\mathbf{k}}}{\partial\varepsilon_{\mathbf{k}}}\frac{\partial}{\partial r}\Psi_k^*\right], \tag{3.44}$$

$$\int_{V_{\text{res}}}|\varphi_{\mathbf{k}}|^2\,d^3r=\int_{V_{\text{res}}}\varphi_k^*\frac{\partial\mathbf{W}}{\partial\varepsilon_{\mathbf{k}}}\varphi_{\mathbf{k}}\,d^3r-\oint_{V_{\text{res}}}dS\left[\varphi_k^*\frac{\partial}{\partial r}\frac{\partial\varphi_{\mathbf{k}}}{\partial\varepsilon_{\mathbf{k}}}-\frac{\partial\varphi_{\mathbf{k}}}{\partial\varepsilon_{\mathbf{k}}}\frac{\partial}{\partial r}\varphi_k^*\right]. \tag{3.45}$$

In carrying out these transforms we utilize Gauss' theorem as well as Eqs. (3.21) and (3.26). The second integral on the right-hand side of Eq. (3.45) coincides with the integral on the right-hand side of Eq. (3.44) since φ and Ψ are identical outside the core.

Subtracting Eq. (3.45) from Eq. (3.44) we obtain

$$\int|\Psi_{\mathbf{k}}|^2\,d^3r-\int|\varphi_{\mathbf{k}}|^2\,d^3r=-\int_{V_{\text{res}}}d^3r\,\varphi_k^*\frac{\partial W}{\partial\varepsilon_{\mathbf{k}}}\varphi_{\mathbf{k}}. \tag{3.46}$$

Summing Eq. (3.46) over all $\mathbf{k}<\mathbf{k}_F$,

$$Z_d=-\sum_{\mathbf{k}}\int_{V_{\text{res}}}d^3\mathbf{r}\,\varphi_k^*(\mathbf{r})\frac{\partial W}{\partial\varepsilon_{\mathbf{k}}}\varphi_{\mathbf{k}}(\mathbf{r}), \tag{3.47}$$

where Z_d is the so-called depletion charge.

Therefore, the true charge density in the vicinity of the core diminishes. This is indicated by the presence of the "orthogonalization hole" whose formation is accounted for by introducing the effective valency Z^* in place of the valency Z:

$$Z^*=Z+Z_d=Z\left(1+\frac{Z_d}{Z}\right)=Z(1+\alpha). \tag{3.48}$$

It is clear from Eq. (3.46) that Z_d is nonzero for all energy-dependent potentials. The value of the depletion charge may exceed 30% of the valency, although this quantity, as a rule, is lower in transition metals than in simple metals. We note that α may also be less than zero (e.g., in beryllium).

3.3. Model pseudopotentials

In order to construct a pseudopotential it is necessary to find the wave functions and the atomic electron energies. The atomic electron density can be found by the wave functions and the crystalline pseudopotential is then formulated as the superposition of the atomic pseudopotentials. The resulting pseudopotential is a first principles pseudopotential. This is a very difficult program to implement. Hence, empirical or model pseudopotentials are most commonly used. In the empirical pseudopotential method the Fourier components of the pseudopotential are selected so that the calculation results for a certain characteristic of a metal are identical to experiment. The derived Fourier components of the pseudopotential are then used to calculate the other properties of this metal.

In the model pseudopotential method the pseudopotential is based on certain physical mechanisms, while the parameters entering into the analytical expressions are treated as fitting parameters. Either the atomic (ionic) characteristics or solid state characteristics are used, as noted above, as the experimental data for the fit. Occasionally these two approaches are combined.

Most common data on the energy levels of free electrons are used as the atomic (ionic) characteristics; the characteristics of solids commonly used include the conditions[254,255,286,296]

$$\left.\frac{\partial U_{\text{tot}}}{\partial \Omega}\right|_{r_s=rs\,\exp} = -P=0; \quad c_{44}(\Omega)=c_{44}^{\exp} \tag{3.49}$$

($C_{44}^{\exp}$ is the shear modulus at zero temperature and pressure), the phonon spectra, the electrical conductivity of liquid metals, the Fermi surface characteristics, etc.

Model pseudopotentials that are fit to atomic (ionic) properties may be used to a greater extent compared to those fit by solid state properties to formulate *a priori* atomistic models of solids. At the same time, calculation results on the physical properties using pseudopotentials fit to the properties of solids may be closer to the data since the intractable features of solids can be implicitly accounted for in this fit. This means that both types of fit have their advantages and drawbacks. We will therefore briefly discuss the primary concepts that must be followed in selecting the model pseudopotential. (a) It is advisable to determine the parameters of the pseudopotential based on experimental characteristics that are not directly related to those that must be investigated with the pseudopotential under analysis in the present study; (b) it is desirable to use pseudopotentials without nonphysical oscillations for $q > 2k_F$ otherwise a damping factor should be used; and (c) it would be desirable to either directly or indirectly account for the pseudopotential nonlocality effect.

We consider some of the more commonly used model pseudopotentials.

3.3.1. The Ashcroft pseudopotential (empty core, local pseudopotential)

This potential takes the form[287]

$$W(r)=\begin{cases} 0, & r<R_M, \\ -2z/r, & r>R_M. \end{cases} \tag{3.50}$$

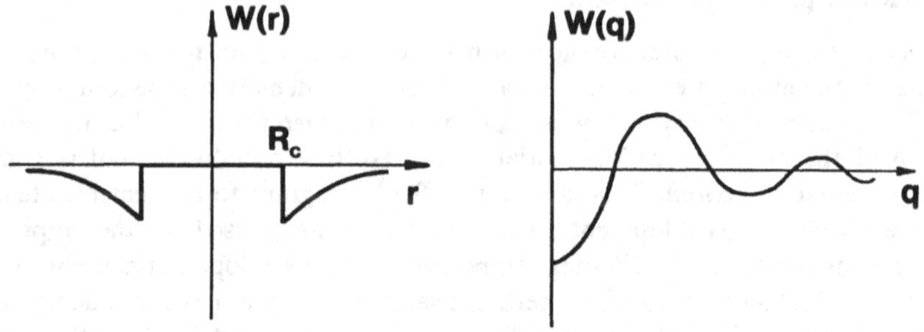

FIG. 3.6. General form of the Ashcroft pseudopotential and its form factor (pseudopotential: unscreened; form factor: screened).

The radius R_M is found by fitting the calculated values to experiment for such factors as the electrical resistance of liquid metals, data on the Fermi surface, and the equilibrium volume. A linear relation between R_M and r_c is observed for alkaline, as well as certain di- and trivalent metals, which may also provide the basis of the criterion for selecting R_M although there are no strict criteria.

We define the form factor of the pseudopotential:

$$\langle \mathbf{k}+\mathbf{q}|W|\mathbf{k}\rangle = \frac{1}{\Omega}\int d^3r\, e^{-i(\mathbf{k}+\mathbf{q})\mathbf{r}}W(\mathbf{r})e^{i\mathbf{k}\mathbf{r}}. \tag{3.51}$$

Since for a local pseudopotential which includes the Ashcroft pseudopotential

$$\langle \mathbf{k}+\mathbf{q}|W|\mathbf{k}\rangle = W(\mathbf{q}) = \frac{1}{\Omega}\int d^3\mathbf{r}\, e^{-i\mathbf{q}\mathbf{r}}W(\mathbf{r}), \tag{3.52}$$

the form factor of the Ashcroft pseudopotential takes the form

$$W(\mathbf{q}) = \frac{1}{\Omega}\int_{R_M}^{\infty} d^3\mathbf{r}\left(-\frac{2z}{r}\right)e^{-i\mathbf{q}\mathbf{r}} = -\frac{8\pi z}{\Omega r^2}\cos qR_M. \tag{3.53}$$

The Ashcroft pseudopotential and its form factor are shown in Fig. 3.6.

Oscillations of $W(q)$ with large q are physically meaningless and are caused by the fact that $W(r)$ in r space has a discontinuity at $r = R_M$. This complicates summing the series in inverse space. In order to improve the convergence, the Ashcroft pseudopotential is commonly multiplied by a damping multiplier of the type $\exp[-\alpha(q/2k_F)^4]$ (α is selected to achieve rapid convergence of the series in inverse space).

3.3.2. The Heine–Abarenkov pseudopotential (nonlocal pseudopotential)

This potential takes the form[288,289]

$$W(r) = \begin{cases} -\sum\limits_{l=0}^{l_{\max}} A_l P_l, & \text{where } r<R_M, \\ -2z/r, & \text{where } r>R_M. \end{cases} \tag{3.54}$$

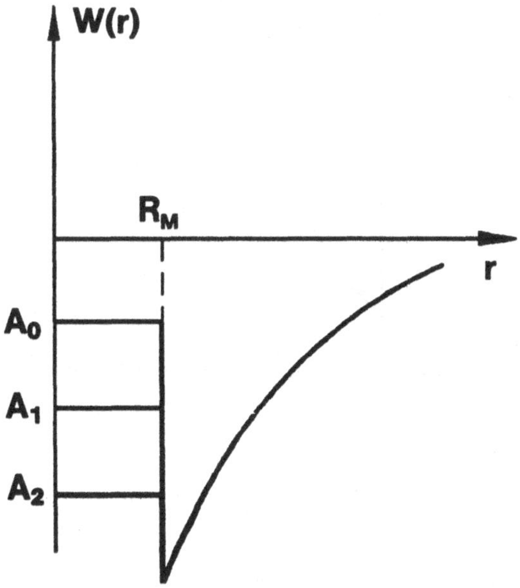

FIG. 3.7. Nonlocal Heine–Abarenkov pseudopotential.

Ordinarily the series in l breaks off at $l = 2$ if we assume $A_l = A_2$ for $l \leqslant 2$ (occasionally $l_{\max} = 3$ is used for transition metals).

We introduce the Heaviside function θ:

$$\theta(x) = \begin{cases} 1, & x \geqslant 0, \\ 0, & x < 0. \end{cases} \tag{3.55}$$

Then the Heine–Abarenkov pseudopotential can be recast as

$$W(r) = -2z/r - \theta(R_M - r)\{(A_2 - 2z/r) + (A_0 - A_2)\hat{P}_0 + (A_1 - A_2)\hat{P}_1\}, \tag{3.56}$$

where $\hat{P}_l$ are defined as Eq. (3.28).

The parameters A_l are obtained from spectroscopic data on the ionic energy levels: the energy states of a free ion of charge $z - 1$ are determined experimentally. The energy levels are calculated for pseudopotential (3.54). The parameters A_l with which the calculated energies coincide with experiment are assumed to be the true parameters.

The parameters $A_l(E)$ for energies that are not identical to the term values can be determined by interpolation or extrapolation. The result of this calculation indicates a weak energy dependence of A_l. Hence $A_l(E) = A_l(E_F)$ is commonly assumed. The Heine–Abarenkov pseudopotential is shown in Fig. 3.7.

3.3.3. *The Heine–Abarenkov pseudopotential (local pseudopotential)*

This potential is given as

$$W(r) = \begin{cases} -A, & r \leqslant R_M, \\ -2z/r, & r > R_M. \end{cases} \tag{3.57}$$

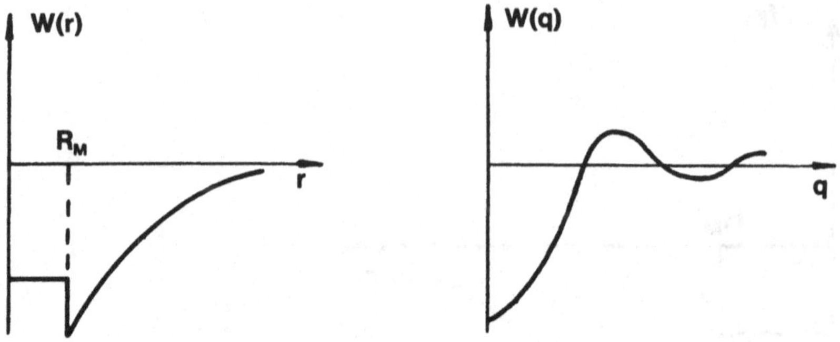

FIG. 3.8. Local Heine–Abarenkov pseudopotential and its form factor (pseudopotential: unscreened; form factor: screened).

This pseudopotential is a two-parameter pseudopotential (parameters A and R_M). Its form factor takes the form

$$W(q) = -\frac{8\pi z}{\Omega q^2}\left[\left(1 - \frac{A(R_M)}{z}\right)\cos(qR_M) + \frac{A}{zqR_M}\sin(qR_M)\right]. \tag{3.58}$$

The local Heine–Abarenkov pseudopotential is shown in Fig. 3.8. The pseudopotentials (3.53) and (3.56), similar to the Ashcroft pseudopotential, are discontinuous functions in r space which cause oscillations in q space.

3.3.4. *The Krasko–Gurskiy pseudopotential*

This potential takes the form

$$W(r) = 2z\left(\frac{e^{-r/r_c}}{r} + \frac{a}{r_c}e^{-r/r_c}\right) - \frac{2z}{r}. \tag{3.59}$$

This is a local two-parameter pseudopotential with parameters r_c and a. The form factor $W(r)$ takes the form

$$W(q) = \frac{8\pi z[(2a-1)q^2r_c^2 - 1]}{\Omega q^2[q^2r_c^2 + 1]^2}. \tag{3.60}$$

The parameters r_c and a can be determined from, for example, the conditions

$$W(q_0) = 0, \tag{3.61}$$

$$\partial U/\partial\Omega\,|_{\Omega_{\text{exp}}} = 0, \tag{3.62}$$

where q_0 is the point corresponding to the zero of $W(q)$ (q_0 is found from Fermi surface data or the electrical resistance data of the liquid metals) and Ω_{exp} is the experimental atomic volume. Equation (3.61) is the equation of state of the crystal at a pressure $P = 0$. Unlike the preceding pseudopotentials the form factor of the Krasko–Gurskiy pseudopotential does not oscillate with large q.

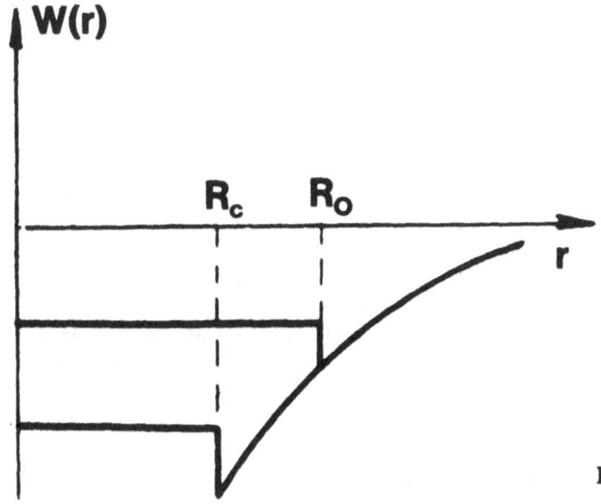

FIG. 3.9. Russell–Taylor pseudopotential.

3.3.5. *The Rasolt–Taylor pseudopotential (nonlocal[291])*

This potential is written as

$$W(r)=\begin{cases}\{-2z/r-[A_0(E)-2z/r]\theta(R_0-r)\}\hat{P}_0, & l=0,\\ \{-2z/r-[A_1(E)-2z/r]\theta(R_1-r)\}\hat{P}_1, & l=1,\\ \{-2z/r-[A_2(E)-2z/r]\theta(R_2-r)\}\hat{P}_2, & l=2.\end{cases} \tag{3.63}$$

Figure 3.9 shows $W(r)$ ($R_l = R$ for all $l > 2$).

The dependence of R_l on l differentiates pseudopotential (3.63) from Eq. (3.54). The parameters A_0, A_1, R_0, and R_1 are selected based on the coincidence of the scattering phase shifts at the Fermi energy in pseudopotential (3.63) and the atomic Hartree–Fock potential accurate to the integer F. Other methods of finding these parameters are also possible.

3.3.6. *The Shaw pseudopotential[292]*

This potential is written by

$$W(r)=\sum_{l=0}^{l_0}[A_l(E)\theta(R_l-r)-(2z/r)\theta(r-R_l)]\hat{P}_l-\sum_{l=l_0+1}^{\infty}\frac{2z}{r}\hat{P}_l. \tag{3.64}$$

For the Shaw pseudopotential (Fig. 3.10) the ionic potential is substituted by A_l for those values of l for which bound states exist (l_0 is the maximum value of l at which bound states will still exist). This pseudopotential is smoother than the Heine–Abarenkov one.

The relation between A_l and R_l is given by the condition

$$A_l=2z/R_l. \tag{3.65}$$

There is also a large number of other model pseudopotentials.

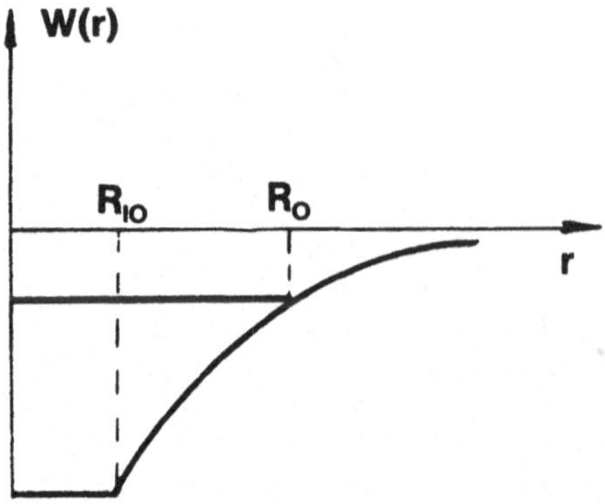

FIG. 3.10. Shaw pseudopotential.

3.3.7. Factorization procedure

Assume the potential field acting on the electron is described by ionic pseudopotentials $W(\mathbf{r})$ periodically distributed throughout the crystalline lattice. Then the crystal potential takes the form

$$V(\mathbf{r})=\sum_{\nu} W(\mathbf{r}-\mathbf{t}_{\nu}), \tag{3.66}$$

where $\mathbf{t}_\nu$ is the radius vector through the ν sites of the lattice and $W(\mathbf{r}-\mathbf{t}_\nu)$ is the pseudopotential of the ion at the ν site acting on the electron at point $\mathbf{r}$.

Let $W(\mathbf{r})$ be the local pseudopotential. We find the Fourier component $V(\mathbf{r})$:

$$\begin{aligned} V(\mathbf{q}) &= \frac{1}{\Omega}\int d^3\mathbf{r} \sum_{\nu} W(\mathbf{r}-\mathbf{t}_\nu)e^{-i\mathbf{qr}} \\ &= \frac{1}{\Omega}\sum_{\nu} e^{-i\mathbf{q}\mathbf{t}_\nu}\int d^3\mathbf{r}\, W(\mathbf{r}-\mathbf{t}_\nu)e^{i\mathbf{q}(\mathbf{t}_\nu-\mathbf{r})} \\ &= S(\mathbf{q})\,W(\mathbf{q}). \end{aligned} \tag{3.67}$$

Here

$$S(\mathbf{q})=\frac{1}{N}\sum_{\nu} e^{-i\mathbf{q}\mathbf{t}_\nu}, \tag{3.68}$$

$$W(\mathbf{q})=\frac{1}{\Omega_0}\int d^3\mathbf{r}\, W(\mathbf{r}-\mathbf{t}_\nu)e^{i\mathbf{q}(\mathbf{r}-\mathbf{t}_\nu)}. \tag{3.69}$$

$S(\mathbf{q})$ is called the lattice structure factor. For the Bravais lattices

$$S(\mathbf{q})=\begin{cases} 1, & \mathbf{q}=\mathbf{g}_n, \\ 0, & \mathbf{q}\neq\mathbf{g}_n, \end{cases} \tag{3.70}$$

where g_n is the reciprocal lattice vector.

Therefore, the form factor of the crystallization pseudopotential can be represented as the product of the structurally dependent quantity $S(\mathbf{q})$ and the structurally independent form factor of the pseudopotential of the ion. Such a representation is called factorization. Factorization in the form of Eq. (3.69) is rigorously defined only for ordered systems. In a disordered system the potential $V(\mathbf{r})$ has no translational symmetry. Its form factor can be divided into translationally invariant and fluctuating parts:

$$V(\mathbf{q}) = S(\mathbf{q})\,\overline{W}(\mathbf{q}) + \langle \mathbf{k}+\mathbf{q}|\delta W|\mathbf{k}\rangle. \qquad (3.71)$$

In this case $\overline{W}(\mathbf{q})$ is the averaged form factor of the ionic pseudopotential consisting of the form factors of the ionic pseudopotentials of the crystal components multiplied by the corresponding fractions of ions of the given type [it is necessary to go over to the averaged volume per ion in calculating $W(\mathbf{q})$], δW is the difference of the component pseudopotentials.

3.4. Temperature dependence of the pseudopotential form factor

In 1966 G. P. Motulevich derived a rather simple expression for the pseudopotential form factor in which the effect of thermal oscillations of the lattice ions was accounted for by analogy to atomic vibrational effects in x-ray scattering theory.

We outline this approach based on Yastrebov and Katsnelson's work.[255] For a lattice consisting of fixed, periodically distributed atoms,

$$V(\mathbf{q}) = W(\mathbf{q})S(\mathbf{q}) = W(\mathbf{q})\delta_{\mathbf{q}\mathbf{g}_n}, \qquad (3.72)$$

$$V(\mathbf{g}_n) = W(\mathbf{g}_n). \qquad (3.73)$$

Consequently, for the squared modulus of the matrix element of the crystal, we have

$$|V(\mathbf{g}_n)|^2 = |W(\mathbf{g}_n)|^2. \qquad (3.74)$$

It is possible to account for atomic vibrations in the crystal by representing the radius vector of the instantaneous position of the atom as the vector $\mathbf{R}_\nu = \mathbf{t}_\nu + \mathbf{u}_\nu$, where $\mathbf{t}_\nu$ is the radius vector of the lattice site and $\mathbf{u}_\nu$ is the atomic displacement vector from this site. In this case assuming the fixed $W(\mathbf{q})$ and neglecting anharmonic effects it is possible to reduce all variations in $|V(\mathbf{g}_n)|^2$ to a temperature change in the structure factor

$$|V(\mathbf{g}_n)|^2 = |W(\mathbf{g}_n)|^2 \sum_j \sum_{j'} \langle e^{i\mathbf{q}(\mathbf{u}_\nu - \mathbf{u}_{\nu'})}\rangle. \qquad (3.75)$$

Averaging the latter expression assuming that atomic shifts are small we obtain

$$|V(|\mathbf{g}_n|)|^2 = |W(\mathbf{g}_n)|^2 \delta_{\mathbf{q}\mathbf{g}_n} \exp(-\tfrac{1}{2}\mathbf{g}^2\overline{\mathbf{u}^2}\mathbf{g}_n), \qquad (3.76)$$

where $\overline{u^2_{\mathbf{g}_n}}$ is the average squared projection of the atomic shift onto the direction of the scattering vector $\mathbf{g}_n$. Since

$$\overline{u^2_{\mathbf{g}_n}} = \tfrac{1}{3}\overline{u^2}, \qquad (3.77)$$

where $\overline{u^2}$ is the average squared atomic shift, we find

$$|V(\mathbf{g}_n)|^2 = |W(\mathbf{g}_n)|^2 \delta_{\mathbf{q}\mathbf{g}_n} e^{-(1/6)g_n^2\overline{u^2}}. \qquad (3.78)$$

The average squared $\overline{u^2}$ is determined by the interatomic interaction forces and this can be found by means of crystal lattice dynamics. This requires representing the shift of the νth atom $\overline{u_\nu^2}$ as a superposition of the contributions of all elastic waves propagating through the crystalline lattice

$$u_\nu = \sum_{l=1}^{N} \sum_{i=1}^{3} a_{l_i} e_{l_i} \cos\{\omega_{l_i} t - \mathbf{k}_{l_i} \mathbf{r}_\nu - 2\pi\delta_{l_i}\}, \tag{3.79}$$

where a_{l_i} is the amplitude of the l wave of i polarization, e_{l_i} is the direction of atomic shift in the l_i wave, $\mathbf{k}_{l_i}$ is the wave vector, and δ_{l_i} is the phase of the l_i elastic wave. If we represent each elastic wave as an oscillator, within the framework of quantum statistics

$$\overline{u_\nu^2} = \frac{\hbar}{mN} \sum_l \sum_i \left\{\frac{1}{e^{\hbar\omega/k_BT} - 1} + \frac{1}{2}\right\} \frac{1}{\omega_{li}}. \tag{3.80}$$

Summation in Eq. (3.80) is possible if we know the wave distribution relative to the frequencies in the vibrational spectrum. In the Debye approximation we can obtain from Eq. (3.80),

$$\overline{u^2} = \frac{9\hbar^2}{mk_B\theta} \left(\frac{T}{\theta}\right)^2 \int_0^x d\xi \left[\frac{\xi}{e^\xi - 1} + \frac{\xi}{2}\right], \tag{3.81}$$

where θ is the Debye characteristic temperature, m is the atomic mass, k_B is the Boltzmann constant, $x = \theta/T$, and $\xi = \hbar\omega/k_BT$.

The exponent in Eq. (3.78) takes the form

$$\frac{1}{6} g^2 \overline{u^2} = 2M = \frac{3}{2} \frac{g^2\hbar^2}{mk_B\theta} \left(\frac{T}{\theta}\right)^2 \int_0^x \left[\frac{\xi\, d\xi}{e^\xi - 1} + \frac{\xi\, d\xi}{2}\right]. \tag{3.82}$$

It is clear that $\bar{u}^2$ grows with increasing temperature so that $|V(\mathbf{g}_n)|^2$ will with increasing temperature decay as

$$|V(\mathbf{g}_n)|^2 = |W(\mathbf{g}_n)|^2 \delta_{\mathbf{q}\mathbf{g}_n} e^{-2M} = |W(\mathbf{g}_n)|^2 \delta_{\mathbf{q}\mathbf{g}_n} e^{-(1/6)g_n^2\overline{u^2}}. \tag{3.83}$$

3.5. The pseudopotentials of the d metals

Calculations of the band structure of transition metals revealed that they contain narrow d bands that intersect and hybridize with the band of near-free electrons. Hence such systems cannot be described within the framework of the free electron approximation.

From the viewpoint of scattering theory the d-phase shift in the d metals ($l = 2$) reveals the strong energy dependence in the narrow range of Γ near energy E. This is illustrated in Fig. 3.11. According to quantum scattering theory the tangent of the d-phase shift takes the form

$$\tan \eta_2 = \frac{\Gamma/2}{E_d - E} + \tan \eta_2'(E), \tag{3.84}$$

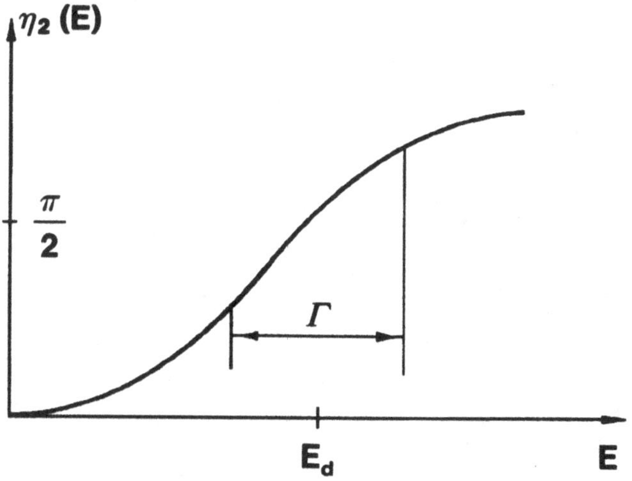

Fig. 3.11. *d*-phase shifts plotted as a function of energy.

where η_2' is the phase shift at energies far from E_d. The mechanism behind the onset of the resonance levels in *d* metals is as follows. The Schrödinger equation for the radial wave function in the field $V(\mathbf{r})$ includes the centrifugal potential $l(l+1)/\mathbf{r}^2$:

$$-\frac{1}{r^2}\frac{\partial}{\partial r}\left(r^2\frac{\partial}{\partial r}R_{lm}(\mathbf{r})\right)+\left[V(\mathbf{r})+\frac{l(l+1)}{r^2}-E\right]R_{lm}(\mathbf{r})=0, \tag{3.85}$$

which behaves like a repulsion potential. If the centrifugal potential is added to the potential $V(\mathbf{r})$ we obtain the effective potential shown in Fig. 3.12. The potential $V_{\text{eff}}(\mathbf{r})$ has a minimum and a maximum (for $l\geqslant 2$) which results in bound states at energies above the reference zero. The resulting bound state has a finite width Γ. The electron has a finite probability of tunneling through the potential barrier into the conduction band (probability of $\sim\Gamma/\hbar$). Hence such bound states are called virtual states. The electron lifetime of this state is inversely proportional to $\Gamma(\sim\hbar/\tau)$. If electron energy E is far from energy E_d (the virtual level energy) ordinary weak scattering will occur; this is characterized by phase shifts $\eta_2'(E)$. Resonance scattering can be represented schematically as follows: when the electron energy closes on E_d it is trapped at the virtual level and leaves the level after time $\hbar/\Gamma$. In certain cases it must be remembered that the virtual state is spin-nondegenerate, that is, electrons with different spin projections scatter differently.

If the potential contains a resonance state of energy E_d then the energy distribution of the electrons will vary sharply near E_d and additional oscillations in the electron density will occur in real space[293,294]:

$$\Delta\rho(r)=\sum_l 3\left(\frac{k_F^3}{3\pi^2}\right)(-1)^l\frac{(2l+1)}{(k_Fr)^2}\frac{\Gamma}{k_F}\cos[2k_dr+2\eta_l(k_d)], \tag{3.86}$$

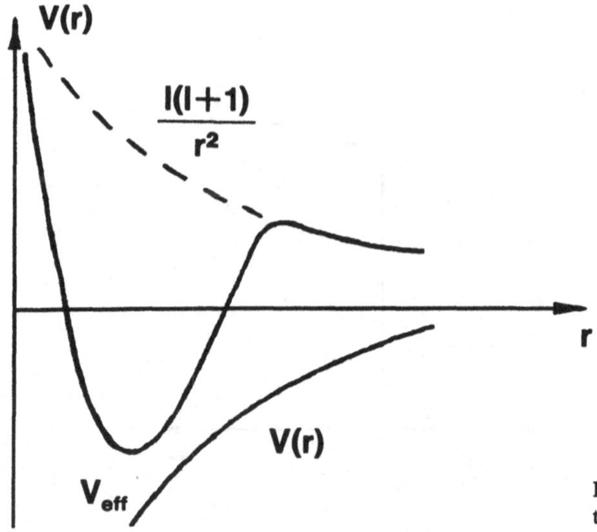

FIG. 3.12. Effective potential illustrating the bound states.

where $k_d = \sqrt{E_d}$.

Clearly the behavior of $\Delta\rho(r)$ differs from the Friedel law.

Therefore, in calculating the physical properties of d metals it is necessary to account for resonance scattering of the conduction electrons. The application of the standard pseudopotential method in this case will not yield satisfactory results. The reason can be traced to the special role of the d states which cannot be classified as core states nor as states in the conduction band: the d electrons are localized to a greater degree than the conduction electrons and are substantially different from the core electrons.

The primary concepts of the pseudopotential method for transition metals were developed by Harrison.[286] The approach can be described as follows. We introduce the basis set in explicit form as d wave functions $|d\rangle$. Then the wave function of the valence electron will take the form

$$\Psi = |\varphi\rangle + \sum_\alpha a_\alpha |\alpha\rangle + \sum_d a_d |d\rangle. \tag{3.87}$$

Substituting Eq. (3.86) into the Schrödinger equation we obtain

$$[-\nabla^2 + V(\mathbf{r})]\,|d\rangle = E_d |d\rangle - \Delta|d\rangle, \tag{3.88}$$

where we use the following designations:

$$E_d = \langle d| - \nabla^2 + V|d\rangle, \tag{3.89}$$

$$\Delta|d\rangle = \delta V|d\rangle - |d\rangle\langle d|\delta V|d\rangle. \tag{3.90}$$

δV is the deviation of the free ion potential from the ion potential in the metal. The quantity E_d represents the average Hamiltonian in the metal over the d states.

After some simple transforms (for details see Harrison's book[286]) we arrive at the equation

$$\nabla^2|\varphi\rangle + W|\varphi\rangle - \sum_d \frac{\Delta|d\rangle\langle d|\Delta|\varphi\rangle}{E_d - E} = E|\varphi\rangle, \tag{3.91}$$

where the operator W is defined as

$$W|\varphi\rangle = V|\varphi\rangle - \sum_\alpha (E_\alpha - E)|\alpha\rangle\langle\alpha|\varphi\rangle + \sum_d (E - E_d)|d\rangle\langle d|\varphi\rangle$$
$$+ \sum_d (|d\rangle\langle d|\Delta|\varphi\rangle + \Delta|d\rangle\langle d|\varphi\rangle). \tag{3.92}$$

The operator W is the pseudopotential of the d metal.

The additive term

$$\sum_d \frac{\Delta|d\rangle\langle d|\Delta|\varphi\rangle}{E_d - E}$$

(the hybridization potential) appears in Eq. (3.91) due to the hybridization of the d state with the free electron states. The hybridization potential is a high quantity for $E \to E_d$. Hence Eq. (3.91) cannot be solved by perturbation theory.

The Harrison approach is too cumbersome and hence is rarely used. Model pseudopotentials, one of which includes the HAA pseudopotential, are commonly used in calculating the properties of d metals. This pseudopotential takes the form[295]

$$W(r) = \begin{cases} -c - (A_0 - c)\hat{P}_0 - (A_1 - c)\hat{P}_1 - (A_2 - c)\hat{P}_2, & r < R_M, \\ -2z/r, & r > R_M, \end{cases} \tag{3.93}$$

where $c = 2z/R_M$. For $A_2 = C$, Eq. (3.93) becomes the pseudopotential of simple metals.

The form of Eq. (3.91) suggests that the pseudopotential of the d metal can be represented as a weak pseudopotential and a certain term that accounts for resonance scattering [this can easily be observed if Eq. (3.91) is treated as an equation with the potential $W - \Sigma_d$]. This approach was developed in Dagens' study.[280] The author proposed the nonlocal energy-dependent potential

$$W(\mathbf{r}) = W_{\mathrm{HA}} + W_{\mathrm{res}}, \tag{3.94}$$

where W_{HA} is a Heine–Abarenkov-type pseudopotential describing weak scattering and W_{res} is the resonance potential describing the hybridization effects of the s and d electrons:

$$W_{\mathrm{res}} = \frac{4\pi \Sigma_m Y_{lm} Y^*_{lm} \gamma(r)\gamma(r')}{E - E_d}. \tag{3.95}$$

The numerator of Eq. (3.95) contains a Hermitian operator affecting the d electrons [$\gamma(r)$ is the fitting parameter].

Occasionally the following equation is used for W_{res} (Refs. 281and 325):

$$W_{\mathrm{res}}(k,q) = \frac{5}{4\pi} P_2(\cos\theta)\beta(k)\beta(|\mathbf{k}+\mathbf{q}|)\frac{k^2 - E_d}{(k^2 - E_d)^2 + \Gamma^2/2}, \tag{3.96}$$

where $\beta(k) = A_d\Omega_0^{-1/2}(k/k_d)^2 \exp[(k_d^2 - k^2)/k_0^2]$, A_d and Γ are the fitting parameters related to the position and width of the d band.

3.6. Pseudopotential screening

The field produced by any ion polarizes the surrounding medium causing a redistribution of electron density that will reduce the effect of the bare field (the screening phenomenon).

It was demonstrated in Sec. 3.2 that this effect can be taken into account by dividing the Fourier components of the bare potential by $\varepsilon^*(\mathbf{q})$. Let the bare pseudopotential be $W^b(\mathbf{r})$. The Fourier component of the screened pseudopotential $W(\mathbf{q})$ takes the form

$$W(\mathbf{q}) = \frac{W^b(\mathbf{q})}{\varepsilon^*(\mathbf{q})}, \tag{3.97}$$

where $\varepsilon^*(\mathbf{q})$ is defined by Harrison[286] (see also Refs. 285 and 296). We analyze the effect of screening for the simplest case $\varepsilon^*(\mathbf{q}) = \varepsilon_{\mathrm{RPA}}(\mathbf{q})$ (Ref. 286) (random phase approximation), in which

$$e_{\mathrm{RPA}}(q) = 1 + \frac{4k_F}{\pi q^2}\left(\frac{1}{2} + \frac{4k_F^2 - q^2}{8k_F q} \ln\left|\frac{q + 2k_F}{q - 2k_F}\right|\right). \tag{3.98}$$

We can easily show that for small $\mathbf{q}$:

$$\lim_{q\to 0} \varepsilon(q) \sim \frac{1}{q^2} \sim \infty \tag{3.99a}$$

and for large $\mathbf{q}$:

$$\lim_{q\to\infty} \varepsilon(q) \sim 1 + \frac{\text{const}}{q^2} \approx 1. \tag{3.99b}$$

This means that the long-wavelength ($\mathbf{q}\to 0$) components of the pseudopotential are screened rather well while the short-wavelength components are screened poorly.

For $q = 2k_F$, $\varepsilon(q)$ has a singularity which results in oscillations of the screened pseudopotential with large r. We will demonstrate this. Going from q to r space,

$$W(r) = \frac{\Omega}{2\pi^2} \int_0^\infty dq\, q^2 \frac{W^b(q)}{\varepsilon(q)} \frac{\sin qr}{gr} = \frac{\Omega}{2\pi^2 r} \int_0^\infty dq\, q \sin qr \frac{W^b(q)}{\varepsilon(q)}. \tag{3.100}$$

After double integration by parts we find

$$W(r) = \frac{\Omega}{2\pi^2 r}\left\{ -\frac{\cos qr}{r} \frac{qW^b(q)}{\varepsilon(q)}\bigg|_0^\infty + \frac{\sin qr}{r^2} \frac{d}{dq}\left[\frac{qW^b(q)}{\varepsilon(q)}\right]_0^\infty \right.$$
$$\left. -\frac{1}{r^2} \int dq \sin qr \frac{d^2}{dr^2}\left[\frac{q(W^b(q))}{\varepsilon(q)}\right]\right\}. \tag{3.101}$$

The terms corresponding to points near the singularity $q = 2k_F$ in the first two terms of Eq. (3.101) cancel. Moreover,

$$W^b_{(q)} \quad \text{and} \quad \frac{d}{dq} W^b_{(q)}$$

go to zero as $q \to \infty$ and the first two terms here also vanish.

Finally we obtain the expression

$$W(r) = -\frac{\Omega}{2\pi^2 r^3} \int_0^\infty dq \sin qr \frac{d^2}{dq^2} \left[\frac{qW^b(q)}{\varepsilon(q)} \right]. \tag{3.102}$$

Due to the rapid oscillations of sin qr the integrals vanish. However, the second derivative of $\varepsilon(q)$ has an infinite jump at $q = 2k_F$. It is necessary to account for the most singular terms in Eq. (3.102) using the explicit form of $\varepsilon(q)$ after which we can easily find $d^2\varepsilon/dg^2$. The most singular term takes the form

$$\left(\frac{d^2\varepsilon(q)}{dq^2} \right)_{\text{sing}} = -\frac{1}{4\pi k_F^2(2k_F - q)}, \tag{3.103}$$

the most singular term in $d^2/dq^2[qW^b(q)/\varepsilon(q)]$ is

$$\frac{d^2}{dq^2} \left[\frac{qW^b(q)}{\varepsilon(q)} \right]_{\text{sing}} = -\frac{2k_F W^b(2k_F)}{\varepsilon^2(2k_F)} \left(\frac{d^2\varepsilon}{dq^2} \right)_{\text{sing}}. \tag{3.104}$$

It is not necessary to account for the other terms since they are slowly varying functions of q and their contribution to the integral vanishes with large r.

Substituting Eqs. (3.103) and (3.104) into Eq. (3.102) we obtain

$$W(r) \simeq -\frac{\Omega W^b(2k_F)}{4\pi^3 k_F \varepsilon^2(2k_F)^3} \int_0^\infty dq \frac{\sin qr}{2k_F - q}. \tag{3.105}$$

The integral in Eq. (3.105) is easily defined:

$$W(r) \simeq \frac{\Omega W^b(2k_F)}{4\pi^2 k_F \varepsilon^2(2k_F)} \frac{\cos(2k_F r)}{r^3}. \tag{3.106}$$

Thus, unlike the Thomas–Fermi approximation, oscillations of the screened potential are obtained with large r. However, such oscillations are called Friedel oscillations and were discovered by J. Friedel in 1958.

The physical cause of Friedel oscillations is the sharp step in the Fermi distribution at $q = 2k_F$. Harrison[286] proposed that at finite temperatures the long-range nature of the screened pseudopotential may change significantly. This idea was developed in Khannanov.[297] The author represented $\varepsilon(q)$ as

$$\varepsilon(q) = 1 - \frac{4}{\pi q^2} \int d^3k \frac{f(k)}{k^2 - |k+q|^2}, \tag{3.107}$$

where $f(k)$ is the Fermi–Dirac distribution function for the degenerate electron gas:

$$f(k)=\frac{1}{\exp[(k^2-\mu)/T]+1}. \tag{3.108}$$

In Eq. (3.108) T is temperature in energy units and μ is the chemical potential.

We introduce the function $y(q)$:

$$I(q)=\int d^3k \frac{f(k)}{k^2-|k+q|^2}. \tag{3.109}$$

Angle integration in Eq. (3.109) yields the expression

$$I(q)=\frac{\pi}{q}\int_0^\infty dk\, f(k)k\ln\left|\frac{k-q/2}{k+q/2}\right|=\frac{\pi}{q}\Psi(q), \tag{3.110}$$

where

$$\Psi(q)=\frac{1}{2}\int_{-\infty}^{\infty} dk\, f(k)k\ln\left|\frac{k-q/2}{k+q/2}\right|. \tag{3.111}$$

The author of this study has demonstrated that the function $\Psi(k)$ has derivatives of any order:

$$\Psi^{n+1}(q)=\frac{1}{2}\int_{-\infty}^{\infty} dk[f(k)k]^n\left(\frac{1}{k-q/2}+\frac{1}{k+q/2}\right). \tag{3.112}$$

In terms of $\Psi(k)$ the dielectric function $\varepsilon(q)$ takes the form

$$\varepsilon(q)=1-\frac{4}{\pi^2}\frac{\Psi(q)}{q^2}. \tag{3.113}$$

We can easily see that incorporating the temperature spread of the Fermi distribution causes the divergence of the derivatives of $\varepsilon(q)$ of any order to vanish and thereby significantly smooths over the oscillations in the potential [the author assumes a complete vanishing of the oscillations due to the singularities in $\varepsilon(q)$]. Moreover, he does not exclude pseudopotential oscillations associated with the singularities of $W(q)$. It should be noted that there are few theoretical studies devoted to the effect of temperature on the features of electron–ion interaction.

In concluding this section we focus on how to account for nonlinear effects in screening theory.[298,299]

Assume we introduce into an electron gas with an average electron density ρ a scatterer which distorts the electron density in its immediate vicinity. We characterize the scatterer by the electron density $\rho_{ex}(\mathbf{r})$. The deviation of the total (accounting for screening) electron density ρ_{tot} from ρ is obviously the induced density $\rho_{ind}(\mathbf{r})$. We represent $\rho_{ex}(\mathbf{r})$ and $\rho_{ind}(\mathbf{r})$ as

$$\rho_{ext}(\mathbf{r})=\sum_{\mathbf{q}}{}' \rho_q^0 \exp(i\mathbf{qr}), \tag{3.114}$$

$$\rho_{cp}(\mathbf{r})=\sum_{\mathbf{q}}{}' \rho_q \exp(i\mathbf{qr}). \tag{3.115}$$

Here the relation between ρ_q and ρ_q^0 takes the form

$$\rho_{\mathbf{q}}=\alpha_{\mathbf{q}}\rho_{\mathbf{q}}^{0}+\sum_{\mathbf{q}_1+\mathbf{q}_2=\mathbf{q}}\alpha_{\mathbf{q}_1\mathbf{q}_2}\rho_{\mathbf{q}_1}^{0}\rho_{\mathbf{q}_2}^{0}, \tag{3.116}$$

where α_q is the polarizability of the electron gas (α_q is linear and $\alpha_{q_1q_2}$ is quadratic). Using Eqs. (3.114) and (3.115) we rewrite expansion (3.116) as

$$V_{\text{tot}}(\bar{\mathbf{q}})=V_{\text{ext}}(\bar{\mathbf{q}})-\alpha_{\mathbf{q}}V_{\text{ext}}(\mathbf{q})+\frac{8\pi}{q^2}\sum_{\mathbf{q}_1+\mathbf{q}_2=\mathbf{q}}\alpha_{\mathbf{q}_1\mathbf{q}_2}\frac{q_1^2q_2^2}{8\pi}V_{\text{ext}}(\mathbf{q}_1)V_{\text{ext}}(\mathbf{q}_2)+\cdots. \tag{3.117}$$

If we discard the third term in this equation we obtain

$$V_{\text{tot}}(\mathbf{q})=V_{\text{ext}}(\mathbf{q})(1-\alpha_{\mathbf{q}})=\frac{V_{\text{ext}}(\mathbf{q})}{\varepsilon_{\text{RPA}}(\mathbf{q})}, \tag{3.118}$$

where

$$\varepsilon_{\text{RPA}}(\mathbf{q})=\frac{1}{1-\alpha_{\mathbf{q}}}. \tag{3.119}$$

Therefore we arrive at results (3.97).

The electron density $\rho_{\text{ind}}(\mathbf{q})$ can be expanded as

$$\rho_{\text{cp}}(\mathbf{q})=\chi^0(\mathbf{q})V_{\text{tot}}(\mathbf{q})+\sum_{\mathbf{q}_1+\mathbf{q}_2=\mathbf{q}}\chi_{\mathbf{q}_1\mathbf{q}_2}^{0}V_{\text{tot}}(\mathbf{q}_1)V_{\text{tot}}(\mathbf{q}_2)+\cdots. \tag{3.120}$$

We can show[299] that for a system of noninteracting electrons,

$$\rho_{\text{tot}}(\mathbf{r})=\frac{k_P^3}{3\pi^2}-\sum_{\mathbf{q}}{}' e^{i\mathbf{qr}}\frac{k_Fm}{\pi^2\hbar^2}V_{\text{tot}}(q)\varphi_1(\mathbf{q},k_F)+V_{\text{tot}}(\mathbf{q})V_{\text{tot}}(\mathbf{q}')\frac{m}{\hbar^2}\varphi_2(\mathbf{q},\mathbf{q}',k_F), \tag{3.121}$$

where the function $\varphi_1(\mathbf{q}_1,k_F)$ identically coincides with the previous function $f_0(\mathbf{q})$, while the quadratic function is as follows[298]:

$$\varphi_2(\mathbf{q},\mathbf{q}',k_F)=\frac{k_F}{qq'\sin^2\theta}\left[(q+q'\cos\theta)\ln\left|\frac{q'+2k_F}{q'-2k_F}\right|\right]$$
$$+(q'+q\cos\theta)\ln\left|\frac{q+2k_F}{q-2k_F}\right|$$
$$-|\mathbf{q}+\mathbf{q}''|\cos\theta\ln\left|\frac{|\mathbf{q}+\mathbf{q}''|+2k_F}{|\mathbf{q}+\mathbf{q}'|-2k_F}\right|+T(q,q',k_F),$$

(3.122)

$$T(q,q',k_F)=\begin{cases}\sqrt{R}\ln\left|\dfrac{(2k_F)^3+2k_Fqq'\cos\theta-2k_F(q+q')+qq'\sqrt{R}}{(2k_F)^3+2k_Fqq'\cos\theta-2k_F(q+q')-qq'\sqrt{R}}\right|, \\ R\geqslant 0, \\ -2\sqrt{-R}\arctan\dfrac{qq'\sqrt{-R}}{(2k_F)^3+2k_Fqq'\cos\theta-2k_F(q+q')^2}, \\ R<0,\end{cases}$$

(3.123)

where $R=(\mathbf{q}+\mathbf{q}')^2-(2k_F)^2\sin\theta$. There is a simpler equation than Eq. (3.122) for φ_2, which is an approximate equation and takes the form

$$\varphi_2(q,q',E_F)=\frac{k_F}{\sqrt{q^2+q'^2+qq'}}\ln\left|\frac{\sqrt{q^2+q'^2+qq'}+2k_F}{\sqrt{q^2+q'^2+qq'}-2k_F}\right|.$$

(3.124)

Using Eq. (3.121) we can obtain

$$V_{\text{tot}}(q)=V_{\text{ext}}(q)-\frac{k_{\text{TF}}^2}{q^2}\varphi_1(q,E_F)V_{\text{tot}}(q)$$
$$+\frac{k_{\text{TF}}^2}{q^24E_F}\sum_{\mathbf{q}_1+\mathbf{q}_2=\mathbf{q}}\varphi_2(\mathbf{q}_1,\mathbf{q}_2,E_F)V_{\text{tot}}(\mathbf{q}_1)V_{\text{tot}}(\mathbf{q}_2). \quad (3.125)$$

Equation (3.125) can be solved together with Eq. (3.117) if we limit the analysis to a linear approximation in Eq. (3.117) and substitute this into Eq. (3.115). We can obtain [discarding linear terms in Eq. (3.125)]

$$V_{\text{ext}}(\mathbf{q})(1-\alpha_{\mathbf{q}})=V_{\text{ext}}(\mathbf{q})-\frac{k_{\text{TF}}^2}{q^2}V_{\text{ext}}(\mathbf{q})(1-\alpha_{\mathbf{q}})\varphi_1(\mathbf{q},k_{\text{TF}}),$$

(3.126)

from which

$$\frac{1}{1-\alpha_{\mathbf{q}}}=1+\frac{k_{\text{TF}}^2}{q^2}\varphi_1=\varepsilon_{\text{RPA}}. \quad (3.127)$$

If we now account for quadratic terms in Eq. (3.117) and substitute $V_{\text{tot}}(\mathbf{q})$ from Eq. (3.117) into Eq. (3.125) we obtain

$$\sum_{\mathbf{q}_1+\mathbf{q}_2=\mathbf{q}} V_{\text{ext}}(\mathbf{q}_1)V_{\text{ext}}(\mathbf{q}_2)\left\{\frac{8\pi}{q^2}\frac{q_1^2q_2^2}{(8\pi)^2}\alpha_{\mathbf{q}_1\mathbf{q}_2}\left(1+\frac{k_{\text{TF}}^2}{q^2}\varphi_1\right)\right.$$

$$\left.-\frac{k_{\text{TF}}^2}{q^2}\frac{\varphi_2}{4E_F}(1-\alpha_{\mathbf{q}_1})(1-\alpha_{\mathbf{q}_2})\right\}. \tag{3.128}$$

From here

$$\alpha_{\mathbf{q}_1\mathbf{q}_2}=\frac{k_{\text{TF}}^2 8\pi}{\varepsilon_{\text{RPA}}(\mathbf{q})4E_F}\frac{\varphi_2(\mathbf{q}_1,\mathbf{q}_2,E_F)}{\varepsilon_{\text{RPA}}(\mathbf{q}_1)\varepsilon_{\text{RPA}}(\mathbf{q}_2)q_1^2q_2^2}. \tag{3.129}$$

Therefore, the following equation is valid for the potential $V_{\text{tot}}(\mathbf{q})$ screened accounting for linear effects:

$$V_{\text{tot}}(\mathbf{q})=\frac{V_{\text{ext}}(\mathbf{q})}{\varepsilon_{\text{RPA}}(\mathbf{q})}+\frac{k_{\text{TF}}}{q^2\varepsilon_{\text{RPA}}(\mathbf{q})4E_F}\sum_{\mathbf{q}_1+\mathbf{q}_2=\mathbf{q}}\varphi_2(\mathbf{q}_1,\mathbf{q}_2)\frac{V_{\text{ext}}(\mathbf{q}_1)V_{\text{ext}}(\mathbf{q}_2)}{\varepsilon_{\text{RPA}}(\mathbf{q}_1)\varepsilon_{\text{RPA}}(\mathbf{q}_2)}. \tag{3.130}$$

We consider as our example the contribution of nonlinear effects to the screening of a point charge in an electron gas. The Fourier component of the potential takes the form

$$V_{\text{ext}}(\mathbf{q})=-\frac{8\pi z}{q^2}. \tag{3.131}$$

Using Eq. (3.130) for potential (3.131) yields the result[298]

$$V_{\text{tot}}(\mathbf{q})=\frac{8\pi z}{q^2\varepsilon(\mathbf{q})}[1-A(\mathbf{q})], \tag{3.132}$$

where

$$A(q)=\frac{k_{\text{TF}}^2 z}{\pi E_F}\int_0^\infty dq'\int_{-1}^{1}d\xi\,\frac{q'^2\varphi_2(\mathbf{q}',\mathbf{q}-\mathbf{q}',E_F)}{\mathbf{q}'^2\varepsilon(\mathbf{q}')(\mathbf{q}-\mathbf{q}')^2\varepsilon(\mathbf{q}-\mathbf{q}')}, \tag{3.133}$$

where $\xi=\cos\theta$ and θ is the angle between $\mathbf{q}$ and $\mathbf{q}'$. As $\mathbf{q}\to\infty$,

$$A(0)=z\left(\frac{\pi}{4k_Fa_0}\right)^{1/2}\frac{1}{1+\pi ka_0}. \tag{3.134}$$

For this case

$$\lim_{q\to 0}V_{\text{tot}}(\mathbf{q})=-\frac{8\pi z^*}{q^3+k_{\text{TF}}^2}, \tag{3.135}$$

where

$$z^*=z(1-A_0). \tag{3.136}$$

Going over to $\mathbf{r}$ space we obtain

$$\lim_{r\to\infty} V(r) = -\frac{z^*}{r} e^{-k_{TF} r}. \tag{3.137}$$

It is clear that compared to linear response theory [in which $A(0)=0$, $z^*=z$] the valency z is replaced by the effective value z^*. For $A(0)>1$, z^* may change sign. This represents a new result compared to linear screening theory. A similar effect was identified by Almbladh[300] where the author analyzed proton screening in an electron gas. Specifically it was determined that for $r_s > 1.9$ hydrogen is in the H^- state. Equations (3.134) and (3.136) predict this trend.

3.7. Binding energy of metals

The binding energy of a metal (alloy) is determined by the sum of the electron–ion and electron–electron interaction energies as well as the interaction energies between ions themselves and includes the free electron gas energy, the electrostatic energy, and the energy of the band structure*:

$$u_{es} = \frac{1}{2}\frac{z^2 e^2 \alpha}{r_0}, \tag{3.138}$$

where $r_0 = (3\Omega_0/4\pi)^{1/3}$. The quantity α is called the Madelung constant. It is calculated for many types of crystalline lattices (e.g., for bcc, $\alpha = -1.791\,86$, for fcc $\alpha = -1.79\,175$, for hcp $\alpha = -1.791\,68$). The fact that $\alpha < -1$ suggests that systems consisting of separate ions comprising the crystal and clusters of isolated molecules will be more stable due to the electrostatic energy of crystals, metals, and semiconductors.

In addition to direct Coulomb interaction between ions there is also indirect interaction through the conduction electrons which defines the energy of the band structure U_{bs}. This is due to the difference between the energy spectrum of the electron gas in the crystal and the energy spectrum of the free electrons.

The expression for the band structure energy is obtained by summing the contributions of first order perturbation theory accounting for electron–electron interactions U_{EL-EL}^{IN}:

$$U_{bs} = \frac{2\Omega}{(2\pi)^3} \sum_{\mathbf{k}}{}' \int d^3k \frac{|\langle \mathbf{k}+\mathbf{q}|V|\mathbf{k}\rangle|^2}{k^2 - |\mathbf{k}+\mathbf{q}|^2} - U_{EL-EL}^{IN}. \tag{3.139}$$

We introduce the function $\chi(q)$ called the Lindhard function:

$$\chi(q) = \frac{2\Omega}{(2\pi)^3 N} \int \frac{d^3k}{k^2 - |\mathbf{k}+\mathbf{q}|^2} = -\frac{z}{4}\left(\frac{2}{3}E_F\right)^{-1}\left\{1 + \frac{1-x^2}{2x}\ln\left|\frac{1+x}{1-x}\right|\right\}. \tag{3.140}$$

Then for the local pseudopotentials we have

$$U_{bs} = \sum_{\mathbf{g}}{}' |S(\mathbf{g})|^2 |V(\mathbf{g})|^2 \chi(g) - U_{EL-EL}^{IN}. \tag{3.141}$$

*Here the binding energy is defined as the difference between the energy of individual ions and electrons. The literature also utilizes other definitions; the neutral atom energy can also be used as the reference level (Ref. 321).

The electron–electron interaction energy $U^{\text{IN}}_{\text{EL-EL}}$ can be defined as the interaction energy of the screened potential V_{ind} and the induced electron density ρ_{ind}:

$$U^{\text{IN}}_{\text{EL-EL}}=\frac{1}{2\Omega}\int d^3r\,\rho_{\text{ind}}(\mathbf{r})\,V_{\text{ind}}(\mathbf{r}), \tag{3.142}$$

or, going over to the Fourier components (accounting for Parseval's theorem)

$$U^{\text{IN}}_{\text{EL-EL}}=\frac{\Omega}{2}\sum_q{}'\rho_{\text{ind}}(\mathbf{b})\,V_{\text{ind}}(\mathbf{q}). \tag{3.143}$$

We find $\rho_{\text{ind}}(\mathbf{q})$, $V_{\text{ind}}(\mathbf{q})$.

We write the wave function of the electron in first order perturbation theory as

$$\Psi_{\mathbf{k}}(\mathbf{r})=|\mathbf{k}\rangle+\sum_{q\neq 0}{}'\frac{|\langle\mathbf{k}+\mathbf{q}|V|\mathbf{k}\rangle|}{k^2-|\mathbf{k}+\mathbf{q}|^2}|\mathbf{k}+\mathbf{q}\rangle. \tag{3.144}$$

The induced density represents the difference between the entire electron density and the average density of the homogeneous electron gas:

$$\rho_{\text{ind}}(\mathbf{r})=\rho(\mathbf{r})-\rho_0$$
$$=\frac{2}{(2\pi)^3}\int_{k<k_F}d^3k\sum\left[\frac{\langle\mathbf{k}+\mathbf{q}|V|\mathbf{k}\rangle}{\mathbf{k}^2-|\mathbf{k}+\mathbf{q}|^2}e^{i\mathbf{k}\mathbf{r}}+\frac{\langle\mathbf{k}|V|\mathbf{k}+\mathbf{q}\rangle}{\mathbf{k}^2-|\mathbf{k}+\mathbf{q}|^2}e^{i\mathbf{k}\mathbf{r}}\right]. \tag{3.145}$$

Going over to Fourier components,

$$\rho_{\text{ind}}(\mathbf{q})=\frac{2}{(2\pi)^3}\left[\int_{k<k_F}\frac{d^3k\langle\mathbf{k}+\mathbf{q}|V|\mathbf{k}\rangle}{\mathbf{k}^2-|\mathbf{k}+\mathbf{q}|^2}-\int_{k<k_F}\frac{d^3\mathbf{k}\langle\mathbf{k}|V|\mathbf{k}+\mathbf{q}\rangle}{\mathbf{k}^2-|\mathbf{k}+\mathbf{q}|^2}d^3\mathbf{k}\right].$$

Since integration in each of the integrals is carried out with respect to all **k**, we can replace **k** by $-\mathbf{k}$ in the second integral and the second term will be the complex conjugate of the first term. As a result the terms in braces will be identical from which

$$\rho_{\text{ind}}(\mathbf{q})=\frac{4}{(2\pi)^3}\int_{k<k_F}d^3\mathbf{k}\frac{\langle\mathbf{k}+\mathbf{q}|V|\mathbf{k}\rangle}{\mathbf{k}^2-|\mathbf{k}+\mathbf{q}|^2}=\frac{4}{(2\pi)^3}V(\mathbf{q})\int\frac{d^3\mathbf{k}}{\mathbf{k}^3-|\mathbf{k}+\mathbf{q}^2}. \tag{3.146}$$

The Poisson equation in q space takes the form

$$V_{\text{ind}}(\mathbf{q})=\frac{8\pi}{q^2}\rho_{\text{ind}}(\mathbf{q}). \tag{3.147}$$

Substituting Eq. (3.146) into Eq. (3.147) and accounting for Eq. (3.140) we obtain

$$V_{\text{ind}}(\mathbf{q})=\frac{8\pi}{\Omega q^2}V(\mathbf{q})\chi(\mathbf{q}). \tag{3.148}$$

Using Eqs. (3.147), (3.148), and (3.139) we have

$$U_{\text{bs}}=\sum_{\mathbf{g}}{}'|S(\mathbf{g})|^2|W(\mathbf{g})|^2\chi(\mathbf{g})\varepsilon(\mathbf{g}). \tag{3.149}$$

The last two multipliers in Eq. (3.149) are independent of the crystal structure and their product is called the characteristic function of the energy of the band structure:

$$\Phi_{bs}(\mathbf{g}) = |W(\mathbf{g})|^2 \chi(\mathbf{g}) \varepsilon(\mathbf{g}). \tag{3.150}$$

We rewrite Eq. (3.149) as

$$U_{bs} = \frac{1}{N^2} \sum_{\nu,\mu,m} \Phi_{bs}(\mathbf{q}_m) e^{i\mathbf{q}_m(t_\nu - t_\mu)}. \tag{3.151}$$

We define the function $\varphi(\mathbf{r})$ as the Fourier transform $\Phi_{bs}(\mathbf{q})$:

$$\varphi_{bs}(\mathbf{r}) = \frac{2\Omega}{(2\pi)^3} \int d^3\mathbf{q}\, \Phi_{bs}(\mathbf{q}) e^{i\mathbf{q}\mathbf{r}}. \tag{3.152}$$

It is clear from Eqs. (3.151) and (3.152) that the energy of the band structure can be represented as the sum of the energy of the pair interatomic interactions:

$$U_{bs} = \frac{1}{2N} \sum_{\nu,\mu} \varphi_{bs}(\mathbf{r}_\nu - \mathbf{r}_\mu). \tag{3.153}$$

The potential φ_{bs} describes pair ion interaction through the conduction electron gas, that is, this is the indirect interaction potential. Adding the Coulomb interaction potential to Eq. (3.152) we obtain the effective interatomic pair potential:

$$\varphi(r) = \frac{z^2 e^2}{r} + \frac{\Omega}{\pi^2} \int_0^\infty dq\, \Phi_{bs}(q) \frac{\sin qr}{qr}. \tag{3.154}$$

Rewriting Eq. (3.154),

$$\varphi(r) = \frac{(ze)^2}{r} \left\{ 1 - \frac{2}{\pi} \int_0^\infty dq\, [1 - \varepsilon^{-1}(q)] M^2(q) \frac{\sin qr}{q} \right\}, \tag{3.155}$$

where

$$\frac{4\pi e^2}{\Omega q^2} M(\mathbf{q}) = \langle \mathbf{k} + \mathbf{q} | W | \mathbf{k} \rangle. \tag{3.156}$$

We can then easily obtain for $\varphi(r)$.

$$\varphi(r) = \frac{(ze)^2}{r} \left\{ 1 - \frac{2}{\pi} \int_0^\infty dq \frac{4\pi e^2 \Pi_{eff} M^2(q) \sin qr}{[q^2 + 4\pi e^2 \Pi_{eff}(q) q]} \right\}.$$

In order to write the expression for the total energy we must still account for the interaction of a uniformly distributed electron with the non-Coulomb part of the pseudopotential:

$$U_1 = \lim_{q \to 0} z \left[\frac{4\pi z}{\Omega q^2} + W(\mathbf{q}) \right]. \tag{3.157}$$

Therefore the binding energy U takes the form

$$U = U_0 + U_1 + U_{bs} + U_{es} + \cdots. \tag{3.158}$$

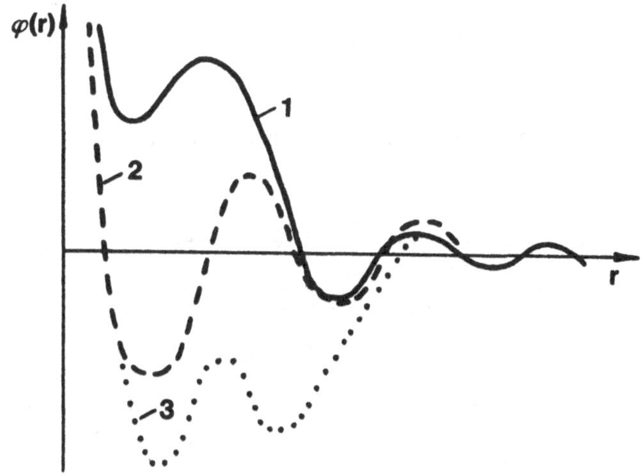

FIG. 3.13. Pair potential plotted as a function of screening method.

Knowing the form factor of the pseudopotential $W(\mathbf{q})$, the dielectric function $\varepsilon^*(\mathbf{q})$, the structure factor $S(\mathbf{q})$, and the parameter α we can calculate the binding energy of the metal. The results of these calculations can be used for, for example, predicting stable crystalline modifications.[256]

We note that in each specific case it is necessary to rigorously analyze the effect of the approximations used on the calculation results. For example, a calculation of pair potential (3.156) with different $\varepsilon^*(\mathbf{q})$ revealed a strong dependence of the potential shape, the position of the first maximum, and its quantity on the form of $\varepsilon^*(\mathbf{q})$. Figure 3.13 shows the potential $\varphi(\mathbf{r})$ for Al with different $\varepsilon^*(\mathbf{q})$. We can show using the equation for $\phi_{es}(\mathbf{q})$ and by carrying out double integration by parts in Eq. (2.155) that the asymptotic form of the effective pair potential takes the form

$$\varphi(r) \sim \frac{9\pi z^2[W(2k_F)]^2}{E_F} \frac{\cos 2k_F r}{(2k_F r)^3}. \tag{3.159}$$

Oscillations of $\Psi(\mathbf{r})$ can be attributed to the singularity in $\varepsilon(\mathbf{q})$ for $\mathbf{q} = 2k_F$ which was described previously.

3.8. The pseudopotential method and electron density functional theory, binding energy p

The relations of electron density-functional theory were reported earlier. In order to simplify the presentation we will again provide the Kohn–Sham equations

$$[-\nabla^2 + W^0(\mathbf{r}) + W^H(\mathbf{r}) + W^{exc}(\mathbf{r})]\Psi_k(\mathbf{r}) = E_k\Psi_k(\mathbf{r}), \tag{3.160}$$

where $W^0(\mathbf{r})$, $W^H(\mathbf{r})$, and $W^{exc}(\mathbf{r})$ are the external, Hartree, and exchange-correlation potentials, respectively. In pseudopotential theory the potential $W^0(\mathbf{r})$ is replaced by a pseudopotential. Expanding the wave function $\Psi_\mathbf{k}(\mathbf{r})$ in plane waves,

$$\Psi_{\mathbf{k}}(\mathbf{r})=\sum_{\mathbf{g}}{}' a_{\mathbf{k}}(\mathbf{g})\,|\mathbf{k}+\mathbf{g}\rangle, \tag{3.161}$$

$$|\mathbf{k}+\mathbf{g}\rangle=\Omega^{-1/2}\exp[-i(\mathbf{k}+\mathbf{g})\mathbf{r}]. \tag{3.162}$$

Substituting Eq. (3.161) into Eq. (3.160) and carrying out the standard transforms we obtain

$$\sum_{\mathbf{g}'}[(E_{\mathbf{k}}-E^0_{\mathbf{k}+\mathbf{g}'})\delta_{\mathbf{g}\mathbf{g}'}-W(\mathbf{g}-\mathbf{g}')]a_{\mathbf{k}}(\mathbf{g}')=0, \tag{3.163}$$

$$E^0_{\mathbf{k}+\mathbf{g}}=|\mathbf{k}+\mathbf{g}|^2, \tag{3.164}$$

$$W(\mathbf{g})=W^0(\mathbf{g})+W^H(\mathbf{g})+W^{\mathrm{exc}}(\mathbf{g}). \tag{3.165}$$

Solving the secular equation for each $\mathbf{k}$ it is possible to find the band structure energy. This approach is well known. A new idea was proposed in 1981 (Ref. 301) for determining the band structure energy which is as follows. We introduce the coefficients $\widetilde{W}_{\mathbf{k}}(\mathbf{q})$:

$$\widetilde{W}_{\mathbf{k}}(\mathbf{g})=(E_{\mathbf{k}}-E^0_{\mathbf{k}+\mathbf{g}})a_{\mathbf{k}}(\mathbf{g}). \tag{3.166}$$

Then secular equation (3.163) takes the form

$$\sum_{\mathbf{g}'}\left[\delta_{\mathbf{g}\mathbf{g}'}-\frac{W(\mathbf{g}-\mathbf{g}')}{E_{\mathbf{k}}-E^0_{\mathbf{k}+\mathbf{g}}}\right]\widetilde{W}_{\mathbf{k}}(\mathbf{g})=0, \tag{3.167}$$

$$\Psi_{\mathbf{k}}(\mathbf{r})=\sum_{\mathbf{g}}\frac{\widetilde{W}_{\mathbf{k}}(\mathbf{g})\,|\mathbf{k}+\mathbf{g}\rangle}{E_{\mathbf{k}}-E^0_{\mathbf{k}+\mathbf{g}}}. \tag{3.168}$$

Averaging $\widetilde{W}_{\mathbf{k}}(\mathbf{q})$ over $\mathbf{k}$ we obtain certain energy-independent coefficients $\widetilde{W}(\mathbf{g})$. We require $W(\mathbf{g})$ to satisfy the equation

$$\sum_{\mathbf{g}'}\left[\delta_{\mathbf{g}\mathbf{g}'}-W(\mathbf{g}-\mathbf{g}')\left\langle\frac{1}{E_{\mathbf{k}}-E^0_{\mathbf{k}+\mathbf{g}}}\right\rangle\right]\widetilde{W}(\mathbf{g}')=0. \tag{3.169}$$

We introduce the function $F(\mathbf{g})$:

$$F(\mathbf{g})=\left\langle\frac{1}{E_{\mathbf{k}}-E^0_{\mathbf{k}+\mathbf{g}}}\right\rangle=\frac{1}{Nz}\sum_{\mathbf{k}}\frac{1}{(E_{\mathbf{k}}-E^0_{\mathbf{k}+\mathbf{g}})}. \tag{3.170}$$

Equation (3.169) takes the form

$$\sum_{\mathbf{g}'}[\delta_{\mathbf{g}\mathbf{g}'}-W(\mathbf{g}-\mathbf{g}')F(\mathbf{g}')]\widetilde{W}(\mathbf{g}')=0. \tag{3.171}$$

The electron density distribution in the system can be written as

$$\rho(\mathbf{r})=c^2\frac{z}{\Omega_0}\left\{1+2\,\mathrm{Re}\sum_{\mathbf{g}}{}'\widetilde{W}(\mathbf{g})F(\mathbf{g})e^{i\mathbf{g}\mathbf{r}}\right.$$

$$\left.+\sum_{\mathbf{g}\mathbf{g}'}{}'\widetilde{W}(\mathbf{g})W^*(\mathbf{g}')\Phi(\mathbf{g},\mathbf{g}')e^{i(\mathbf{g}-\mathbf{g}')\mathbf{r}}\right\}. \tag{3.172}$$

We find C from conservation of particle number

$$c^2=\left[1+\sum_{\mathbf{g}'}{}' \,|\,\widetilde{W}(\mathbf{g})\,|^2\Phi(\mathbf{g},\mathbf{g}')\right]^{-1}, \tag{3.173}$$

$$\Phi(\mathbf{g},\mathbf{g}')=\left\langle\frac{1}{(E_{\mathbf{k}}-E^0_{\mathbf{k}+\mathbf{g}})(E_{\mathbf{k}}-E^0_{\mathbf{k}+\mathbf{g}})}\right\rangle. \tag{3.174}$$

Carrying out the Fourier transforms of the function $\rho(\mathbf{r})$ we obtain

$$\rho(\mathbf{g})=\begin{cases} z/\Omega_0 \quad \text{where } g=0, \\ \dfrac{z}{\Omega_0}c^2\left[2\,\mathrm{Re}\,\widetilde{W}(\mathbf{g})F(\mathbf{g})+\sum_{\mathbf{g}\neq\mathbf{g}'}\widetilde{W}(\mathbf{g}-\mathbf{g}')W^*(\mathbf{g}')\varphi(\mathbf{g}-\mathbf{g}',\mathbf{g}')\right] \\ \quad \text{where } \mathbf{g}\neq 0. \end{cases} \tag{3.175}$$

The total crystal energy within the framework of electron density-functional theory takes the form

$$E=\sum_{\mathbf{k}}E_k-\frac{e^2}{2}\int d^3r\,dr'\,\frac{\rho(\mathbf{r})\rho(\mathbf{r}')}{|\mathbf{r}-\mathbf{r}'|}+E_{\mathrm{exc}}[\rho]-\int d^3r\,W^{\mathrm{exc}}(\mathbf{r})\rho(\mathbf{r})+E_{\mathrm{es}}. \tag{3.176}$$

We recast this equation in the momentum representation (calculated per ion)

$$E_N=\frac{1}{N}\sum_{\mathbf{k}}E_{\mathbf{k}}-\frac{e^2}{2}\Omega_0\sum_{\mathbf{g}}{}'\frac{4\pi}{g^2}|\rho(\mathbf{g})|^2$$

$$+\Omega_0\sum_{\mathbf{g}}\left[\varepsilon_{\mathrm{exc}}(\mathbf{g})-W^{\mathrm{exc}}(\mathbf{g})\right]\rho(\mathbf{g})+E_{\mathrm{es}}. \tag{3.177}$$

The first term in Eq. (3.177) accounting for averaging over $\mathbf{k}$ takes the form

$$u_{\mathrm{Coul}}=\frac{1}{N}\sum_{\mathbf{k}}E_{\mathbf{k}}=\frac{3}{10}e^2zk_F^2+zW(\mathbf{g}=0)+z\sum_{\mathbf{g}}\widetilde{W}(\mathbf{g})W^*(\mathbf{g})F(\mathbf{g}), \tag{3.178}$$

$$W(g=0)=\lim_{\mathbf{g}\to 0}\left[W^0(\mathbf{g})+\frac{4\pi ze^2}{\Omega_0 g^2}\right]+W^{\mathrm{exc}}(\mathbf{g}=0). \tag{3.179}$$

Kreske[301] has proposed (using for the exchange-correlated Nozières–Pines energy) an approximation by Chebyshev polynomials:

$$E_{\mathrm{exc}}=\Omega_0\sum_{\mathbf{g}}\left[\varepsilon_{\mathrm{exc}}(\mathbf{g})-W^{\mathrm{exc}}(\mathbf{g})\right]\rho(\mathbf{g})$$

$$=zE^0_{\mathrm{exc}}+\Omega_0^2\frac{0.916}{r_s}\left(-\frac{4}{3}A_2+\frac{1}{3}B_2\right)\sum_{\mathbf{g}}{}'\,|\rho(\mathbf{g})|^2, \tag{3.180}$$

$$E^0_{\mathrm{exc}}=-\frac{0.916}{r_s}-0.115+0.031\ln r_s. \tag{3.181}$$

$$W^{\text{exc}}(g) = -\frac{4}{3}\frac{0.458e^2}{r_s} \times \begin{cases} A_0 + \left(\dfrac{z}{\Omega_0}\right)^{-2} A_2 \sum_{\mathbf{g}'}{}' |\rho(\mathbf{g}')|^2 & \text{where } g=0, \\ \left(\dfrac{z}{\Omega_0}\right)^{-1} A_1\rho(\mathbf{g}) + \left(\dfrac{z}{\Omega_0}\right)^{-2} A_2 \sum_{\mathbf{g}'}{}' \rho(\mathbf{g}'-\mathbf{g})\rho(\mathbf{g}') & \text{where } g\neq 0. \end{cases}$$

Here

$$A_0=0.99\,287+0.0904r_s-0.0938r_s^2-0.000\,52r_s^3;$$
$$A_1=0.3049+0.007\,51r_s;$$
$$A_2=-0.03132-0.001r_s;$$
$$B_2=0.12496.$$

The technique described above makes it possible to calculate total crystal energy without using perturbation theory. This method was implemented in research by Fuchs *et al.*[302]

Chapter 4

Application of electronic theory to calculation of physical properties

4.1. Calculation results for the binding energy and their comparison to experiment

4.1.1. Binding energy, pseudopotential method

The equations required for calculating the binding energy by the pseudopotential method are provided in Chap. 3. This chapter noted that a necessary condition for obtaining the correct data is satisfaction of equilibrium condition (3.2). A consistent calculation of all binding energy components (3.1); specifically the Ewald energy U_{es}, the free electron gas energy U_0, the energy corresponding to the first (U_1), second ($U_2 = U_{bs}$), and third (U_3) order perturbation theory for some pseudopotentials will satisfy condition (3.2) and hence an additional routine was proposed[260]; a fit based on the equilibrium condition of the quantity U_1. This routine was useful in, for example, employing the Heine–Abarenkov–Animalu pseudopotential (see Chap. 3).

The greatest contribution to U_{tot} derives from the Ewald energy $U_{es} = (1/2)ze^2/r\alpha$, where α is the Madelung constant. For densely packed structures $\alpha \sim -1.792$ and its quantities for bcc, fcc, and hcp structures differ by less than 0.01%. It is interesting that due to the Ewald energy the most stable structure was not a densely packed fcc or hcp but rather was a bcc structure ($\alpha = -1.791\,361$). This phenomenon is due to the effect of the contribution of the second (primarily) neighbors which are substantially closer to the first neighbors for the bcc compared to the fcc and hcp. An analogous situation occurs with the hcp family of structure since their minimum value was $\alpha = -1.791\,676\,90$ for a structure with a c/a ratio of 1.6356 rather than for a structure with an ideal c/a ratio of 1.6333 ($\alpha = -1.791\,676\,24$).

The data for a number of simple metals shown in Table 4.1 represent a good example of the close correlation between calculation and experiment. Table 4.1 provides data obtained both by means of first principles pseudopotentials (rows 2–4) and by means of model pseudopotentials. It is clear that with the exception of L_i the discrepancy between calculation and experiment for both groups of pseudopotentials rarely exceeds 1%. For L_i an equally good agreement was obtained only in the last studies (rows 4, 9, and 10).

TABLE 4.1. Results of bonding energy ($-u$) calculations for a number of simple metals compared to experiment.

No.	Source (Ref)	Li	Na	K	Al	Rb
1	266	0.2585	0.2305	0.194	2.08	0.1855
2	303	0.2697	0.2413	0.1916	2.0926	0.1721
3	304	0.275	0.235	0.1905	2.1265	0.183
4	305	0.2590	0.2305	0.197	···	0.190
5	266	0.2743	0.231	0.192	···	0.1819
6	306	0.272	0.2315	0.1945	2.067	···
7	260	0.282	0.227	0.1915	2.115	0.172
8	307	0.2802	0.2308	0.1871	2.071	···
9	308	0.256	0.230	0.195	2.077	···
10	309	0.2549	0.2301	0.1943	2.078	···

Overall we can reliably claim that both first principles and model pseudopotentials, at least those that have been developed recently,[305,308] make it possible, with a high degree of accuracy (up to 1%), to calculate the binding energy of simple metals. This confirms the promise of using the pseudopotential for calculating the characteristics of metals and alloys that determine the binding energies. At the same time this accuracy will hardly always be sufficient for calculating stable crystalline structures since, as noted in Chap. 3, the structure-dependent part of the energy accounts for only 2%–4% of the total energy. This means that correct prediction of stable structures requires special research for each broad group of objects. It is no coincidence that the authors of many studies[261,303] focused special attention on the selection of pseudopotentials, the refinement of their parameters, the selection of screening functions, etc. An even more extreme problem in this sense is the situation regarding the use of pseudopotentials to analyze the stability of transition metal structures. For these metals, as demonstrated by Katsnelson *et al.*,[262] it is advisable to go beyond the second order perturbation theory and to also account for the pseudopotential parameters. It is also necessary to go from Ω_i to $\overline{\Omega}$ and from z_j to z (Ref. 255) and carry out the corresponding transformation in the form factors of the component pseudopotentials for these alloys. Moreover, as demonstrated in, for example, Katsnelson *et al.*,[271] it is necessary to account for ordering effects and, in principle, atomic shifts.[257,258]

4.1.2. The electron density functional method

Following Egorushkin and Kohn[310] we write the total energy of the metal as

$$E[\rho]=\frac{1}{2}\sum_{i,j}\frac{z_i z_j}{|\mathbf{R}_i-\mathbf{R}_j|}-\int dr\sum_i\frac{z_i}{|\mathbf{r}-\mathbf{R}_i|}\rho(\mathbf{r})+\frac{1}{2}\int dr\,dr'\frac{\rho(\mathbf{r})\rho(\mathbf{r}')}{|\mathbf{r}-\mathbf{r}'|}, \tag{4.1}$$

where z_i and z_j are the nuclear charges, $\mathbf{R}_i$ and $\mathbf{R}_j$ are their coordinates or

$$E[\rho]=E_q[\rho]+G[\rho]. \tag{4.2}$$

Here E_q is the electrostatic interaction energy of the nuclei and the electrons.

We give $\rho(\mathbf{r})$ as a sum of electron densities of all atoms $\rho_{am}(\mathbf{r})$:

$$\rho(\mathbf{r})=\sum_i \rho_{am}(|\mathbf{r}-\mathbf{R}_i|). \tag{4.3}$$

Then the electrostatic energy of the metal can be given as

$$E_q=\sum_p n_p \left\{\frac{z^2}{2\mathbf{R}_p}-z\int d\mathbf{r}\frac{\rho_{am}(|\mathbf{r}-\mathbf{R}_p|)}{\mathbf{r}}\right.$$

$$\left.+\frac{1}{2}\int d\mathbf{r}\,d\mathbf{r}'\,\rho_{am}(\mathbf{r})\frac{\rho_{am}(|\mathbf{r}-\mathbf{R}_p|)}{|\mathbf{r}-\mathbf{r}'|}\right., \tag{4.4}$$

where n_p is the number of atoms in the sphere labeled p and $\mathbf{R}_p$ is the radius of the same sphere.

For simplicity we introduce the functions $\gamma(\mathbf{R}_p)$ and $\Gamma(\mathbf{R}_p)$:

$$\int d\mathbf{r}\frac{\rho_{am}(|\mathbf{r}-\mathbf{R}_p|)}{\mathbf{r}}=\frac{z}{R_p}-\gamma(\mathbf{R}_p), \tag{4.5}$$

$$\int d\mathbf{r}\,d\mathbf{r}'\frac{\rho_{am}(|\mathbf{r}-\mathbf{R}_{p'}|)}{|\mathbf{r}-\mathbf{r}'|}\rho_{am}(\mathbf{r}')=\frac{z^2}{\mathbf{R}_p}-\Gamma(\mathbf{R}_p). \tag{4.6}$$

Subject to Eqs. (4.5) and (4.6), Eq. (4.4) takes the form

$$E_{\mathbf{q}}=E_{\mathbf{q}}^a+\sum_{p\neq 0} n_p\left[z\gamma(\mathbf{R}_p)-\frac{1}{2}\Gamma(\mathbf{R}_p)\right], \tag{4.7}$$

$$E_{\mathbf{q}}^a=-z\int d\mathbf{r}\frac{\rho_{am}}{\mathbf{r}}+\frac{1}{2}\int d\mathbf{r}\,d\mathbf{r}'\frac{\rho_{am}(\mathbf{r})\rho_{am}(\mathbf{r}')}{|\mathbf{r}-\mathbf{r}'|}. \tag{4.8}$$

A local approximation can be used to calculate $G[\rho]$. Therefore, the resulting formulas make it possible to determine the total metal energy.

The equation for total energy can be recast as

$$E=\sum_i E_i+\frac{1}{2}\sum_{i,j}{}' E_{ij}+\delta E, \tag{4.9}$$

where E_i is the atomic energy at the i site, e_{ij} is the pair interaction energy, and δE is the contribution of ternary, etc., interactions:

$$E_i=E[\rho_a(\mathbf{r}-\mathbf{R}_i)], \tag{4.10}$$

$$E_{ij}=E[\rho_a(\mathbf{r}-\mathbf{R}_i)+\rho_a(\mathbf{r}+\mathbf{R}_j)]-E[\rho_a(\mathbf{r}-\mathbf{R}_i)]-E[\rho_a(\mathbf{r}-\mathbf{R}_j)]. \tag{4.11}$$

Knowing the energy E it is possible to find the binding energy

$$U=E[\rho]-E[\rho_a]. \tag{4.12}$$

Table 4.2 provides the binding energies calculated by Katsnelson *et al.*[311] for a simple metal (Al), transition metals (Ti, Ni) and a noble metal (Cu). It is clear that the discrepancy between calculation and experiment in the binding energy is

TABLE 4.2. Binding energy, calculated by TFED.

Metal	$-U$ (exp.)	$-U$ (calc.)
Al	2.08	2.092
Ti	0.931	0.949
Ni	0.443	0.395
Cu	0.412	0.422

0.5% for Al, 2% for Ti, 2.5% for Cu, and 10% for Ni. The similarity of the results from calculation and experiment for Al was somewhat poorer than from calculations by the model pseudopotential method. However, this should not cast a shadow on the density-functional method which is still in the initial stage of its development. The agreement between the density-functional calculation and experiment should be treated as a satisfactory agreement although in purely applied aspects the model pseudopotential is preferred for simple metals.

4.2. Methods of calculating the elastic properties and dynamical matrix

4.2.1. Elastic properties

The shape of solids is altered under external forces. If the stresses are sufficiently weak the strains will be proportional to the stresses and in this case the generalized Hooke's law holds according to which the strain tensor components ε_{ik} at each point are linear functions of the stress components σ_{lm}. The components of these tensors are related by means of the components of the elasticity moduli tensors C_{iklm} or their inverse constant elasticity tensor S_{iklm}.

The free energy per unit of volume of the strained body as a function of the strain tensor component (in Voight notation) under minor strain can be given as the series

$$F=F_0+\sum_{p,q}\frac{\partial^2 F}{\partial\varepsilon_p\partial\varepsilon_q}\varepsilon_p\varepsilon_q, \qquad (4.13)$$

where F_0 is the free energy per unit volume of an unstressed body; the derivatives in Eq. (4.13) are evaluated for $\varepsilon_p=0$, that is, for a body under no stress. The second derivatives are the second order elasticity moduli. For the ground state (all $\varepsilon_p=0$, $T=0$) we write

$$c_{pq}=\frac{\partial^2 U}{\partial\varepsilon_p\partial\varepsilon_q}, \qquad (4.14)$$

where U is the total energy whose calculation techniques were discussed previously. A variety of calculation schemes are used to calculate the modulii of elasticity; we consider only two such schemes. In one of these schemes[312] the study applies the parameters ε_i, γ_i, and V as the strain parameters describing the transition from an unstrained state to a strained state; these parameters relate the particle coordinates before and after strain by the relations

$$\begin{aligned} x' &= x(1+\varepsilon_1), & x' &= x+\gamma_1 y, & x' &= V^{1/3}x, \\ y' &= y(1+\varepsilon_1)^{-1}, & y' &= y, & y' &= V^{1/3}y, \\ z' &= z, & z' &= z, & z' &= V^{1/3}z. \end{aligned} \tag{4.15}$$

The parameter V describes the change in volume and the remaining parameters characterize the shear strain components. In this notation we can obtain the following for the moduli of elasticity:

$$\frac{\partial^2 U}{\partial V^2} = \tfrac{1}{3}(c_{11}+2c_{12}) = B, \tag{4.16}$$

$$\frac{\partial^2 U}{\partial \gamma_1^2} = c_{44}, \quad \frac{1}{4}\frac{\partial^2 U}{\partial \varepsilon_1^2} = \frac{1}{2}(c_{11}-c_{12}) = c'. \tag{4.17}$$

Rather simple calculations of the moduli of elasticity can be carried out for crystals of cubic symmetry if we know the interatomic interaction pair potentials $\varphi(\mathbf{r})$. In this case,[257]

$$c_{11} = \frac{1}{6\Omega}\sum_{l,\alpha}{}' X_\alpha^4(l)\left[\frac{\varphi''}{R^2}-\frac{\varphi'}{R^3}\right]_{R=R_l}, \tag{4.18}$$

$$c_{12} = \frac{1}{6\Omega}\sum_{l,\alpha}{}' X_\alpha^2(l)X_\beta^2(l)\left[\frac{\varphi''}{R^2}-\frac{\varphi'}{R^3}\right]_{R=R_l}, \tag{4.19}$$

where l characterizes the triple defining the position of the site in the lattice, X_α is the projection of the vector of the lattice site $\mathbf{R}_l$ onto the corresponding axis α, and Ω is the atomic volume. We can then use Eqs. (4.16) and (4.17) to find the bulk modulus of elasticity B and the shear moduli. The elasticity moduli calculated using these or similar equations by means of the pseudopotential method are similar to the experimental values not only after fitting the pseudopotential parameters by means of the atomic properties (e.g., equilibrium conditions), but also often following fitting by electronic characteristics (Fermi surface, etc.). In order to enhance the agreement between calculation and experiment it is necessary to fit not only the form factors of the pseudopotential but also to rigorously select the other fitting parameters (e.g., the volumetric and correlation fitting parameters).

Thus, according to Baks and Trefilov[261] a fit of the local Heine–Abarenkov pseudopotential by means of conditions (3.49) and the selection (after appropriate testing) of exchange and correlation corrections in the Vashishta–Singwi form makes it possible to calculate the moduli B of alkali metals accurate to better than 3%. However, this is of course a record result; the ordinary discrepancy between calculation and experiment is greater. However, since the most important element is not the calculation of individual values but rather the calculation of the dependence of specific values on different factors (such as pressure or temperature) such a deviation between calculation and experiment is not critical.

4.2.2. Dynamical matrix

The atoms in a crystalline lattice shift from equilibrium due to thermal vibrations. Let $U^j_{\nu l}$ be the j component of the displacement vector $U_{\nu l}$ of the atom at the νth site of the lth cell. The vector $U^j_{\nu l}$ can be given as

$$U^j_{\nu l}=e^{i\mathbf{Q}1}U^j_{\nu 0}(t)=e^{i\mathbf{Q}1}U^j_{\nu\mathbf{Q}}(t), \tag{4.20}$$

where $\mathbf{Q}$ is the wave vector. It is possible to show that $U^j_{\nu\mathbf{Q}}$ satisfies the equations

$$\sum_{j'\nu'}[G^{jj'}_{\nu\nu'}(\mathbf{Q})-\omega^2M_\nu\delta_{\nu\nu'}\delta_{jj'}]U^{j'}_{\nu'\mathbf{Q}}=0, \tag{4.21}$$

where $\omega(\mathbf{Q})$ is the vibrational frequency spectrum, $G^{jj'}_{\nu\nu'}$ is the dynamical matrix, and M_ν is the ion mass at site ν. For any nontrivial solution of system (4.21) the determinant of the matrix of coefficients of amplitudes will be equal to zero. This makes it possible to find the frequency spectrum $\omega(\mathbf{Q})$.

One of the studies devoted to calculating the phonon spectra of pure metals[295] has calculated the phonon spectra of the 3*d*, 4*d*, and 5*d* series metals. Since the unit cell of pure metals contains only a single atom the matrix in Eq. (4.21) is a (3×3) matrix. For crystals in the cubic system the matrix can be written as follows in the principal directions:

$$\sum_{j'}[M\omega^2\delta_{jj'}-G^{jj'}(\mathbf{Q})]U^{j'}_{\mathbf{Q}}=0, \tag{4.22}$$

where $G^{jj'}$ is a dynamical matrix that can be represented as the sum of three components describing the contributions of electrostatic energy, the repulsive Born–Mayer term, and the band structure energy, respectively. These can also be written for the squared frequencies as follows:

$$\omega^2=\omega^2_{\text{es}}+\omega^2_{\text{b}\mu}+\omega^2_{\text{bs}}, \tag{4.23}$$

where, for example, $\omega^2_{\text{es}}=G_{\text{es}}M^{-1}$, etc.

For the contribution of the band structure energy:

$$[\omega^2_{\text{bs}}(\mathbf{Q})]^{jj'}=-\left(\frac{4\pi z^2}{M\Omega}\right)\left\{\sum_n\frac{(\mathbf{Q}+\mathbf{g}_n)^{(j)}(\mathbf{Q}+\mathbf{g}_n)^{(j')}}{|\mathbf{Q}+\mathbf{g}_n|^2}\Phi(\mathbf{Q}+\mathbf{g}_n)\right.$$
$$\left.-\sum_{n\neq 0}\frac{g^j_n g^{j'}_n}{|\mathbf{g}_n|^2}\Phi(\mathbf{g}_n)\right\}, \tag{4.24}$$

where

$$\Phi(\mathbf{q})=\left[\frac{4\pi z(1+\alpha ef)}{\Omega q^2}\right]^{-2}|W^b(\mathbf{q})|^2\frac{\varepsilon(q)-1}{\varepsilon(q)[1-f(q)]}. \tag{4.25}$$

The calculation results are close to the experimental data.[255]

4.3. Free energy and the equation of state of metals

The free energy is one of the most important thermodynamic functions and is related to internal energy U and entropy S by the equation

$$F=U-TS. \tag{4.26}$$

In turn, in order to find the equation of state it is necessary to carry out volume differentiation of the free energy at a constant temperature:

$$P=-\left(\frac{\partial F}{\partial v}\right)_T. \tag{4.27}$$

The free energy can be given[313] as the sums

$$F=U_0+k_BT\sum_{j=1}^{3N}\ln(1-e^{-\hbar v_j/k_BT}), \tag{4.28}$$

from which it is possible to find the following for the equation of state:

$$P=-\left(\frac{\partial U_0}{\partial v}\right)_T-k_BT\sum_{j=1}^{3N}\frac{e^{-\hbar v_j/k_BT}}{1-e^{-\hbar v_j/k_BT}}\left(\frac{\hbar}{k_BT}\right)\frac{\partial v_j}{\partial v}. \tag{4.29}$$

Here U_0 is the sum of the potential energy of the crystal in equilibrium and the energy of the null vibrations. The change in phonon frequencies with changing volume is characterized by the Grüneisen parameter

$$\gamma=-\frac{d\ln v_i}{d\ln v}. \tag{4.30}$$

Equation (4.29) takes the form

$$P=-\left(\frac{\partial U_0}{\partial v}\right)_T+\gamma\frac{k_BT}{V}\sum_j\left(\frac{\hbar v_j}{k_BT}\right)\frac{e^{-\hbar v_j/k_BT}}{1-e^{-\hbar v_j/k_BT}}. \tag{4.31}$$

In order to find the average Grüneisen parameter it is necessary to first define the second moment of the frequency spectrum $\overline{v^2}$:

$$\gamma=\frac{1}{2}\frac{\partial\ln\overline{v^2}}{\partial\ln v}. \tag{4.32}$$

Using the equations and the work of Portnoy *et al.*,[257] expressing the relation between $\overline{v^2}$ and the derivatives of the interatomic interaction potential φ', φ'', and φ''' it is possible to obtain

$$\gamma=-\frac{1}{6}\frac{\Sigma_i c_i[-(2/R)\varphi'(R)+2\varphi''(R)+R\varphi'''(R)]_{R=R_i}}{\Sigma_i c_i[(2/R)\varphi'(R)+\varphi''(R)]_{R=R_i}}, \tag{4.33}$$

where C_i is the coordination number.

The calculations carried out have demonstrated[255] that the calculated and measured values of γ are in satisfactory agreement. Knowing the free energy and its temperature dependence it is possible to determine many important characteristics of solids such as the polymorphic transformation temperature (based on the intersection points between the temperature dependences of the free energies of the competing phases), the binding strength (based on the function $3p\Omega=-\partial U/\partial x$

at the minimum where $x = \ln a/a_0$), the cohesive energy, and the contribution of various electron groups (s,p,d) to cohesive energy formation, etc.[314]

4.4. Calculation of electrical properties

4.4.1. The electrical resistivity of metals; general relations

Transport properties which include electrical resistance and thermovoltage are normally calculated assuming that the conduction electron forms an electron gas that interacts with an ensemble of regularly (for crystals) or irregularly (for liquids and amorphous solids) distributed ions and that is scattered by the ions of this ensemble. Such scattering determines the electrical resistance of the medium (or its inverse quantity: the electrical conductance) and can be calculated by means of perturbation theory or more general techniques. The initial equation obtained for calculating the electrical conductance G in an approximation of the relaxation time τ is the relation[255]

$$G=\tfrac{1}{3}e^2 v_F^2 \tau N(E_F). \tag{4.34}$$

The relaxation time τ (the time between two successive electron scattering events) is given by the equation

$$\frac{1}{\tau}=\int dS'\, P(\mathbf{k},\mathbf{k}')(1-\cos\theta), \tag{4.35}$$

where $P(\mathbf{k}, \mathbf{k}')$ is the probability of electron scattering from state $\mathbf{k}$ to $\mathbf{k}'$ (per unit of time), θ is the angle between $\mathbf{k}$ and $\mathbf{k}'$, and S is the surface formed by all possible vectors $\mathbf{k}'$. If electron scattering is treated as an elastic process, then $|\mathbf{k}| = |\mathbf{k}'|$. The Wiedemann–Franz law—the criterion of absolutely elastic scattering—holds at a temperature $T > \theta_D$ (the Debye temperature).

The problem of finding τ reduces to determining $P(\mathbf{k},\mathbf{k}')$. This can be carried out both within the framework of pseudopotential theory and by means of the scattering phase shifts. For a weak ionic potential the probability $P(\mathbf{k},\mathbf{k}')$ is given through the form factor of the pseudopotential

$$P(\mathbf{k},\mathbf{k}')=\frac{2\pi}{\hbar}|S(\mathbf{q})|^2|V(\mathbf{q})|^2\delta(\varepsilon_{\mathbf{k}'}-\varepsilon_{\mathbf{k}}). \tag{4.36}$$

Substituting Eq. (4.36) into Eq. (4.35) and carrying out integration with respect to the angles we obtain

$$\rho=\frac{3\pi m N\Omega}{8\hbar e^2 E_F}\int_0^{2k_F} d\left(\frac{q}{k_F}\right)|S(\mathbf{q})|^2|V(\mathbf{q})|^2\left(\frac{q}{k_F}\right)^3. \tag{4.37}$$

This formula is used in electrical resistance calculations of metals including both crystalline and noncrystalline metals. In a number of cases it is possible to obtain satisfactory agreement with experiments. The primary difficulties in using Eq. (4.37) include determining the structure factor and selecting the pseudopotential.

For transition and noble metals, electron–ion interaction is occasionally more easily described in terms of phase shifts by means of a t matrix:

$$t(E_F,q) = -\frac{2\pi\hbar^3}{m(2mE_F)^{1/2}}\left(\frac{1}{\Omega}\right)\sum_l (2l+1)\sin\eta_l \exp(i\eta_l)P_l\cos(\theta). \tag{4.38}$$

In this case the equation for ρ takes the form

$$\rho = \frac{12\pi\Omega}{\hbar e^2 V_F^2}\int_0^1 \left(\frac{q}{2k_F}\right)^3 d\left(\frac{q}{2k_F}\right)|t(E_F,\mathbf{q})|^2|S(\mathbf{q})|^2. \tag{4.39}$$

Using Eq. (4.39) for transition and noble metals makes it possible to improve the agreement between the calculations and experiments.

4.4.2. Temperature dependence of ρ for crystalline metals

We consider the temperature dependence of electrical resistance. Rewriting the equation for ρ we get

$$\rho = \frac{\pi\Omega}{2zk_F^3}\sum_{\mathbf{q}(|\mathbf{q}|<2k_F)} \langle S^*(\mathbf{q})S(\mathbf{q})\rangle_0 V^2(\mathbf{q})\mathbf{q}. \tag{4.40}$$

Within the framework of the Einstein model

$$\langle S^*(\mathbf{q})S(\mathbf{q})\rangle_0 = \left[S^*(\mathbf{q})S(\mathbf{q}) - \frac{1}{N}\right]\exp(-2W_0q^2) + \frac{1}{N}, \tag{4.41}$$

where

$$W_0 = \frac{\coth(\theta_E/2T)}{4Mk_B\theta_E} \tag{4.42}$$

and θ_E is the Einstein temperature.

Substituting Eq. (4.41) into Eq. (4.40) we obtain

$$\rho = \rho_1 + \rho_2. \tag{4.43}$$

Here

$$\rho_1 = \frac{\pi\Omega_0}{2zk_F^3}\sum_{\mathbf{q}(|\mathbf{q}_0|<2k_F)} |S_0(\mathbf{q})|^2\exp(-2W_0q^2)V^2(\mathbf{q})\mathbf{q}, \tag{4.44}$$

$$\rho_2 = \frac{\Omega^2}{4\pi zk_F^3}\int_0^{2k_F} d\mathbf{q}[1-\exp(-2W_0q^2)]V^2(\mathbf{q})\mathbf{q}^3. \tag{4.45}$$

The calculations reveal that the temperature dependence of the electrical resistance is nearly entirely determined by the temperature dependence of ρ_2.

4.4.3. Residual resistance

Since impurities and defects are always present in a metal, resistance always has a measurable (nonzero) value even at very low temperatures (here we neglect superconductivity effects which are special in its nature). This is called residual resistivity, which is determined by conduction electron scattering of the metal matrix by the perturbing potential of the impurity or the defect. According to

Mattheissen's rule the total resistance of a metal consists of the sum of the ideal resistance of the pure metal and the residual resistance. We note that significant deviations are often observed from Matthiessen's rule.

We consider the residual resistance due to the substitution point defects. Studying the effect of point defects on electrical resistance is important for evaluating the degree of structural perfection of metals and detecting defects that most substantially alter the properties. Ordinarily the effect of point defects on metal resistance is estimated by the effect of 1 at. %.

In terms of the form factors of an impurity atom and the matrix atoms ρ_{res} takes the form

$$\rho_{\text{res}}=\frac{3\pi m\Omega_M}{800e^2\hbar E_F}\int_0^2 dx\,|W_P(x)-W_M(x)|^2x^3, \tag{4.46}$$

where Ω_M is the atomic volume of the metal matrix, W_P and W_M are the form factors of the screened pseudopotential of the impurity and the matrix, respectively, and $x = q/k_F$.

Equation (4.46) can be modified to account for lattice distortions that occur when impurity atoms are introduced to the metal. Indeed the form factor of a metal containing an impurity can be given as

$$W(\mathbf{q})=S(\mathbf{q})W_M(\mathbf{q})+S'(\mathbf{q})[W_P(\mathbf{q})-W_M(\mathbf{q})], \tag{4.47}$$

where $S(\mathbf{q})$ and $S'(\mathbf{q})$ are the structure factors of the matrix lattice and the impurity, respectively,

$$S(\mathbf{q})=\frac{1}{N}\sum_j e^{-i\mathbf{q}\mathbf{r}_j}, \tag{4.48}$$

$$S'(\mathbf{q})=\frac{1}{N}\sum_{j(\mathbf{r})} e^{-i\mathbf{q}\mathbf{r}_j}. \tag{4.49}$$

Summation is carried out only over ions of the second type in $S'(\mathbf{q})$.

We assume that atomic shifts from equilibrium are small compared to $\mathbf{r}_j$, which permits expanding the structure factors $S(\mathbf{q})$ and $S'(\mathbf{q})$ into a Taylor series expansion in the neighborhood of $\mathbf{r}_j$ (j is the site number):

$$W(\mathbf{q})=[S(\mathbf{q})+\delta S(\mathbf{q})]W_M(\mathbf{q})+[S'(\mathbf{q})+\delta S'(\mathbf{q})][W_P(\mathbf{q})-W_M(\mathbf{q})], \tag{4.50}$$

where

$$\delta S(\mathbf{q})=-\frac{i}{N}\sum \mathbf{q}\delta\mathbf{r}_j\exp[-i\mathbf{q}\mathbf{r}_j]. \tag{4.51}$$

It is clear that the change in the form factor of the metal attributed to the substitution impurity takes the form

$$dW(\mathbf{q})=\delta S(\mathbf{q})W_M(\mathbf{q})+[S'(\mathbf{q})+\delta S'(\mathbf{q})][W_P(\mathbf{q})-W_M(\mathbf{q})]. \tag{4.52}$$

Accounting for Eqs. (4.46), (4.47), and (4.59) and the relation of ρ to the relaxation time we obtain for the residual resistance which is related to the impurity atoms:

$$\rho_{\text{res}}=\frac{3\pi m\Omega_M N}{800\hbar e^2 E_F}\left[\int_0^2 |\delta S(x)|^2 |W_M(x)|^2 x^3\,dx+\int_0^2 dx|S'(x)|^2|W_P(x)\right.$$
$$-W_M(x)|^2x^3+\int_0^2 dx\{\delta S(x)S'(x)+\delta S(x)S'(x)\}W_M(x)\{W_P(x)$$
$$\left.-W_M(x)\}x^3\right]. \tag{4.53}$$

Equation (4.53) discards terms above second order of smallness.

The first two terms in Eq. (4.53) correspond to the contribution to ρ_{res} due to distortions of the metal-matrix lattice and the change in the scatterer potential; the third term—the "interference" term—simultaneously accounts for lattice distortion and the change in potential.

If lattice distortions are neglected we obtain Eq. (4.46) from Eq. (4.53). Here the result must be reduced to a quantity corresponding to a single atomic percent of the impurity.

We note that calculations using Eq. (4.53) do not always make it possible to achieve satisfactory agreement with experiment and are highly dependent on the selection of the specific pseudopotential.

An extensive survey of calculation results for ρ_{res} of dilute alloys based on nontransition metals with alkali and alkaline-earth element impurities compared to experiment is provided.[315,316] The Ashcroft, Heine–Animalu, Krasko–Gurskiy, etc., pseudopotentials were used while the Singwi, Vashishta–Singwi, and Geldart–Vosko exchange and correlation corrections were used. It was determined that with the exception of the Krasko–Gurskiy pseudopotential it was possible for the remaining cases to select pseudopotential and screening parameters to achieve a discrepancy between calculation and experiment of less than 30%.

While there are only a few studies devoted to calculating ρ_{res} for condensed alloys, primary attention here will not be devoted to the quantitative agreement between calculation and experiment but rather will focus on the dependence of the calculated quantities of various physical factors such as ordering, static atomic shifts, etc. From this viewpoint the calculation of Katsnelson *et al.*[317] is of interest; this study calculated ρ_{res} of a Cu–Zn alloy containing 15 at. % of Zn by the equation

$$\rho_{\text{res}}=\text{const}\int_0^2 dx|W(q)^2|x^3, \tag{4.54}$$

where, taking account of the short-range order effects, the square of the pseudopotential matrix element of the crystal takes the form

$$|W(\mathbf{q})|^2=|c(\mathbf{q})|^2|\Delta\overline{W}(\mathbf{q})-\overline{W}(\mathbf{q})\mathbf{q}\mathbf{A}_{\mathbf{Q}}|^2, \tag{4.55}$$

$$|c(\mathbf{q})|^2=\frac{c_A c_B}{N}\sum_j \alpha(\rho_i)\exp(i\mathbf{q}\boldsymbol{\rho}_j), \tag{4.56}$$

where $\overline{W}(\mathbf{q})$ and $\Delta W(\mathbf{q})$ is the average form factor and the difference of the form factors of the component pseudopotentials, respectively. A_Q is the function introduced by Krivoglaz[318] characterizing the dimensional effect, related to the volumetric concentration derivative $(1/v)dv/dc$ and $\alpha(\rho_i)$ are the short-range order parameters. Substitution of Eqs. (4.55) and (4.56) into Eq. (4.54) makes it possible to represent ρ_{res} as the sum of three terms: ρ_D, ρ_{SRO}, and ρ_{SAS} which represent the following contributions: the disordered configuration of atoms of different types; the change in ρ_D due to the onset of short-range order and the change in ρ_D due to static atomic shifts. Calculations that used the HAA pseudopotential for Zn and the Yurev–Vatolin pseudopotential[281] for Cu have demonstrated that $\rho_D = 2.8\ \mu\Omega$ cm, $\rho_{SRO} = 0.6\ \mu\Omega$ cm, $\rho_{SAS} = -0.6\ \mu\Omega$ cm, $\rho_{res} = 1.6\ \mu\Omega$ cm, and $\rho_{exp} = 2.6\ \mu\Omega$ cm. Therefore, the discrepancy between the results of the calculation and experiment was less than 50% which is not at all bad. However, the most interesting aspect is that not only short-range order effects but also static atomic shift effects make a negative contribution to ρ_{res}. The latter effect is due to the influence of the interference term noted above.

The residual resistance can be calculated within the framework of the phase shift method by the equation[218,320]

$$\rho_{res} = \frac{4\pi c\hbar}{z^M c^2 k_F} \sum_{l=0} (2l+1) \sin^2(\Delta\eta_{l+1}(E_F) - \Delta\eta_l(E_F)), \qquad (4.57)$$

where

$$\Delta\eta_{l(l+1)} = \eta^P_{l(l+1)}(E_F) - \eta^M_{l(l+1)}(E_F) \qquad (4.58)$$

is the difference of the phase shifts of the impurity and the matrix.

The residual resistivity of several (25) dilute alloys of noble metals containing a transition impurity have been calculated by Stepanyuk *et al.*[218] and Katsnelson *et al.*[320] The phase shifts were found by solving the Schrödinger equation for the potentials of the matrix and the impurity. The potential was derived in MT form by the Mattheis[197] method. The resulting potential was shifted within the MT sphere by a quantity to satisfy the Friedel sum rules for $\eta_l^{I(M)}$. The same approach was used for all alloys. In most cases it was possible to obtain a satisfactory agreement with experiments.

In conclusion we will briefly focus on the possibilities for calculating the vacancy induced residual resistance.

The presence of a vacancy can be described within the framework of the pseudopotential method by the potential of an ion of opposite sign remote from the metal. The problem reduces to a calculation of the structure factor, determination of the pseudopotential of the remote ion, and incorporating lattice distortions.

If an ion is removed from site $\mathbf{r}_i$ the structure factor $|S(\mathbf{q})| \sim N^{-1}$ while the formula for the residual resistance due to the vacancy takes the form

$$\rho_{res}^{vac} = \frac{1}{N} \frac{3\pi m\Omega}{800 e^2 \hbar E_F} \int_0^2 dx\, x^3 |W(x)|^2. \qquad (4.59)$$

At low defect concentrations the resistance associated with different vacancies is additive. Hence the multiplier $1/N$ must be replaced by the atomic valency con-

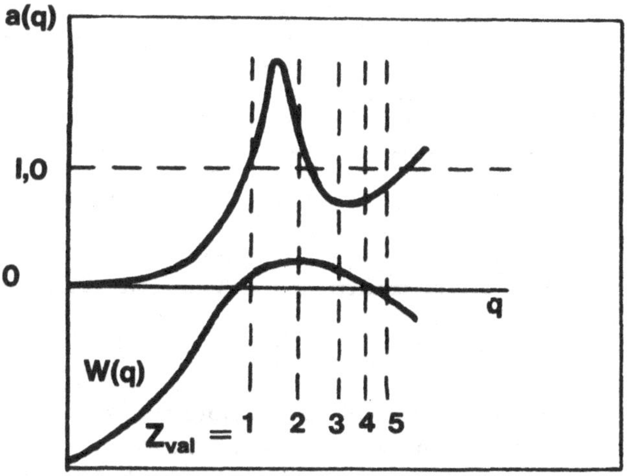

FIG. 4.1. General form of $a(\mathbf{q})$ and $W(\mathbf{q})$ and the integration limits in Eq. (4.37) plotted as a function of metal valency.

centration. A calculation by Eq. (4.59) yields diminished values compared to experiment and this is due to neglecting lattice distortions.

It is also possible to calculate the residual resistance of the vacancy within the framework of the phase shift formalism:

$$\rho_{\text{res}}^{\text{vac}}=\frac{4\pi\hbar}{z^M e^2 k_F}\sum_{l=1}^{\infty} l\sin^2(\eta_l^P(E_F)-\eta_{l+1}^M(E_F)). \tag{4.60}$$

The phase shifts in the potential can be found by the self-consistent Kohn–Sham method using the "vacancy in jellium" model.

4.4.4. Electrical resistance of noncrystalline metals

The initial relations for calculating the electrical resistance of liquid and amorphous metals are Eqs. (4.34), (4.37), and (4.39) which use the following equation as the structure factor:

$$a(\mathbf{q})=|S(\mathbf{q})|^2=\frac{1}{N}\sum_j\sum_{j'} e^{i\mathbf{q}(\mathbf{R}_j-\mathbf{R}_{j'})}, \tag{4.61}$$

where j and j' run the coordinates of all irregularly distributed atoms.

Contrary to the analogous equation for crystals, $a(\mathbf{q})$ for noncrystalline matter consists of terms characterizing smeared spots whose value decreases, while the extent of smearing grows with $\mathbf{q}$, since only a part of $a(\mathbf{q})$ for $\mathbf{q} < 2k_F$ makes a contribution to the electrical resistivity between the integration limits in Eq. (4.30). The range of $a(\mathbf{q})$ responsible for the principal peak of this function and its wings (see Fig. 4.1) is the most significant element for calculating electrical resistance as demonstrated by analysis. The figure shows that for monovalent metals the upper

TABLE 4.3. Calculated electronic resistivity of metallic melts and their comparison with experiment.

		Reference						Exp.
Element	T (°C)	323	324	324	325	326	222	327
Cu	1150	26.7	19.1	10.9	20.9	21.6	23	21.5
Ag	1000	28.5	35.2	39.0	...	28.2	15	18.1
Au	1150	...	34.1	49.0	...	52.1	35	32.0
Ni	1500	54.9	...	...	48.8	...	96	85

integration limit lies on the left wing of the structure maximum while the upper limit of divalent and polyvalent metals lies on the right wing.

The dependence of the product $|V(\eta)|^2 a(\eta)$ $(\eta = q/2k_F)$ in the integral

$$\rho = \frac{12\pi\Omega_0}{\hbar e^2 v_F^2} \int_0^1 d\eta\, \eta^3 |V(\eta)|^2 a(\eta) \tag{4.62}$$

on η is nonmonotonic.[255] Two regions ($q < q_0$ and $q > q_0$) are clearly differentiated here [q_0 is the value of q where $v(q) = 0$]. The first of these regions corresponds to plasma electrical resistance ρ_L^{PL} corresponding to scattering by long-wavelength density fluctuations while the second region corresponds to the structural electrical resistivity ρ_L^{STR} where the total resistance is the sum of these parts.

For the first region assuming for simplicity that $\langle |v(q)|^2 \rangle = (1/2)|V(0)|^2 = 1/2[(2/3E_F)]^2$, $a(q) = a(0)$, $q_0/2k_F \simeq 0.8 \simeq \eta_0$, we obtain

$$\rho_L^{PL} = \frac{12\pi}{\hbar e^2} \frac{1}{v_F^2} \frac{1}{N} \int_0^{\eta_0} \frac{1}{2} \left(\frac{2}{3} E_F \right)^2 a(0) \eta^3\, d\eta \simeq \frac{0.3\, a(0)}{\hbar e^2} \frac{E_F^2}{N} \frac{}{v_F^2} \approx \frac{0.3\pi a(0) m}{2\hbar e^2 N}. \tag{4.63}$$

Since $a(c)$ must grow with rising temperature (since the wings of the structure maximum in Fig. 4.1 will grow with increasing temperature), ρ_L^{PL} is greater the higher the temperature. Judging by Fig. 4.1, ρ_L^{PL} makes the primary contribution to the electrical resistance of the monovalent liquid (or amorphous) metals which will grow with temperature. The equation for the structure part of the electrical resistance can be given as

$$\rho_L^{STR} = \frac{12\pi}{\hbar e^2} \frac{1}{v_F^2} \frac{1}{N} \int_{\eta_0}^1 d\eta |V(\eta)|^2 \eta^3 a(\eta). \tag{4.64}$$

It is obvious that a calculation of ρ_L^{STR} is possible if the structure factor $a(\eta)$ for $q < 2k_F$ is known in addition to $|v(\eta)|^2$.

Equation (4.62) and appropriate pseudopotentials can also be used to calculate the numerical values of ρ_L and in a number of cases have been successful. Silonov *et al.*[322] calculated the electrical resistance of Cu, Ag, Au, and Ni liquid metals by means of a model resonance pseudopotential[281] whose results are shown in Table 4.3. It is clear from the table that only the pseudopotential calculation, in accordance with Yurev *et al.*,[281] made it possible to obtain overall satisfactory agreement with experiments. In all other studies satisfactory agreement, if observed, occurs only for isolated metals.

However, a qualitative analysis of the temperature dependence of ρ_L^{STR} and ρ_L^{TOT} is also possible without such a calculation. At the same time, based on even the most general considerations, it is clear that with increasing temperature the maximum of $a(\mathbf{q})$ will diminish while the wings of the maximum will grow (since the degree of disorder of the liquid or amorphous solid will grow with increasing temperature), although in this case the degree of modulation of $a(\mathbf{q})$ will diminish. It therefore follows from an analysis of Fig. 4.1 that ρ^{STR} will noticeably diminish with increasing temperature for divalent metals and will have a small (more likely a negative) magnitude for polyvalent metals. Therefore, ρ_L^{TOT} for divalent metals will diminish with increasing temperature, while for polyvalent metals it will vary little with variations in temperature. These predictions are consistent with the experimental data.[327]

It is necessary to derive an equation for the structure factor $a(\mathbf{q})$ based on structural concepts of the corresponding condensed matter in order to formulate a rigorous quantitative theory of electrical resistance. Therefore, Meisel and Cote[328] derived equations for the structure factors of amorphous metals with a Debye phonon spectrum that are used in x-ray and neutron $S^x(\mathbf{q})$ scattering theory as well as electrical resistance $S^p(\mathbf{q})$ theory. The dynamical structure factors were averaged over the frequencies of the elastic waves in order to obtain these expressions; the structure factors in the case of scattering of penetrating radiations and electrical resistance take the form

$$S^x(\mathbf{q}) = \int_{-\infty}^{\infty} d\omega\, S(\mathbf{q},\omega), \tag{4.65}$$

$$S^p(\mathbf{q}) = \int_{-\infty}^{\infty} d\omega\, S(\mathbf{q},\omega) x n(x). \tag{4.66}$$

Here $x = \hbar\omega/k_B T$, $n(x) = (e^x - 1)^{-1}$. Consistent with van Hove the dynamical structure factor can be represented as the series

$$S(\mathbf{q},\omega) = S_0(\mathbf{q},\omega) + S_1(\mathbf{q},\omega) + \cdots, \tag{4.67}$$

where in the harmonic approximation

$$S_0(\mathbf{q},\omega) = a(q) e^{-2M(q)} \delta(\omega), \tag{4.68}$$

$$S_1(\mathbf{q},\omega) = e^{-2M(q)} \frac{n(-\omega)}{-\omega} \sum_{\varphi} \frac{\hbar q^2}{2M} [a(\mathbf{q}+\boldsymbol{\varphi})\delta(\omega+\omega_\varphi) + a(\mathbf{q}-\boldsymbol{\varphi})\delta(\omega-\omega_\varphi)], \tag{4.69}$$

$$a(\mathbf{q}+\boldsymbol{\varphi}) = \frac{1}{N} \sum_m \sum_n \exp[i(\mathbf{q}+\boldsymbol{\varphi})(\mathbf{r}_m - \mathbf{r}_n)]. \tag{4.70}$$

Here $\boldsymbol{\varphi} = (\varphi, j)$, j and φ is the polarization index and the wave vector of the elastic wave, respectively.

The first term in $S(\mathbf{q},\omega)$ corresponds to elastic scattering and hence $S_0^x(\mathbf{q}) \sim S_0^p(\mathbf{q})$.

For $S_1^x(\mathbf{q})$ and $S_1^p(\mathbf{q})$ we can write

$$S_1^x(\mathbf{q})=\alpha(\mathbf{q}) \int_0^1 d\left(\frac{\varphi}{\varphi_p}\right) \frac{\varphi}{\varphi_p}\left[n(x)+\frac{1}{2}\right] \int \frac{d\Omega}{4\pi} a(\mathbf{q}+\boldsymbol{\varphi}), \tag{4.71}$$

$$S_1^p(\mathbf{q})=\alpha(\mathbf{q})(\theta/T) \int_0^1 d\left(\frac{\varphi}{\varphi_p}\right) (\varphi/\varphi_p)^2 n(x) n(-x) \int \frac{d\Omega}{4\pi} a(\mathbf{q}+\boldsymbol{\varphi}), \tag{4.72}$$

Here $x = \hbar\omega_\varphi/k_BT = (\hbar\omega_p/k_BT)/(\varphi/\varphi_D) = (\theta/T)(\varphi/\varphi_D)$; φ_D is the wave vector of an elastic wave corresponding to the radius of the Debye sphere and θ_D is the Debye characteristic temperature.

Accounting for multiphonon effects we have

$$S^p(\mathbf{q})=S_0^p(\mathbf{q})+S_1^p(\mathbf{q})+\{1-[1+2M(\mathbf{q})]e^{-2M(\mathbf{q})}\}. \tag{4.73}$$

where the last term at moderate temperatures will not exceed $2M^2(\mathbf{q})$.

We can see from the resulting equation for $S_0^p(\mathbf{q})$ and $S_1^p(\mathbf{q})$ that with increasing temperature $S_0^p(\mathbf{q})$ decays in proportion to the Debye–Waller factor $e^{2M(\mathbf{q})}$, and $S_1^p(\mathbf{q})$ grows in proportion to $Te^{2M(\mathbf{q})}$. As demonstrated by analysis[329] the primary properties of the experimentally observed temperature dependence of ρ can be fully explained based on the concepts outlined above.

4.4.5. The thermopower of noncrystalline metals

The thermopower Q is related to temperature T and the energy dependence of the electrons in charge transfer $\rho(E)$ by the equation

$$Q=-\frac{\pi^2 k_B^2 T}{3eE_F}\left[\frac{\partial \ln \rho(E)}{\partial \ln (E)}\right]_{E=E_F} \tag{4.74}$$

which holds for crystalline and noncrystalline metals and alloys.

In the near-free electron model $\rho(E) = A_0[\sigma_n(E)/E]$, where σ_n is the electron scattering cross section of the ion:

$$\sigma_n \sim \int_0^1 d\eta\, a(\eta)\,|V(\eta)|^2 \eta^3. \tag{4.75}$$

We denote the equation in brackets in Eq. (4.74) by $-\xi$. Then

$$\xi=\left[\frac{\partial \ln E}{\partial \ln E}-\frac{\partial \ln \sigma_n(E)}{\partial \ln E}\right]_{E=E_F}=1-r, \tag{4.76}$$

where

$$r=\left.\frac{\partial \ln \sigma_n(E)}{\partial \ln E}\right|_{E-E_F}. \tag{4.77}$$

We introduce

$$g_0 = \frac{a(2k_F)\,|V(2k_F)|^2}{\langle a\omega^2 \rangle}, \tag{4.78}$$

where

$$\langle a\omega^2 \rangle = \frac{1}{4k_F^4} \int^{2k_F} d\eta\, a(\eta)\,|V(\eta)|^2 \eta^3. \tag{4.79}$$

It can be demonstrated that $r = -2 + 2g_0$ and $\xi = 3 - 2g_0$. The parameter $g_0 = 0$ if $|\gamma(2k_F)|^2 = 0$. In all remaining cases $g_0 > 0$. If $0 < 2g_0 < 1.5$, the thermopower has a normal sign. If $a(2k_F)$ and $|v(2k_F)|^2$ are simultaneously large, g_0 may be ~ 1.5 and the sign of the thermopower becomes anomalous. This condition corresponds to the case in which the maxima of $|\omega(\mathbf{g})|^2$ and $a(\mathbf{q})$ are very close and their positions correspond to $q \sim 2k_F$. It follows that the anomalous sign on the thermopower will coexist with the large electrical resistances. These theoretical predictions are in satisfactory agreement with experimental data.

It is easily demonstrated that the theory of the electrical properties of noncrystalline metals can also be applied to metallic noncrystalline alloys.[255,327]

4.5. Calculation of x-ray spectra

The x-ray spectroscopy is one of the fundamental methods of investigating the electronic structure of condensed matter. The basic advantages of x-ray spectroscopy include the capability to obtain information on energy states across the entire width of the valence bands and to analyze the energy distribution of electrons of different symmetry,[330] which permits an effective determination of the nature of the chemical bond in solids.[331] An especially useful technique is to analyze the electronic structure by means of combined x-ray and photoelectron spectroscopy techniques[332] which makes it possible to combine on a common energy scale the spectra of different components of multicomponent systems. Significant progress has been made recently in analyzing a variety of disordered systems by means of x-ray spectroscopy.[332,333]

In order to interpret x-ray spectra, calculations "from first principles" of these spectra are carried out based on their electronic structure which can be obtained by any one-electron technique. Figure 4.2 shows the x-ray band scheme: emission and absorption. The x-ray line intensity in the one-electron approximation takes the form

$$I(\omega) \sim \omega^3 \sum_{nk} |\langle \varphi_{nk}|H'|\varphi_c\rangle|^2 \delta(\omega - \varepsilon_n + \varepsilon_c), \tag{4.80}$$

where φ_c is the wave function of the core state and $H' = \exp(-i\mathbf{qr})\mathbf{A}\cdot p$ ($\mathbf{q}$ is the photon wave vector, $\mathbf{A}$ is the polarization vector, and p is the momentum operator). In the dipole approximation $\exp(i\mathbf{qr}) \approx 1$ and we then have

$$I(\omega) \approx \omega^3 \sum_n \oint dS \frac{|\langle \varphi_{nk}|\nabla|\varphi_c\rangle|^2}{|\nabla_k \varepsilon_n(k)|}. \tag{4.81}$$

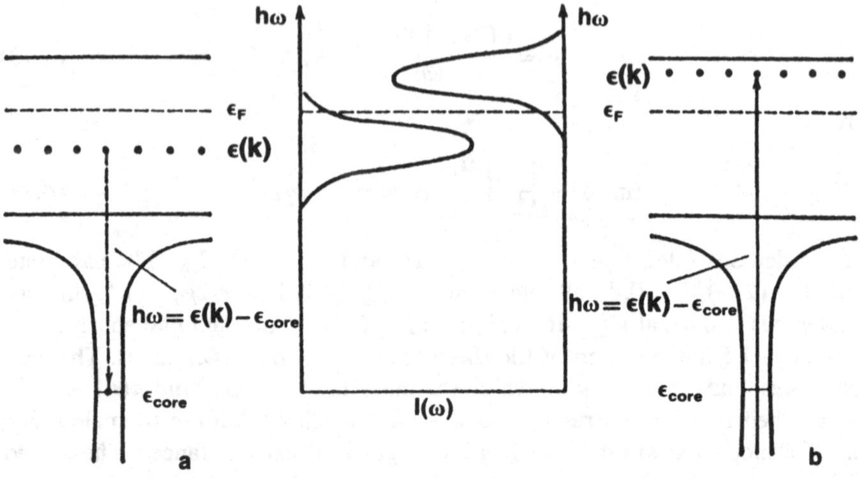

FIG. 4.2. X-ray excitation scheme.

If we use, for example, the LAPW method described above it is possible to obtain calculation equations for the emission and absorption spectra:

$$I^s_{nl}(\hbar\omega) \sim (\hbar\omega\varepsilon^s_{nl})^3 \left\{ \frac{l}{2l+1} [M^s(nl,l-1,\hbar\omega)]^2 n^s_{l-1}(\hbar\omega) + \frac{l+1}{2l+3} [M^s(nl,l+1,\hbar\omega)]^2 n^s_{l+1}(\hbar\omega) \right\}, \tag{4.82}$$

$$\mu^s_{nl}(\omega) \sim (\hbar\omega - \varepsilon^s_{nl})^2 \left\{ \frac{l}{2l+1} [M^s(nl,l-1,\hbar\omega)]^2 n^s_{l-1}(\hbar\omega) + \frac{l+1}{2l+3} [M^s(nl,l+1,\hbar\omega)]^2 n^2_{l+1}(\hbar\omega) \right\}.$$

Here $M^s_{nl}(nl,l \pm 1,\varepsilon)$ is the transition probability matrix element and $n^s_l(S)$ is the partial density of states

$$n^s_l(\varepsilon) = \sum_n \oint dS \frac{Q^s_{kl}}{|\nabla_k \varepsilon_n(k)|}. \tag{4.83}$$

Q^s_{kl} is the partial charge [see Eq. (2.149)] and nl characterizes the core levels of the Sth atom in the unit cell.

It is possible to use one of the three forms of the transition probability matrix element in the dipole approximation[334] $\langle\varphi_{nk}|\hat{O}|\varphi_c\rangle$ where $\hat{O} = r, \nabla, \nabla\gamma_{\text{eff}}(r)$ to calculate the x-ray spectra. These matrix elements are related by the following equations:

$$\langle\varphi_{nk}|\nabla|\varphi_c\rangle = \tfrac{1}{2}(\varepsilon - \varepsilon_c)\langle\varphi_{nk}|r|\varphi_c\rangle, \tag{4.84}$$

$$\langle \varphi_{nk} | \nabla V_{\text{eff}}(r) | \varphi_c \rangle = (\varepsilon - \varepsilon_c) \langle \varphi_{nk} | \nabla | \varphi_c \rangle, \tag{4.85}$$

which hold exactly with integration over all space. However, since the MT approximation is primarily used in calculating the x-ray spectra, the relation between the matrix elements and the different operators is broken. As demonstrated by Nemoshkalenko and Aleshin,[334] the best approximation in this situation is a matrix element with the operator $\nabla v_{\text{eff}}(\mathbf{r})$, that is,

$$M(nl,l\pm 1,\varepsilon) = \int_0^{R_c} dr\, P^s_{nl}(r) P^s_{l\pm 1}(\varepsilon,r) \nabla V_{\text{eff}}(r). \tag{4.86}$$

Here $P^s_{nl}(r)$ is the radial part of the wave function of the core electron calculated with the effective self-consistent crystal potential and $P^s_{l\pm 1}(\varepsilon,r)$ is the radial part of the wave function of the electron in the valence band (the conduction band).

The theoretical intensity is ordinarily smeared by the width of the core level and the instrumental resolution, as well as by accounting for the Auger effect by means of the following equation[335]:

$$I'(\varepsilon) \sim \frac{1}{4} \int_{-\infty}^{\infty} d\varepsilon' \frac{I(\varepsilon')\Gamma(\varepsilon')/2}{(\varepsilon - \varepsilon')^2 + [\Gamma(\varepsilon')/2]^2}, \tag{4.87}$$

where

$$\Gamma(\varepsilon) = \Gamma_c + \Gamma_{\text{inst}} + \Gamma_0(\varepsilon), \quad \Gamma_0(\varepsilon) = \omega_0 = \left[1 - \frac{\varepsilon - \varepsilon_0}{\varepsilon_F - \varepsilon_0} \right]$$

(ω_0 is the parameter and ε_0 is the energy of the bottom of the valence band).

The formula for calculating the intensity of the emission spectra of disordered alloys was derived by Györffy and Stocks.[338] Using Green's function apparatus for a two-component alloy[336] the following equation can be obtained:

$$I^s(\hbar\omega - \varepsilon_c) \sim |M^s|^2 \operatorname{Im} \left\{ G^{(s)}_{00}(\hbar\omega) + S \sum_{\rho} [\overline{G^{(s)}_{\rho 0}(\hbar\omega)} + G^{(s)}_{0\rho}(\hbar\omega)] \right.$$
$$\left. + S^2 \sum_{\rho_1\rho_2} G_{\rho_1\rho_2}(\hbar\omega) \right\}, \tag{4.88}$$

where ρ is the radius vector of the nearest neighbor site, S is the overlap integral of the wave functions, and $G^s_{0\rho}$ is Green's function averaged over all atoms surrounding the Sth atom. Using the designations

$$\overline{G}^{(A)}_{00} \equiv g_A, \quad \overline{G}^{(B)}_{00} \equiv g_B, \quad G^{(A)}_{0\rho} \equiv c_A g_{AA} + c_B g_{AB}, \tag{4.89}$$

$$\overline{G}^{(B)}_{0\rho} \equiv c_A g_{BA} + c_B g_{BB}, \quad G_{\rho\rho} \equiv c_A g_A + c_B g_B \tag{4.90}$$

(C_A and C_B are the concentration of the alloy components) Eq. (4.88) will take the form

$$I^A(\hbar\omega - \varepsilon^A_c) \sim |M^A|^2 \operatorname{Im}[g_A + 2zs(c_A g_{AA} + c_B g_{AB}) + zs^2(c_A g_A + c_B g_B)], \tag{4.91}$$

$$I^{\mathrm{B}}(\hbar\omega-\varepsilon_c^{\mathrm{B}})\sim|M^{\mathrm{B}}|^2\,\mathrm{Im}[g_m+2zs(c_{\mathrm{B}}g_{\mathrm{BB}}+c_{\mathrm{A}}g_{\mathrm{BA}})+zs^2(c_{\mathrm{A}}g_{\mathrm{A}}+c_{\mathrm{B}}g_{\mathrm{B}})],\tag{4.92}$$

where z is the number of nearest neighbors. The functions g_{AA}, g_{AB}, g_{BA}, and g_{BB} determine the electron transition probability from a site occupied by, for example, a type A atom to a site occupied by a type B atom (the g_{AB} function).

Sonder and Babanoi[337] used their theoretical results to interpret the experimental spectra of silver and palladium in Pd–Ag alloys.

A theoretical analysis of the x-ray spectra based on the coherent potential method was carried out,[338] utilizing the coherent potential method to derive an equation for the x-ray spectral intensity. This equation takes the form

$$I^s(\hbar\omega-\varepsilon_c)\sim\sum_{l'}|M^s(nl,l',\hbar\omega)|^2\,\mathrm{Im}\,\tau_{l'l}^{s00}(\hbar\omega),\tag{4.93}$$

where

$$M^s(nl,l',\hbar\omega)=-\int dr\,Y_{lm}(\hat{r})R_{nl}^s(r)H'(r)R_{l'}^s(\hbar\omega,r)\,Y_{l'm'}\frac{\sqrt{\hbar\omega}}{\sin S_{l'}(\hbar\omega)}.\tag{4.94}$$

4.6. Optical characteristics of condensed systems

Experimental and theoretical analyses of optical properties yield valuable information on the electronic structure of the water. In this section we will examine certain critical elements in the theoretical approach to solid state optics.

The electron–phonon interaction Hamiltonian takes the form[339,340]

$$\hat{H}=\frac{e}{mc}\sum_j\hat{A}(\hat{\mathbf{r}}_j t)\hat{P}_j,\tag{4.95}$$

where the vector potential of the electromagnetic wave will be represented as

$$\hat{A}(\mathbf{r},t)=A\pi\exp[i(\hat{\eta}\mathbf{r}-\omega t)]+\text{complex conjugation}.\tag{4.96}$$

Here e is electron charge, m is electron mass, c is the speed of light in vacuum, $\hat{\mathbf{r}}_j$ is the coordinate operator of the jth electron, p_j is its momentum operator, A is the amplitude of the vector potential, π is the unit polarization vector, η is the complex wave vector, and ω is the electromagnetic wave frequency.

The transition probability from state $|\lambda\rangle$ to state $|\lambda'\rangle$ in first order perturbation theory is given by the equation

$$P_{\lambda\lambda'}=\frac{2\pi}{\hbar}|\langle\lambda'|\hat{L}|\lambda\rangle|\delta(E_\lambda-E_{\lambda'}+\hbar\omega),\tag{4.97}$$

where $\hat{L}e^{-i\omega t}=\hat{H}$ and $\hat{L}$ is time independent. According to Eq. (4.96),

$$\langle\lambda'|\hat{L}|\lambda\rangle=\langle\lambda'|\frac{e}{mc}A\sum_j\exp[i\widehat{\eta r_j}]\hat{P}_j\pi|\lambda\rangle.\tag{4.98}$$

Ordinarily the dependence of the interaction operator on $\mathbf{r}_j$ is neglected entirely, that is, only the first term in the expansion is retained:

$$\exp(i\widehat{\eta r}) = 1 + 1i\widehat{\eta r} + (i^2)\frac{(\eta r)^2}{2!} + \cdots, \qquad (4.99)$$

which corresponds to the dipole approximation. In this approximation the imaginary part of the band-to-band permittivity takes the form

$$\varepsilon_2(\omega) = \frac{4\pi^2 e^2}{m^2\omega^2\hbar} \sum_{\lambda\lambda'} f(E_\lambda)[1 - f(E_{\lambda'})] \,|\langle\lambda'|\sum_j \hat{P}_{j\pi}|\lambda\rangle|^2 \delta(\omega - \omega_{\lambda\lambda'}), \qquad (4.100)$$

where $f(E_\lambda)$ are the occupation numbers and $\omega_{\lambda\lambda'}$ is the transition frequency from state λ to λ'. Ordinarily $\varepsilon_2(\omega)$ is calculated in the one-electron approximation. Moreover, it is assumed that these states are quasistationary and at the same time the lifetime of the conduction band states may be 10^{-15} s.[341] Elementary perturbation theory does not incorporate the local field effects. However, the recent use of the TDLDA approximation (see Chap. 1) which applies a "self-consistent" field as the perturbation has made it possible to accurately describe photoemission spectra from the core states in many types of matter over an energy range 50–150 eV.[83,84] Standard theory yields results that are too high by a factor of 3–4.

Many previous calculations of the optical properties of metals and dielectrics utilized the constant matrix element (ME) approximation. In this case the function $\varepsilon_2(\omega)$ will be proportional to the combined density of states accurate to the multiplier $1/\omega^2$. The singularities in the combined density of states are determined by the critical van Hove points. Accounting for the energy independence of the ME will alter the relative intensities in the different portions of the spectra, that is, it will produce a certain modulation. The issue of the effect of incorporating ME on the optical spectrum has long been analyzed. Empirical pseudopotential calculations for dielectrics[254] are available; these have demonstrated a strong change in the spectrum when the ME are accounted for. For metals the issue of accounting for ME has been more fully developed. The two-band model as applied to alkali metals contains an analogous equation for the ME (Ref. 342):

$$\langle\Psi_k^-|\hat{P}|\Psi_k^+\rangle = GV_G/\omega, \qquad (4.101)$$

where $|\Psi_k^-\rangle$, $|\Psi_k^+\rangle$ are the valence band and conduction band states in the two-band model, G is the vector characterizing the Brillouin zone, and V_G is the interaction matrix element. This model has provided a satisfactory description of the optical spectrum of Na.

One of the first studies to investigate the effect of accounting for ME on the spectrum of copper[343] has noted that if the state is localized, the ME will be determined by the overlap of wave functions at one center and will have a weak dependence on **k**. However, the optical spectrum may change substantially when ME is incorporated due to the different transition probability for the different pairs of states of the valence band and conduction band. Table 4.4 gives the MEs of copper calculated at various points in the first Brillouin zone. This study has demonstrated that incorporating ME has a strong effect on the spectrum of ε_2, yielding a good agreement between it and experiment (Fig. 4.3) while having virtually no effect on the photoemission spectrum, which indicates the weak dependence of the ME on the state for high energy excitations. However, as the angle-

TABLE 4.4. Matrix elements of some transitions (copper).

Transition $\lambda - \lambda'$	$E_{\lambda'} - E_{\lambda}$ (Ry)	$\frac{\|\langle\lambda\|\hat{p}\|\lambda'\rangle\|^2}{2m}$ (Ry) (Ref. 343)	$\frac{\|\langle\lambda\|\hat{p}\|\lambda'\rangle\|^2}{2m}$ (Ry) (Ref. 348)
$\Gamma_{25} - \Gamma_1$	1.976	0.330	...
$L_1 - L_{2'}$	0.344	0.051	0.055
$L_3 - L_{2'}$	0.214	0.012	0.006
$L_3 - L_{2'}$	0.106	0.193	0.070
$X_{2'} - X_1$	0.336	0.721	0.823
$X_1 - X_{4'}$	0.540	0.042	0.051
$X_5 - X_{4'}$	0.289	0.120	0.083
$X_{4'} - X_1$	0.387	0.780	1.039

resolved photoemission was developed it was determined that incorporating ME was significant in this range as well. Even prior to Janak *et al.*[343] the need to account for the ME was noted[344,345] and was confirmed by subsequent research.[346–348]

The relations of the matrix element of the optical transitions along Δ and Λ were provided by Smith.[348] The band structure and wave functions were calculated within the framework of the combination scheme. The Hamiltonian was divided into the sum of terms describing the strongly bound *d* electrons, the free electrons, and the interaction of the free electrons with the *d* electrons. The matrix elements of the Hamiltonian were selected so that the band structure matched the calculated value.[343] For the Δ_1–Δ_2 transition two zeros on the Δ axis were observed in the ME which in principle refutes the concept of a weak change in the ME. The authors attributed this phenomenon to hybridization of the weakly bound *s* states and *d* states. The calculated spectrum of ε_2 (Ref 348) (Fig. 4.4) reproduces the two principle maxima (at 2.3 and 5 eV). Certain singularities at 4.5 eV vanished from

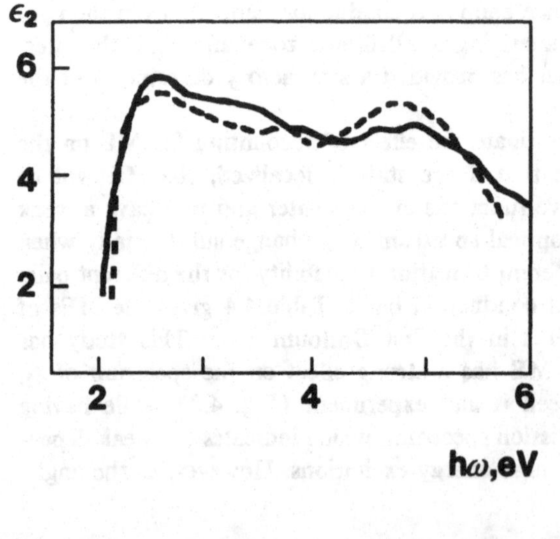

FIG. 4.3. Experimental and theoretical spectra of copper (--- experiment; —theory).

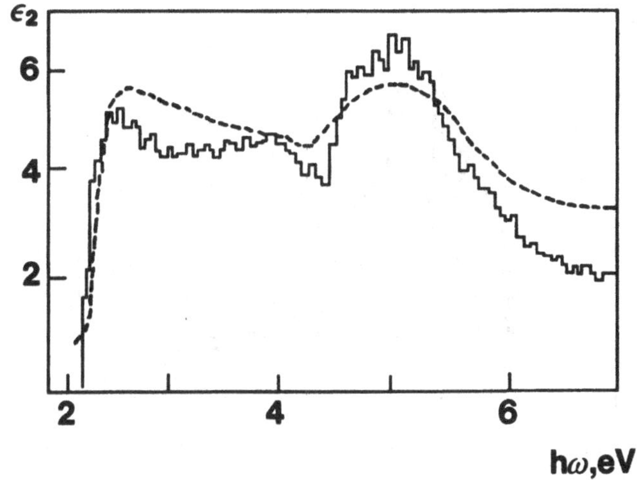

FIG. 4.4. Experimental and theoretical (Ref. 348) spectra of copper.

the spectrum, indicating poorer accuracy of the method compared to Janak *et al.*'s calculations.[343] Overall, the spectrum accurately describes the experimental results.

A number of studies applying more effective LAPW and LMTO techniques to calculate the optical characteristics of metals, dielectrics, and compounds have appeared in recent years. For example, within the framework of the LAPW method the matrix elements of the optical transition are found as follows.[349] We go over to a special coordinate system in which the z axis runs along the wave vector $\mathbf{k}_i$. This makes it possible to eliminate summation over the magnetic quantum number in calculating the matrix element. We give the momentum matrix element as

$$P_{\lambda\lambda'}(\mathbf{k}) = \langle \Psi_{\mathbf{k}\lambda'} | -i\nabla | \Psi_{\mathbf{k}\lambda} \rangle = \sum_{i,j} c^*_{\lambda'}(\mathbf{k}_i) c_\lambda(\mathbf{k}_j) t_{ij}, \tag{4.102}$$

where $C_\lambda(\mathbf{k}_i)$ are the coefficients of the expansion of the electron wave function

$$t_{ij} = \langle \varphi(\mathbf{k}_i r) | -i\nabla | \varphi(\mathbf{k}_j r) \rangle. \tag{4.103}$$

Here and henceforth we use the standard notation of the LAPW method. The contribution to t_{ij} from the region within the MT spheres takes the form

$$t_{x_{i,j}} = B \cos \varphi_{k_j}; \quad t_{y_{i,j}} = B \sin \varphi_{k_j}; \quad t_{z_{i,j}} = c, \tag{4.104}$$

where

$$B = \sum_{l=1}^{\infty} [j_{l-1}(k_i s) j_l(k_j s) A_{i,j}(l-1,l) P_l^1 (\cos \theta_{kj})$$
$$+ j_l(k_i s) j_{l-1}(k_{js}) A_{i,j}(l,l-1) P_{l-1}^1 (\cos \theta_{kj})], \tag{4.105}$$

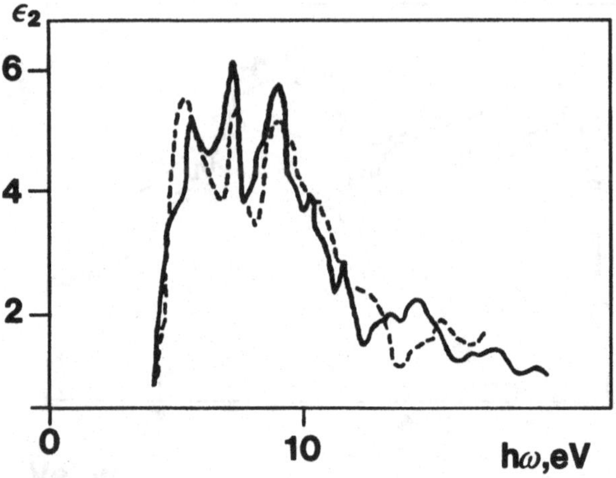

FIG. 4.5. Theoretical and experimental spectrum $\varepsilon_2(\omega)$ of CaS (---- experiment; —theory).

$$c=\sum_{l=1}^{\infty} l[j_{l-1}(k_i s)j_l(k_j s)A_{i,j}(l-1,l)P_l^0(\cos\theta_{kj}) - j_l(k_i s)j_{l-1}(k_j s)A_{ij}(l,l-1)P_{l-1}^0(\cos\theta_{kj})], \quad (4.106)$$

where

$$A_{ij}(l',l)=-2\pi[R_{ij}(l',l)+\tilde{D}_{l'}(k_i^2)S^2-\tilde{D}_l(k_j^2)S^2-S^2],$$

$$R_{ij}(l',l)=\frac{1}{\Phi_{\nu l'}(\tilde{D}_{l'}(k_i^2),S)\Phi_{\nu l}(\tilde{D}_l(k_j^2),s)} \times\Bigg[(E_{\nu l'}+\omega_{\nu l'}^i-E_{\nu l}-\omega_{\nu l}^j)\int_0^S dr\,\Phi_{\nu l}(r)\Phi_{\nu l'}(r)r^3 + (E_{\nu l'}-E_{\nu l}+\omega_{\nu l'}^i)\omega_{\nu l}^j\int_0^S \dot{\Phi}_{\nu l}(r)\Phi_{\nu l'}(r)r^3\,dr + (E_{\nu l'}-E_{\nu l}-\omega_{\nu l}^j)\omega_{\nu l'}^j\int_0^S dr\,\Phi_{\nu l}(r)\dot{\Phi}_{\nu l'}(r)r^3 + (E_{\nu l'}-E_{\nu l})\omega_{\nu l}^j\omega_{\nu l'}^i\int_0^S \dot{\Phi}_{\nu l}(r)\dot{\Phi}_{\nu l'}(r)r^3\,dr\Bigg]. \quad (4.107)$$

The range outside the MT sphere makes the contribution

$$t_{\text{out},j}=k_j\left[\Omega\delta_{ij}-\frac{4\pi s^2}{|\mathbf{k}_i-\mathbf{k}_j|}j_1(|\mathbf{k}_i-\mathbf{k}_j|s)\right]. \quad (4.108)$$

Nemoshkalenko *et al.*[349] apply these formulas to calculate the optical conductivity of an intermetallic CoGa compound with a CsCl structure. The energy po-

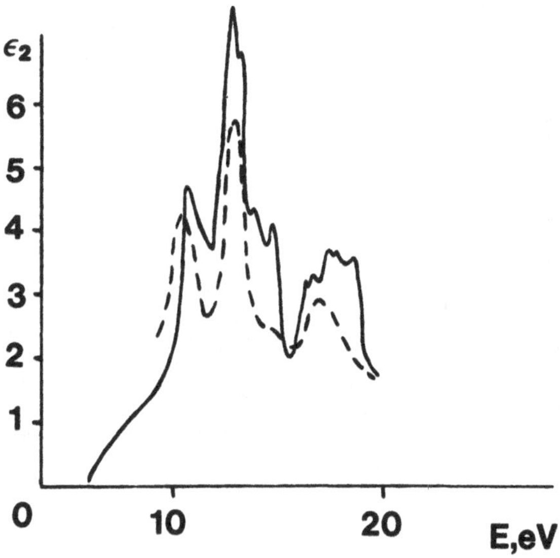

FIG. 4.6. Theoretical and experimental spectrum $\varepsilon_2(\omega)$ of MgO (----- experiment; —theory).

sition of the primary spectral features in the calculation neglecting the matrix element is (on the whole) conveyed properly, although the intensities of the different peaks are incorrect.

We carried out a series of calculations of the optical properties of A^2B^6 compounds (the *c* axis, MgO, SrS, CaO, SrO, BaS, etc.) accounting for both the optical

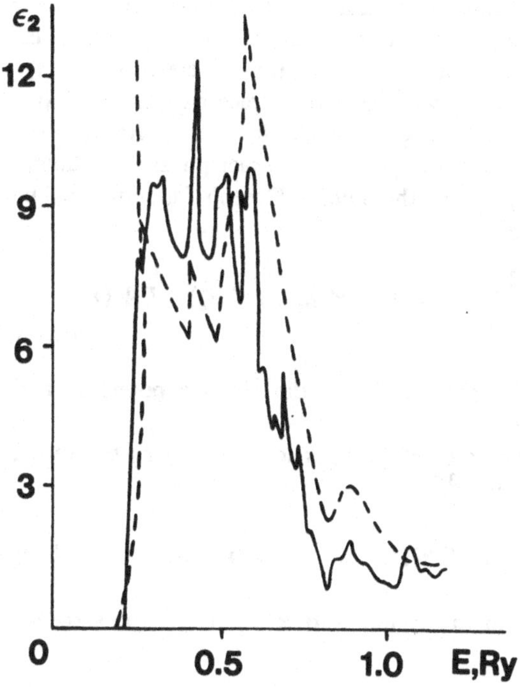

FIG. 4.7. Theoretical and experimental spectrum $\varepsilon_2(\omega)$ of SrS (----- experiment;—theory).

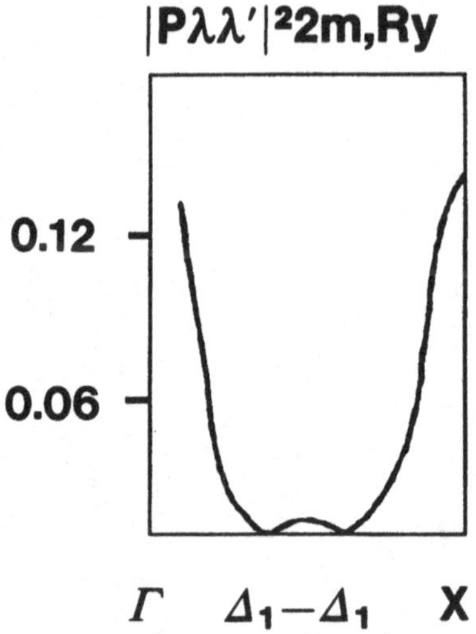

FIG. 4.8. Squared matrix element plotted as a function of **k** for CaS.

transition matrix element and the approximation of its energy independence.[350–353] Figures 4.5–4.7 provide sample frequency dependences of the function ε_2 and available experimental data for the *c* axis, MgO, and SrS. It is clear that satisfactory agreement with experiment is achieved overall which cannot be obtained without the matrix element. Our equations for the matrix element are (on the whole) similar to Eq. (4.107) although a somewhat different derivation method was used. Figure 4.8 provides a sample plot of the squared ME as a function of **k** along the Δ axis for the *G–X* transition.

It should be noted that the calculated spectrum of $\varepsilon_2(\omega)$ is shifted by ΔEg, because the band gap is underestimated, which occurs in all electronic structure calculations of dielectrics based on the local approximation of electron density-functional theory (DFT). From the DFT viewpoint the dielectric gap is the difference between the ionization energy and the electron affinity energy. As we know the eigenvalues in the Kohn–Sham method $(\widetilde{E}_i)$ have no exact physical meaning. The quasiparticle excitation energies given by the poles of Green's function must be determined from the Dyson equation

$$\left[-\frac{\hbar^2\nabla^2}{2m}+V_{\text{ext}}(r)+V_n(r)\right]\Psi_i(r)+\int d^3r'\,\Sigma_{\text{exc}}(r,r',E_c)=E_i\Psi_i(r), \tag{4.109}$$

where $\Sigma_{\text{exc}}(r,r',E)$ is the self-energy operator, V_H is the Hartree potential, and $V_{\text{ext}}(r)$ is the external field potential.

The equation for the self-energy operator of quasiparticles neglecting the vertex function yields the *GW* approximation[354,355]

$$\Sigma_{\text{exc}}(r,r',E)=i\int\frac{d\omega}{2\pi}e^{-i\delta\omega}G(r,r',E-\omega)\,W(r,r',\omega), \tag{4.110}$$

where $G(r,r',\omega)$ is Green's function of the system and $W(r,r',\omega)$ is the screened Coulomb interaction.

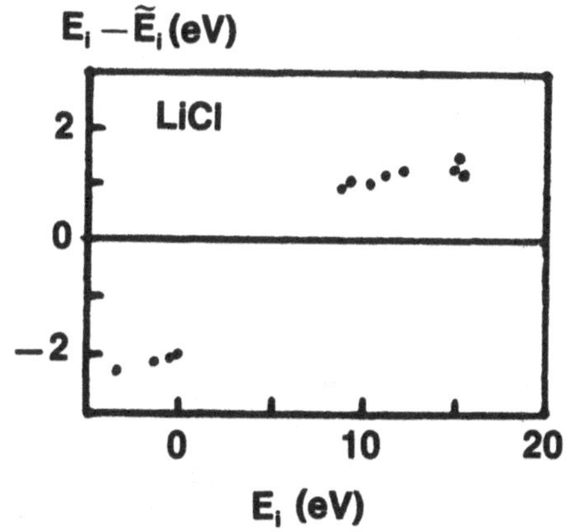

FIG. 4.9. $E_i - \widetilde{E}_i$ correction for LiCl.

The correct value of the dielectric gap was obtained within the framework of this approximation where G and W were calculated by means of DFT wave functions. Figure 4.9 gives the correction $E_i - \widetilde{E}_i$ for LiCl at certain high symmetry points of the Brillouin zone. The jump in the vicinity of the forbidden band and the weak dependence of the correction on the energy outside the forbidden band should be noted. This jump is due to a discontinuity in the functional derivative of the exchange-correlation function[358–360]

$$\left.\frac{\delta E_{\mathrm{exc}}\{n(r)\}}{\delta n(r)}\right|_{N+} - \left.\frac{\delta E_{\mathrm{exc}}\{n(r)\}}{\delta n(r)}\right|_{N-} = c \tag{4.111}$$

which makes it possible (to some degree) to account for the quasiparticle nature of the spectrum by a simple shift in the conduction band.[361–364] This was confirmed by calculation[365] in which the wave function of the quasiparticles coincided with the DFT wave function.

The multiparticle effects were accounted for more rigorously by Kulikov *et al.*[366] The authors of this study calculated the optical conductivity of chromium in good agreement with experiment.

A recently published article[367] derived an approximate equation for the nonlocality correction to the eigenvalues

$$\Delta E_g = \Delta E_g^{\mathrm{HF}} / \varepsilon(0), \tag{4.112}$$

where ΔE_g^{HF} is the difference in the dielectric gap in the Hartree–Fock approximation and the DFT calculation and $\varepsilon(0)$ is the static permittivity.

Calculations using this equation are in good agreement with available theoretical and experimental data.

Research on the optical spectra of metals in the IR range[368] is worthy of mention, because of the deviation of the optical spectra of the transition metals from Drude spectra near 0.1 eV can be attributed to band-to-band transitions. The constant matrix element approximation is not suitable in the IR range. The LMTO method was carried out to calculate[368] the frequency dependence of $\varepsilon_2(\omega)$ for vanadium, niobium, molybdenum, rhodium, and palladium. Figure 4.10 illustrates

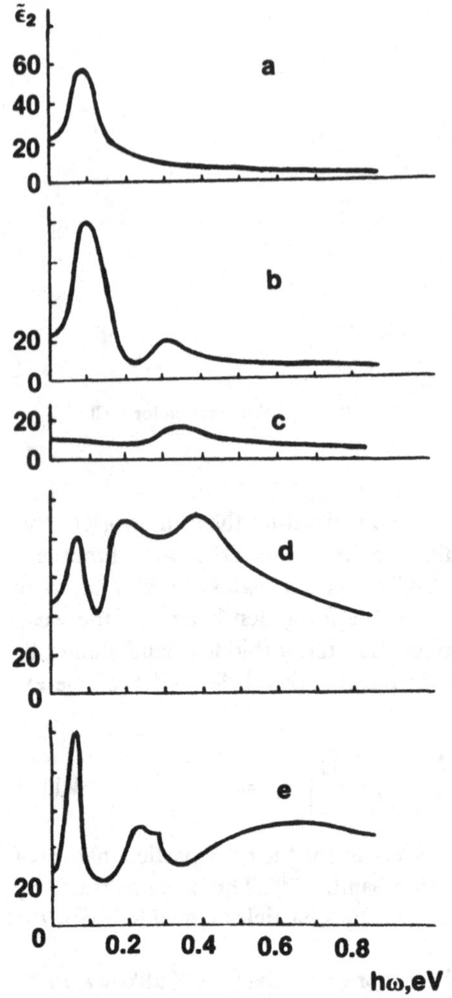

FIG. 4.10. Calculated $\varepsilon_2(\omega)$ relations for (a) vanadium, (b) niobium, (c) molybdenum, (d) rhodium, and (e) palladium.

these results which show that clearly expressed threshold or maxima-type structures appear in the IR. This result refutes the present opinion that there is no structure in the $\varepsilon_2(\omega)$ spectrum in the IR. In order to carry out a comprehensive analysis of the optical characteristic of the metals it is necessary to also account for intraband transitions. The intraband contribution to the imaginary part of the permittivity is given by the Drude formula

$$\varepsilon_2^0(\omega)=\frac{\omega_p}{\gamma\omega(1+\omega^2/\gamma^2)}, \tag{4.113}$$

where γ is the average collision frequency of the conduction electron and ω_p is the plasma frequency

$$\omega_p=(4\pi^2e^2N(E_F)\langle V^2\rangle/3m)^{1/2}/\hbar. \tag{4.114}$$

Here $N(E_F)$ and $\langle v^2 \rangle$ is the density of states and the average squared electron velocity on the Fermi surface.

4.7. Thermodynamic properties of molten alloys

The method given in Chap. 2 for calculating the density of electron states of metallic can be used to find the electron contributions to the entropy of mixing, the enthalpy of formation, and the Gibbs free energy. We will briefly discuss certain specific examples following the works of Gorbunov *et al.*[369–371]

The electron contribution to the entropy of mixing is determined by the equation

$$\Delta S_{\mathrm{EL}} = S - (S_1 + S_2), \tag{4.115}$$

where S is the melt entropy, S_1 is the entropy of the first component while S_2 is the entropy of the second component which can be found through the density of state:

$$S_1 = cS^{\mathrm{A}} = \frac{c\pi^2 k_B^2}{3} T n^{\mathrm{A}}(\varepsilon_F), \tag{4.116}$$

$$S_2 = (1-c)S^{\mathrm{B}} = \frac{c\pi^2 k_B^2}{3} T n^{\mathrm{B}}(\varepsilon_F), \tag{4.117}$$

where c is the concentration of component A.

In order to calculate S it is necessary to know the density of states in the melt $n(E)$ for which the following approximation can be used:

$$n(E) = c n_{\mathrm{A}}(E) + (1-c) n_{\mathrm{B}}(E). \tag{4.118}$$

The calculation begins with formulation of the MT potentials of the pure components. Here it is important to remember that it is necessary to consider the surrounding environment of the given atom by other atoms (A and B) in Eq. (4.118) in order to find $n^{\mathrm{A(B)}}$. The environmental contributions can be found as the averaged characteristics in the virtual crystal approximation

$$V_{\mathrm{Cou}}^{\mathrm{A(B)}} = V_{\mathrm{am}}^{\mathrm{A(B)}} + \frac{1}{2r} \sum_{p=1}^{p_{\max}} \frac{n_p}{R_p} \int_{R_p - r}^{R_p + r} [c V_{\mathrm{am}}^{\mathrm{A}}(t) + (1-c) V_{\mathrm{am}}^{\mathrm{B}}(t)] t \, dt, \tag{4.119}$$

where $V_{\mathrm{Cou}}^{\mathrm{A(B)}}$ is the Coulomb potential of atom A(B) in the melt, $V_{\mathrm{am}}^{\mathrm{A(B)}}$ is the atomic Coulomb potential, R_p is the radius of the pth coordination shell, and n_p is the atomic number of the pth coordination shell.

An analogous relation is also valid for the electron density. Then using the Mattheis[147] scheme we find the total potential as the sum of the Coulomb and exchange terms. Here the potentials of the components must be calculated on the same energy scale. The constants of the MT potentials V_0^{A} and V_0^{B} in the general case will not be equal in the melt.

When V_0^{A} and V_0^{B} are similar the following approximation is possible:

$$V_0^{\mathrm{melts}} = c V_0^{\mathrm{A}} + (1-c) V_0^{\mathrm{B}}. \tag{4.120}$$

This method was applied[371] to investigate the electronic structure and physical properties of aluminide melts of the iron subgroup. Figure 4.11 shows the concen-

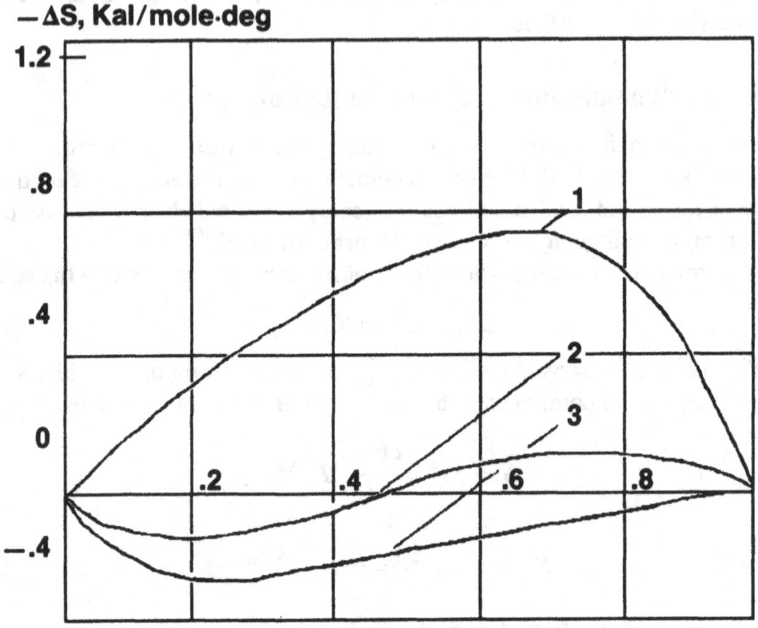

FIG. 4.11. Concentration dependence of the electronic component of the entropy of mixing of Al–Fe (1), Al–Co (2), and Al–Ni (3).

tration dependence of the electronic component of the entropy of mixing the Al–Fe, Al–Co, and Al–Ni melts. The behavior of $\Delta S_{\rm EL}$ is determined by the behavior of the function $n(E)$ of the melts. The total mixing energy for these melts is nearly entirely determined by the electron component $\Delta S_{\rm EL}$. Using methods other than the Anderson–McMillan approach to calculate the concentration dependence of $\Delta S_{\rm EL}$ will yield a relation similar to that shown in Fig. 4.11, although a certain quantitative difference will be observed. The electron contribution to the enthalpy of formation ΔH was calculated for these same melts:

$$\Delta H = \int_0^{E_F} d\varepsilon\, \varepsilon n(\varepsilon) - c \int_0^{E_F^{\rm A}} d\varepsilon\, \varepsilon n_{\rm A}(\varepsilon) - (1-c) \int_0^{E_F^{\rm B}} d\varepsilon\, \varepsilon n_{\rm B}(\varepsilon) - c\frac{\delta}{2} z_{\rm A}^d + (1-c)\frac{\delta}{2} z_{\rm B}^d. \tag{4.121}$$

Here δ is the width of the d band of the melt $z_{\rm A}^d$, and $z_{\rm B}^d$ is the number of component d electrons. Equation (4.121) was obtained assuming that the d electrons make the most significant contribution to enthalpy.

Figure 4.12 demonstrates the dependence of ΔH on the second component concentration in the melts of Al–Fe, Al–Co, and Al–Ni alloys. Similar calculations make it possible to explain the interatomic interactions in the melts.

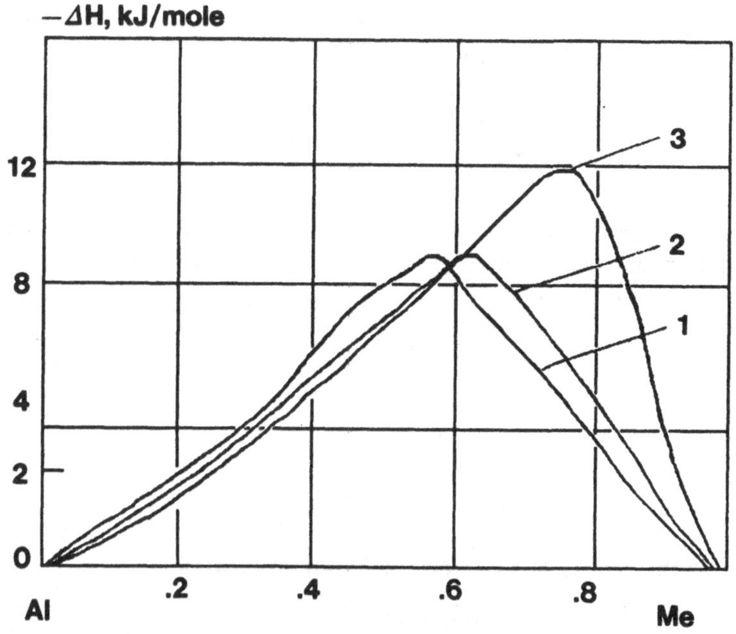

FIG. 4.12. Concentration dependence of the electron contribution to the enthalpy of formation of Al–Fe (1), Al–Co (2), and Al–Ni (3) alloys.

Thus, for example, the absolute value of ΔH will grow in the Al series of melts containing Fe–Co–Ni and will reach a maximum in the nickel alloys which is consistent with experiment and explains the significant thermodynamic stability of the nickel alloys.

Calculations of the entropy of mixing, the enthalpy of mixing, and other thermodynamic characteristics of melts satisfactorily correlate with experiment over a broad range of component concentrations.

4.8. Estimation of the superconducting transition temperature

The modern theory of superconductivity is based on the idea of pairing of electrons with opposite momenta and spin resulting in the formation of an effective Bose particle.[372] Several theoretical and experimental studies have been devoted to explaining the nature of attraction between the pair electrons.[372] The experimental observation of the isotope effect in superconductors served as a clear indication that electron–phonon interaction plays a role in superconductivity.

Electron–phonon interaction screens Coulomb electron repulsion and results in electron renormalization or "dressing." Then the energy of the new "dressed" quasiparticle is determined from the solution of the equation

$$\varepsilon = E_k + \Sigma(k,\varepsilon). \qquad (4.122)$$

Here the self-energy part of $\Sigma(k,\varepsilon)$ describes the phonon contribution to the quasiparticle energy and neglecting the dependence on k in a linear approximation in the energy $\Sigma(\varepsilon)\approx\lambda\varepsilon$ determines the effective electron–phonon interaction constant

$$\lambda=-\frac{\partial}{\partial\varepsilon}\sum(\varepsilon)\bigg|_{\varepsilon=0}=2\int_0^\infty\frac{d\omega\,\alpha^2(\omega)F(\omega)}{\omega},\tag{4.123}$$

where $\alpha^2(\omega)F(\omega)$ is the spectral electron–phonon interaction function. The parameter λ determines the renormalization of electron mass on the Fermi surface from interaction with phonons:

$$\frac{m^*}{m}=\frac{E_k}{\varepsilon}=1+\lambda.\tag{4.124}$$

We can write[372]

$$\int d\omega\,\omega\alpha^2(\omega)F(\omega)=\frac{n(\varepsilon_F)\langle I^2\rangle}{2M},\tag{4.125}$$

where $n(\varepsilon_F)$ is the density of states on the Fermi level, while

$$\langle I^2\rangle=\frac{\int_{S_F}\widehat{dk}\int_{S_F}\widehat{dk'}\,|\langle\Psi_k|e\nabla V|\Psi_k\rangle|^2}{\int_{S_F}\widehat{dk}\int_{S_F}\widehat{dk'}}\tag{4.126}$$

is the average squared electron–phonon coupled matrix element. If we introduce the average squared phonon frequency[373]

$$\langle\omega^2\rangle=\frac{\int d\omega\,\omega\alpha^2(\omega)F(\omega)}{\int(d\omega/\omega)\alpha^2(\omega)F(\omega)},\tag{4.127}$$

the parameter λ can be given as

$$\lambda=\frac{n(\varepsilon_F)\langle I^2\rangle}{M\langle\omega^2\rangle}.\tag{4.128}$$

The parameter λ is an important quantity in the theory of the phonon mechanism of superconductivity since it is used to express the renormalization and energy gap parameters in the Eliashberg equations for the self-energy part of the superconductor. The approximate numerical solution of the Eliashberg equations for the different phonon spectra makes it possible to write the interpolation formula relating the superconducting transition temperature T_c to the effective superconducting parameters.[372] In the case of strong coupling superconductors the McMillan formula[373] is the most commonly used:

$$T_c=\frac{\omega_{\log}}{1.2}\exp\left[-\frac{1.04(1+\lambda)}{\lambda-\mu^*(1+0.62\lambda)}\right],\tag{4.129}$$

as well as the Karakazov equation[375]

$$T_c=\frac{\omega_{\log}}{1.43}\exp\left[-\frac{1+\lambda}{\lambda-\mu^*[1+0.5\lambda/(1+\lambda)]}\right],\tag{4.130}$$

where μ^* is the effective Coulomb pseudopotential introduced by Bogolyubov *et al.* in BCS theory[376]:

$$\mu^* = \frac{V_c n(\varepsilon_F)}{1 + V_c n(\varepsilon_F) \ln(\varepsilon_F/\omega_D)}. \tag{4.131}$$

Here v_c is the matrix element of electron Coulomb interaction, ε_F is the Fermi energy, and ω_D is the Debye phonon frequency.

The "hard MT sphere" approximation[377] is ordinarily used to calculate the electron–phonon interaction constant assuming that the phonon atomic shift from equilibrium will cause a hard shift of the local potential in the MT region while any change in the potential in the external region can be ignored. In the scalar-relativistic approximation we have[378]

$$n(\varepsilon_F)\langle I^2\rangle \equiv \eta$$

$$= \sum_l \eta_l = \frac{\varepsilon_F}{T^2 n(\varepsilon_F)} \sum_l 2(l+1) \left| \int_0^{R_S} dr (P_l^s P_{l+1}^s \right.$$

$$\left. + Q_l^s Q_{l+1}^s) r^2 \frac{dV_{\text{eff}}^s}{dr} \right|^2 \frac{n_l^s(\varepsilon_F) n_{l+1}^s(\varepsilon_F)}{N_l^{(1)}(\varepsilon_F) N_{l+1}^{(1)}(\varepsilon_F)}, \tag{4.132}$$

where P_l^s and Q_l^s are the solution of Eqs. (1.194), while

$$N_l^{(1)}(\varepsilon_F) = \frac{\sqrt{\varepsilon_F}}{\pi} (2l+1) \int_0^{R_S} dr \left\{ [Q_l^s]^2 + \left[1 + \frac{l(l+1)}{(2Mcr)^2} \right] [P_l^s]^2 \right\} r^2. \tag{4.133}$$

Therefore, in order to calculate T_c by Eqs. (4.129) and (4.130) it is necessary to know the partial densities of states on the Fermi level $n_l^s(\varepsilon_F)$ as well as the self-consistent MT potential $v_{\text{eff}}^s(r)$. Information on the phonon spectrum is required to estimate $\langle\omega^2\rangle$.

References

1. M. Born and J. R. Oppenheimer, Ann. Phys. **84**, 457 (1927).
2. B. I. Bartz, Yu. L. Bolotin, E. V. Inopin, and V. Yu. Gonchar, *The Hartree–Fock Method in Nuclear Theory* (Naukova Dumka Press, Kiev, 1982), p. 208.
3. D. Hartree, *Atomic Structure Calculations* (Izd-vo inostr. Int, Moscow, 1968).
4. V. A. Fock, Z. Phys. **61**, 126 (1980).
5. G. Slater, *Self-consistent Field Methods for Molecules and Solids* (Mir Press, Moscow, 1978).
6. M. G. Beselov and L. N. Labzovskiy, *Atomic Theory. Electron Shell Structure* (Nauka Press, Moscow, 1986).
7. *Semiempirical Methods of Calculating Electron Structure*, edited by G. Sigal (Mir Press, Moscow, 1980), Vols. 1 and 2, p. 327.
8. N. O. Lipari, Phys. Status Solidi B **40**, 691 (1970).
9. A. P. Yutzis, Zh. Eksp. Teor. Fiz. **23**, 129 (1952) [Sov. Phys. JETP **23** (1952)].
10. J. C. Slater, Phys. Rev. **81**, 385 (1951).
11. K. Schwarz, Phys. Rev. B **5**, 2466 (1972).
12. K. Schwarz, Theor. Chim. Acta **34**, 225 (1974).
13. A. P. Kovtun, Development of the X_α scattered wave method and its application to analyzing the electronic properties of palladium sulfide-type compounds: abstracts of candidate's dissertations in physics and mathematics, Kiev, 1980.
14. J. Hubbard, Proc. R. Soc. London Ser. A **277**, 237 (1964).
15. S. V. Vlasov and O. V. Farberovskich, Fiz. Tverd. Tel. **24**, 941 (1982) [Sov. Phys. Solid State **24**, 536 (1982)].
16. J. N. Chasalviel, M. Campagna, G. K. Wertheim, and P. Schmidt, Phys. Rev. **14**, 4586 (1976).
17. O. Goscinski, P. T. Pickup, and G. Purvis, Chem. Phys. Lett. **22**, 167 (1973).
18. J. Connoly, in: *Semiempirical Methods of Calculating Electronic Structure*, edited by G. Sigal (Mir Press, Moscow, 1980), Vol. 1, p. 139.
19. C. Lundquist and N. March, *Nonuniform Electron Gas Theory* (Mir Press, Moscow, 1987) (Russian trans.).
20. *Achievements in Electronic Theory of Metals*, edited by P. Ziche and G. Lemman (Mir Press, Moscow, 1984) (Russian trans.).
21. S. Reems, *Multielectron System Theory* (Mir Press, Moscow, 1976) (Russian trans.).
22. D. Pines and F. Nozieres, *Theory of Quantum Liquids* (Mir Press, Moscow, 1967) (Russian trans.).
23. E. P. Wigner, Trans. Faraday Soc. **34**, 678 (1938).
24. D. Bohm and D. Pines, Phys. Rev. **92**, 609 (1953).
25. D. Pines, *Elementary Excitations in Solids* (Mir Press, Moscow, 1965) (Russian trans.).
26. M. Gell-Mann and K. A. Brueckner, Phys. Rev. **106**, 364 (1957).
27. P. Nozieres and D. Pines, Phys. Rev. **111**, 442 (1958).
28. J. Hubbard, Proc. R. Soc. London Ser. A **243**, 336 (1957).
29. T. Gaskell, Proc. Phys. Soc. **77**, 1182 (1961).
30. W. L. McMillan, Phys. Rev. A **138**, 442 (1965).
31. N. Metropolis, A. W. Rosenbluth, M. N. Rosenbluth, and E. Teller, J Chem. Phys. **21**, 1087 (1953).
32. *Monte Carlo Methods in Statistical Physics*, edited by K. Binder (Mir Press, Moscow, 1982) (Russian trans.).
33. D. Ceperley, G. V. Chester, and M. H. Kalos, Phys. Rev. B **15**, 3081 (1977).

34. D. Ceperley, Phys. Rev. B **18**, 3126 (1978).
35. D. L. Freeman, Phys. Rev. B **15**, 5512 (1977).
36. K. S. Singwi, M. P. Tosi, R. H. Land, and A. Sjolander, Phys. **176**, 589 (1968).
37. K. S. Singwi, A. Sjolander, M. P. Tosi, and R. H. Land, Phys. Rev. B **1**, 1044 (1970).
38. J. Hubbard, Phys. Lett. A **25**, 709 (1967).
39. B. D. Gorbachenko and E. G. Maksimov, Usp. Fiz. Nauk **130**, 65 (1980) [Sov. Phys. Usp. **130** (1980)].
40. P. Vashishta and K. S. Singwa, Phys. Rev. B **6**, 875 (1972).
41. U. von Barth and L. Hedin, J. Phys. C **5**, 1200 (1980).
42. S. H. Vosko, L. Wilk, and M. Nusair, Can. J. Phys. **58**, 1200 (1980).
43. I. A. Akhiezer and S. V. Peletminskiy, Zh. Eksp. Teor. Fiz. **38**, 18 (1960) [Sov. Phys. JETP **38** (1960)].
44. A. H. MacDonald and S. H. Vosko, J. Phys. C **12**, 2977 (1979).
45. M. V. Ramana and A. K. Rajagopal, J. Phys. C **14**, 4291 (1981).
46. L. H. Thomas, Proc. Cambridge Philos. Soc. **23**, 542 (1927).
47. E. Fermi, Z. Phys. **48**, 73 (1928).
48. P. Hohenberg and W. Kohn, Phys. Rev. B **136**, 864 (1964).
49. W. Kohn and L. Sham, Phys. Rev. A **140**, 1133 (1965).
50. M. Levy, Proc. Nat. Acad. Sci. U.S.A. **76**, 6062 (1979).
51. O. Gunnarsson and B. I. Lundquist, Phys. Rev. B **13**, 4274 (1976).
52. U. von Barth, in: *The Electronic Structure of Complex System*, edited by P. Pharisean and W. H. Temmerman, NATO, ASI Phys. **113**, 67 (1984).
53. A. K. Rajagopal and J. Callaway, Phys. Rev. B **7**, 1912 (1973).
54. D. C. Langreth and S. H. Vosko, Phys. Rev. Lett. **59**, 497 (1987).
55. D. D. Koelling and B. N. Harman, J. Phys. C **10**, 3107 (1977).
56. O. V. Farberovich, S. V. Vlasov, and G. P. Nizhnikova, Program for self-consistent relativistic calculation of atomic and ionic structures in the local density approximation. Part 1. Numerical techniques and program description, Voronezh, 1983. Manuscript deposited at Voronezh University and the All-Union Institute of Scientific and Technical Information, June 2, 1983.
57. A. N. MacDonald, W. E. Pickett, and D. D. Koelling, J. Phys. C **13**, 2675 (1980).
58. D. C. Langreth and J. P. Perdew, Phys. Rev. B **21**, 54 (1980).
59. O. Gunnarsson, M. Jonson, and B. I. Lundquist, Phys. Lett. A **59**, 177 (1976).
60. O. Gunnarsson, M. Jonson, and B. I. Lundquist, Solid State Phys. **24**, 765 (1977).
61. O. Gunnarsson, M. Jonson, and B. I. Lundquist, Phys. Rev. B **20**, 3136 (1979).
62. O. Gunnarsson and R. O. Jones, Phys. Scr. **21**, 394 (1980).
63. G. W. Pratt, Jr., Phys. Rev. **88**, 1217 (1952).
64. D. C. Langreth and J. P. Perdew, Phys. Rev. B **15**, 2884 (1977).
65. D. C. Langreth and M. J. Mehl, Phys. Rev. Lett. **47**, 446 (1981).
66. N. D. Lang and W. Kohn, Phys. Rev. B **1**, 4555 (1970).
67. D. C. Langreth and M. J. Mehl, Phys. Rev. B **28**, 6112 (1983).
68. J. P. Perdew and A. Zunger, Phys. Rev. B **23**, 5048 (1981).
69. A. K. Theophilou, J. Phys. C **12**, 5419 (1979).
70. U. von Barth, Phys. Rev. A **20**, 1693 (1979).
71. O. B. Farberovich, Phys. Status Solidi B **104**, 265 (1981).
72. C. O. Almbladh and U. von Barth, Phys. Rev. B **31**, 3231 (1985).
73. L. Hedin and S. Lundquist, in: *Solid State Physics*, edited by F. Seitz, D. Turnbull, and H. Ehrenreich (Academic Press, New York, 1969), Vol. 23, p. 1.
74. L. Hedin, Phys. Rev. A **139**, 796 (1965).
75. J. Schwinger, Proc. Natl. Acad. Sci. U.S.A. **37**, 452 (1951).
76. L. J. Sham and W. Kohn, Phys. Rev. **145**, 561 (1966).
77. G. Albman and F. Barth, J. Phys. F **5**, 1155 (1975).
78. R. E. Watson, J. F. Herbst, L. Hodges, B. I. Lundquist, and J. W. Wilkins, Phys. Rev. B **13**, 1463 (1976).
79. A. H. MacDonald, J. Phys. F **10**, 1737 (1980).
80. W. E. Pickett, Comments Solid State Phys. **12**, 57 (1985).

81. S. V. Vlasov, O. V. Farberovich, B. A. Zon, Ye. Yu. Delota, M. German, G. P. Inzhnikova, and V. S. Rostovtsev, Izv. U.S.S.R. Acad. Sci. Ser. Fiz. **50**, 1336 (1986).
82. V. Peuckert, J. Phys. C **11**, 4945 (1978).
83. A. Zangwill and P. Soven, Phys. Rev. A **21**, 1561 (1980).
84. A. Zangwill and P. Soven, Phys. Rev. Lett. **45**, 204 (1980).
85. S. Lundquist, in: *Nonuniform Electron Gas Theory*, edited by S. Lundquist and H. March (Mir Press, Moscow, 1987), p. 151 (Russian trans.).
86. M. M. German, V. Ya. Kupershmidt, and O. V. Farberovich, Fiz. Tverd. Tela **30**, 1822 (1988).
87. V. V. Nemoshkalenko and Yu. N. Kucherenko, *Computational Physics Techniques in Solid State Theory. Nonideal Crystals* (Naukova Dumka Press, Kiev, 1985).
88. *Theory of Chemisorption*, edited by G. Smith (Mir Press, Moscow, 1983) (Russian trans.).
89. P. W. Anderson, Phys. Rev. **109**, 1492 (1958).
90. *Problems of Vitreous State Physics. Collection of Scientific Proceedings*, edited by Yu. R. Zakis (Riga, 1985).
91. V. L. Bonch-Bruevich and A. G. Mironov, Fiz. Tverd. Tela **3**, 3009 (1961) [Sov. Phys. Solid State **3** (1961)].
92. I. M. Lifshits, S. A. Gradeskul, and L. A. Pastur, *Introduction to Theory of Disordered Systems* (Nauka Press, Moscow, 1982).
93. *Electronic Theory of Disordered Semiconductors*, edited by V. L. Bonch-Bruevich (Nauka Press, Moscow, 1981).
94. N. Mott and E. Davis, *Electronic Processes in Noncrystalline Matter*, Moscow **1**, 664 (1982).
95. N. V. Cohan and M. Weissman, J. Phys. C **10**, 383 (1977).
96. C. Papatriantofellon, E. N. Economon, and T. P. Eggerton, Phys. **18**, 920 (1976).
97. B. Kramer, K. Maschke, and L. D. Lande, Phys. Rev. B **8**, 579 (1973).
98. B. A. Kramer, Phys. State Solids **41**, 649 (1970).
99. A. I. Gubanov, *Quantum Electronic Theory of Amorphous Semiconductors* (Leningrad, 1963).
100. L. D. Laude, B. Kramer, and K. Maschke, Phys. Rev. B **8**, 579 (1973).
101. P. Soven, Phys. Rev. **156**, 809 (1967).
102. P. Soven, Phys. Rev. **178**, 1136 (1969).
103. P. Soven, Phys. Rev. B **2**, 4715 (1970).
104. L. Schwartz, F. Brouers, A. B. Vedyayev, and P. Ehrenreich, Phys. Rev. B **4**, 3383 (1971).
105. A. V. Vedyaev, O. A. Kotelnikova, M. Yu. Nikolaev, and A. V. Stefanovich, *Phase Transitions and the Electronic Structure of Alloys* (MGU, Moscow, 1986).
106. G. Ehrenreich and L. Schwarz, *Electronic Structure of Alloys* (Mir Press, Moscow, 1979) (Russian trans.).
107. J. L. Beeby and S. F. Edwards, Proc. R. Soc. London Ser. A **274**, 395 (1963).
108. J. L. Beeby, Proc. R. Soc. London Ser. A **279**, 82 (1964).
109. P. Lloyd, Proc. Phys. Soc. **90**, 217 (1967).
110. J. M. Ziman, Proc. Phys. Soc. **88**, 387 (1966).
111. D. Weaire, Phys. Rev. Lett. **26**, 1541 (1972).
112. D. Weaire and M. F. Thorpe, Phys. Rev. B **4**, 2508 (1972).
113. D. Weaire and M. F. Thorpe, Phys. Rev. B **4**, 3518 (1972).
114. M. F. Thorpe, D. Weaire, and R. Albee, Phys. Rev. B **7**, 3777 (1973).
115. N. J. Kelly and D. W. Bullett, Solid State Commun. **18**, 593 (1976).
116. D. W. Bullett, Phys. Rev. B **14**, 1683 (1976).
117. D. W. Bullett, J. Phys. C **8**, 2695 (1975).
118. J. J. Keller, Phys. C **4**, 3143 (1971).
119. T. C. McGill and J. Klima, Phys. Rev. B **5**, 1517 (1972).
120. P. Lloyd, Proc. Phys. Soc. **90**, 207 (1967).
121. A. I. Gubanov, Dokl. U.S.S.R. Acad. Sci. **159**, 46 (1964).
122. A. I. Gubanov, Fiz. Tekhn. Poluprovodnikov. **3**, 881 (1969).
123. K. H. Johnson, Surf. Sci. **42**, 341 (1974).
124. J. D. Joannopoulos and F. Indurain, Phys. Rev. B **10**, 51 (1974).
125. W. S. Pollard and J. D. Joannopoulos, Phys. Rev. B **117**, 1770 (1978).
126. J. D. Joannopoulos, J. Non-Cryst. Solids **32**, 241 (1979).
127. J. D. Joannopoulos and M. L. Cohen, Phys. Rev. B **7**, 2644 (1973).

128. J. D. Joannopoulos, M. Schuler, and M. L. Cohen, Phys. Rev. B **11**, 2186 (1975).
129. K. D. Wilson, Usp. Fiz. Nauk **141**, 193 (1983) [Sov. Phys. Usp. **141** (1983)].
130. M. V. Sadovskiy, Usp. Fiz. Nauk **133**, 223 (1981) [Sov. Phys. Usp. **133** (1981)].
131. M. I. Klinger, Zh. Eksp. Teor. Phys. **82**, 1687 (1982) [Sov. Phys. JETP **55**, 976 (1982)].
132. F. N. Ignatev, V. G. Karpov, and M. I. Klinger, Fiz. Tverd. Tela **25**, 1265 (1983) [Sov. Phys. Solid State **25**, 272 (1983)].
133. A. M. Stonham, *Defect Theory in Solids* (Mir Press, Moscow, 1978), Vol. 1, p. 570; Vol. 2, p. 358.
134. S. A. Ivarestov, E. A. Kotomin, and A. I. Ermoshkin, *Molecular Models of Point Defects in Wide Band Gap Solids* (Riga, 1983).
135. V. E. Egorushkin and Yu. A. Khon, *Electron Theory of Transition Metal Alloys* (Nauka Press, Novosibirsk, 1985).
136. F. Ducastelle and F. Cyrot-Lackmann, Phys. Chem. Solids **31**, 1295 (1970).
137. F. Cyrot-Lackmann, F. Ducastelle, and J. Friestee, J. Solid State Commun. **8**, 685 (1970).
138. R. Haydock, V. Heine, and M. Kelly, J. Phys. C **5**, 2845 (1972).
139. R. Haydock, V. Heine, and M. J. Kelly, J. Phys. C **15**, 2891 (1982).
140. S. Frota-Pessaa, Phys. Rev. B **28**, 3753 (1982).
141. A. Rahman, M. J. Mandell, and J. P. McTange, J. Chem. Phys. **64**, 1564 (1976).
142. R. Albern, M. Blume, H. Krakauer, and L. Schwartz, Phys. Rev. B **12**, 4090 (1975).
143. R. Car and M. Parrinello, Phys. Rev. Lett. **55**, 2471 (1985).
144. P. Turchi, F. Ducastelle, and G. Treglia, J. Phys. **15**, 21 (1982).
145. R. Haydock, V. Heine, and M. J. Kelly, J. Phys. C **8**, 295 (1975).
146. V. V. Nemoshkalenko and V. N. Antonov, *Computational Physics Methods in Solid State Theory. Band Theory of Metals* (Nauka Dumka Press, Kiev, 1985).
147. T. L. Loucks, *Augmented Plane Wave Method* (Benjamin Inc., New York, 1967).
148. C. Herring, Phys. Rev. **57**, 1169 (1940).
149. J. Callaway, Phys. Rev. **99**, 500 (1955).
150. R. A. Deegan and W. D. Twose, Phys. Rev. **164**, 993 (1967).
151. O. V. Farberovich, S. I. Kurganskii, and E. Domashevskaya, Phys. Status Solidi B **94**, 51 (1979).
152. O. V. Farberovich, S. I. Kurganskii, and E. Domashevskaya, Phys. Status Solidi B **97**, 631 (1980).
153. S. I. Kurganskiy, V. B. Farberovich, and E. P. Domashevskaya, Fiz. Tekhnik. Poluprovodnikov. **14**, 1315 (1980).
154. S. I. Kurganskiy, V. B. Farberovich, and E. P. Domashevskaya, Fiz. Tekhnik. Poluprovodnikov. **14**, 1412 (1980).
155. S. I. Kurganskiy, V. B. Farberovich, and E. P. Domashevskaya, in: *Methods of Calculating the Energy Structure and Physical Properties of Crystals* (Naukova Dumka Press, Kiev, 1982), p. 152.
156. J. S. Faulkner, Prog. Mater. Sci. **27**, 1 (1962).
157. J. C. Slater and G. F. Koster, Phys. Rev. **94**, 1498 (1954).
158. P. W. Anderson, Phys. Rev. **181**, 25 (1969).
159. W. H. Adams, J. Chem. Phys. **34**, 82 (1961); **37**, 2 (1967).
160. D. W. Bullett, Solid State Phys. **35**, 129 (1980); R. Haydock, *ibid.* **35**, 216 (1980); V. Heine, *ibid.* **35**, 1 (1980); and M. J. Kelly, *ibid.* **35**, 296 (1980).
161. R. Richter, H. Eschrig, and B. Velicky, in: Proc. of the 16th Ann. Int. Symp., "Electronic Structures of Metals and Alloys," Dresden, 1986, p. 24.
162. M. Richter and H. Eschrig, Ref. 161, p. 132.
163. O. V. Farberovich, Dissertation for doctorate degree in physics and mathematics, Voronezh, 1984.
164. V. I. Rezer and V. P. Shirokovskiy, Fizika Metall. Metall. **32**, 934 (1971).
165. B. I. Reser and V. V. Dyakin, Phys. Status Solidi B **87**, 41 (1978).
166. M. D. Girardson, J. Math. Phys. **12**, 165 (1971).
167. Z. A. Gurekiy, Dissertation for doctorate degree in physics and mathematics, Kiev, 1984.
168. B. A. Gurekiy and Z. A. Gurekiy, A certain class of pseudopotentials formulated in the fully ortogonalized plane wave basis, Preprint of the Academy of Sciences of the Ukrainian SSR, Institute of Theoretical Physics, 1975, 21 pp.
169. W. Kohn and N. Rostocker, Phys. Rev. **94**, 1111 (1954).
170. J. Korringa, Physica **13**, 392 (1947).
171. B. A. Gurekiy and Z. A. Gurekiy, Ukrainskiy Fiz. Zh. **23**, 19 (1978).
172. P. M. Marcus, Int. J. Quantum Chem. Symp. **1**, 567 (1967).

173. O. K. Andersen, Phys. Rev. B **12**, 864 (1975).
174. D. D. Koelling and G. O. Arbman, J. Phys. F **5**, 204 (1975).
175. B. S. Garbow, J. M. Boyle, J. J. Dongarra, and C. B. Moler, in: *Lecture Notes in Computer Science*, edited by G. Goos and J. Hartman (Springer, Berlin, 1977), Vol. 51.
176. O. V. Farbersovich, S. V. Vlasov, and G. P. Nizhnikova, Program for self-consistent relativistic calculation of atomic and ionic structures in the local density approximation. No. 1, Numerical methods and program description, Voronezh, 1983, No. 2953-83.
177. A. H. MacDonald, W. E. Pickett, and D. D. Koelling, J. Phys. C **13**, 2675 (1980).
178. M. Weinert, J. Math. Phys. **22**, 2433 (1981).
179. M. M. Sobol', *Numerical Monte Carlo Methods* (Nauka Press, Moscow, 1973).
180. N. Elyashar and D. D. Koelling, Phys. Rev. B **13**, 530 (1976).
181. K. S. Sonn, D. G. Dempsy, and L. Kleinman, Phys. Rev. B **13**, 15 (1976).
182. D. G. Dempsy, L. Kleinman, and E. Caruthers, Phys. Rev. B **8**, 2932 (1975).
183. D. G. Dempsy, L. Kleinman, and E. Caruthers, Phys. Rev. B **4**, 1489 (1976).
184. O. Bisi and C. Calandra, Surf. Sci. **83**, 83 (1979).
185. J. R. Smith, J. G. Gay, and F. J. Arlinghaus, Phys. Rev. B **21**, 2201 (1980).
186. J. G. Gay, J. R. Smith, F. J. Arlinghaus, and T. W. Capehart, Phys. Rev. B **23**, 1552 (1981).
187. F. J. Arlinghaus, J. G. Gay, and J. R. Smith, Phys. Rev. B **23**, 5152 (1981).
188. J. G. Gay, J. R. Smith, and F. J. Arlinghaus, Phys. Rev. B **25**, 643 (1982).
189. W. Kohn, Phys. Rev. B **11**, 3756 (1975).
190. N. Kar and P. Soven, Phys. Rev. B **11**, 3761 (1975).
191. K. Kambe, Surf. Sci. **117**, 443 (1982).
192. H. Krakauer and B. K. Cooper, Phys. Rev. B **16**, 605 (1977).
193. O. Fepsen, J. Madsen, and O. K. Andersen, Phys. Rev. B **18**, 605 (1978).
194. H. Krakauer, M. Pasternak, and A. J. Freeman, Phys. Rev. B **1**, 1706 (1979).
195. D. R. Hamasin, L. F. Mattheiss, and H. S. Greenside, Phys. Rev. B **24**, 6151 (1981).
196. M. Pasternak, H. Krakauer, and A. J. Freeman, Phys. Rev. B **2**, 5601 (1980).
197. O. I. Dubrovskiy, S. I. Kurganskiy, O. V. Farberovich, and E. P. Domashevskaya, Poverkhnost' No. 2, 28 (1988).
198. V. N. Tomilenko, The electronic surface structure of certain 4*d*- and 5*d*- transition metals. Dissertation for candidate's degree in physics and mathematics, Kiev, 1987.
199. G. V. Wolf and L. A. Korapanova, Poverkhnost', No. 4, 27 (1985).
200. E. Wimmer, A. Krokauer, and M. Weiner, Phys. Rev. B **24**, 864 (1981).
201. D. D. Koelling, Rep. Prog. Phys. **44**, 139 (1981).
202. S. E. Dorfman, I. I. Mazin, A. V. Ruban, and Yu. A. Uspenskiy, Izv. Vuzov. Fiz. **26**, 103 (1983).
203. C. G. Broydin, Math. Comp. **19**, 577 (1965).
204. G. Dennis and R. Shnabel, *Numerical Unconditional Optimization Techniques and Solution of Nonlinear Equations* (Mir Press, Moscow, 1988) (Russian trans.).
205. G. P. Srivastava, J. Phys. A **17**, 317 (1984).
206. D. Singh, A. Krakauer, and C. S. Wang, Phys. Rev. B **34**, 8391 (1986).
207. A. C. Aitken, Proc. R. Soc. Edinburgh **46**, 289 (1926).
208. E. Mrosan, G. Lehmann, and H. Weittenneck, Phys. Status Solidi B **64**, 131 (1984).
209. E. Mrosan, G. Lehmann, and H. Weittenneck, Phys. Status Solidi B **64**, 11 (1974).
210. E. Mrosan and G. Lehmann, Phys. Status Solidi B **71**, 13 (1975).
211. E. Mrosan and G. Lehmann, Phys. Status Solidi B **77**, 607 (1976).
212. E. Mrosan and G. Lehmann, Phys. Status Solidi B **78**, 159 (1976).
213. V. S. Stepanyuk *et al.*, Phys. Status Solidi B **112**, 211 (1982).
214. A. Sh. Chavchanidre, I. S. Shpotin, and V. S. Stepanyuk, Phys. Status Solidi B **110**, 155 (1982).
215. A. A. Katsnelson, V. S. Stepanyuk, I. S. Shpotin, and I. S. Yastrebov, Metallofizika **2**, 28 (1980).
216. A. A. Katsnelson, V. S. Stepanyuk, I. S. Shpotin, and L. I. Yastrebov, Metallofizika **3**, 23 (1981).
217. V. S. Stenanyuk, G. M. Zhidomirov, A. A. Katsnelson, I. S. Shpotin, and L. I. Yastrebov, Zhur. Fiz. Him. **61**, 124 (1982).
218. V. S. Stepanyuk, A. A. Katsnelson, and L. I. Yasterbov, Metallofizika **7**, 114 (1985).
219. R. Zeller and P. H. Dedericks, Phys. Rev. Lett. **42**, 1713 (1979).
220. R. Padlonsky, R. Ziller, and P. H. Dedericks, Phys. Rev. B **22**, 5777 (1980).
221. P. J. Brospenning, K. Zeller, A. Lodder, and P. H. Dedericks, Phys. Rev. B **29**, 703 (1984).

222. K. Zeller, J. Dents, and P. H. Dedericks, Solid State Commun. **44**, 993 (1982).
223. N. Stefanou, P. J. Brespenning, R. Zeller, and P. H. Dedericks, Phys. Rev. B **36**, 6372 (1987).
224. V. V. Koleshenkov, E. V. Poloentsev, and V. V. Vinokurtsev, Self-consistent calculation of the electronic structure of fragments of ideal and defect-containing solids. Manuscript deposited at the All-Union Institute for Scientific and Technical Information, No. 3914-80 Moscow, 1980, 30 pp.
225. A. Fazzio, J. K. Leite, and M. L. Sigueira, J. Phys. C **12**, 513 (1979).
226. R. A. Zvarestov, *Quantum Mechanical Methods in Solid State Theory* (Izd-vo Leningr. Univ., Leningrad, 1982).
227. A. Zunyer and A. Katzir, Phys. Rev. B **11**, 237 (1975).
228. K. H. Johnson, J. Chem. Phys. **45**, 3085 (1966).
229. O. Gunnarson, J. Harris, and R. O. Jones, J. Phys. C **9**, 1739 (1979).
230. A. K. Williams and J. W. Morgan, J. Phys. C **5**, 1293 (1972).
231. N. Rosch, W. G. Klemperer, and K. H. Johnson, Chem. Phys. Lett. **23**, 149 (1973).
232. D. E. Ellis, Int. J. Quantum Chem. Symp. **2**, 35 (1968).
233. D. E. Ellis and G. S. Painter, Phys. Rev. B **2**, 2887 (1970).
234. M. Wolfsberg and L. Helmholz, J. Chem. Phys. B **20**, 8 (1952).
235. G. Klopman and R. Ivens, in: *Semiempirical Techniques for Electronic Structure Calculations*, edited by G. Stala (Mir Press, Moscow, 1980), p. 47.
236. G. Ehrenreich and L. Schwartz, Solid State Phys. **31**, 149 (1976).
237. V. E. Egorushkin and Yu. A. Khon, *Electronic Theory of Transition Metal Alloys* (Nauka Press, Novosibirsk, 1985).
238. V. A. Gubanov, E. Z. Kurmaev, and A. I. Ivanovskiy, *Quantum Chemistry of Solids* (Nauka Press, Moscow, 1984).
239. J. J. Olson, Phys. Rev. B **12**, 2908 (1975).
240. M. Lax, Phys. Rev. **84**, 621 (1952).
241. L. Schwartz, H. Peterson, and A. Bansil, Phys. Rev. B **12**, 31 (1975).
242. F. Cyrot-Lackman, J. Phys. **27**, 627 (1966).
243. P. W. Anderson (*Proc. Inter. School of Physics*, McMillan, New York, 1967), p. 50.
244. L. E. Ballentine, Can. J. Phys. **44**, 2533 (1966).
245. B. L. Gyorffy, Phys. Rev. B **1**, 3290 (1970).
246. L. Schwartz and H. Ehrenreich, Ann. Phys. **64**, 100 (1971).
247. L. M. Roth, Phys. Rev. B **9**, 2476 (1974).
248. P. Lloyd, Proc. Phys. Soc. **90**, 207 (1967).
249. Y. Ishida and F. Yonesawa, Prog. Theor. Phys. **49**, 73 (1973).
250. S. F. Edwards, Proc. R. Soc. London Ser. A **247**, 518 (1962).
251. J. M. Ziman, Adv. Phys. **13**, 89 (1964).
252. A. Rachman, *Numerical Calculations in Classical Liquid Interactions in Condensed Matter*, edited by J. Woods Halley, NATO Adv. Study Series (Plenum Press, New York, 1977), Vol. 35, p. 417.
253. R. Car and M. Parrinello, Phys. Rev. Lett. **55**, 247 (1985).
254. V. Heine, M. Kohn, and D. Weir, *Pseudopotential Theory* (Mir Press, Moscow, 1973) (Russian trans.).
255. L. I. Yastrebov and A. A. Katsnelson, *Principles of One-electron Solid State Theory* (Nauka Press, Moscow, 1981) (in Russian); (Mir Press, Moscow, 1987), 2nd ed. (in English).
256. A. A. Katsnelson and L. I. Yastrebov, *Pseudopotential Theory of Crystalline Structure* (Izd. MGU, Moscow, 1981).
257. K. I. Portnoy, V. I. Bogdanov, and D. L. Fuks, *Calculation of Interaction and Phase Stability* (Metallurgiya, Moscow, 1981).
258. V. E. Panin, Yu. A. Khon, I. K. Naumov *et al.*, *Phase Theory in Alloys*, edited by A. A. Katsnelson (Nauka Press, Novosibirsk, 1984).
259. A. A. Katsnelson, O. M. Tatarinskaya, and M. M. Khrushchov, Fiz. Met. Metall. **6**, 655 (1987).
260. N. W. Ashcroft and D. C. Langreth, Phys. Rev. **155**, 682 (1967).
261. V. G. Baks and A. V. Trefilov, Fiz. Tverd. Tela **19**, 244 (1977) [Sov. Phys. Solid State **19** (1977)].
262. A. A. Katsnelson, O. M. Tatarinskaya, and M. M. Khrushchov, Fiz. Met. Metall. **5**, 890 (1984).
263. A. O. E. Animalu, Proc. R. Soc. London Ser. A. **294**, 376 (1966).
264. V. M. Silonov, *Introduction to the Statistical and Electronic Theory of Metallic Solids* (Izd-vo MGU, Moscow, 1983).

265. A. A. Katsnelson, V. M. Silonov, and A. Khavadzha Farid, Phys. Status Solidi B **91**, 11 (1979).
266. A. A. Katsnelson, A. O. O. Mekhrabov, and V. M. Silonov, Izvest. Vuzov Ser. Fiz. **10**, 103 (1976).
267. A. A. Katsnelson, A. O. O. Mekhrabov, and V. M. Silonov, Fiz. Met. Metall. **4**, 33 (1978).
268. A. A. Katsnelson, O. V. Krisko, and V. M. Silonov, Fiz. Met. Metall. **60**, 243 (1985).
269. A. A. Katsnelson, V. M. Silonov, and A. Khavadzha Farid, Fiz. Met. Metall. **4**, 52 (1980).
270. S. Bencnert, Phys. Status Solidi B **60**, 483 (1975).
271. A. A. Katsnelson, V. M. Silonov, and T. Abbas, Fiz. Met. Metall. **59**, 372 (1985).
272. V. M. Silonov, Abstract for dissertation of doctorate of physics and mathematics, Moscow State University, 1989.
273. A. A. Katsnelson, V. M. Silonov, and A. N. Prozorov, Fiz. Met. Metall. **57**, 985 (1984).
274. A. A. Katsnelson, V. S. Stepanyuk, L. I. Yastrebov, and O. T. Malyuchkov, Metallofizika **4**, 46 (1982).
275. A. A. Katsnelson and N. V. Popova, Dep. at the All-Union Institute for Scientific and Technical Information, No. 627-82.
276. A. A. Katsnelson, V. M. Silonov, and O. V. Krisko, Preprint No. 4/1986, Physics Department, Moscow State University, 1986, pp. 1–5.
277. D. R. Hamann, M. Schlüter, and P. Chiang, Phys. Rev. Lett. **43**, 1494 (1979).
278. E. V. Chulkov, V. M. Silkin, and E. K. Shirykalov, Fiz. Met. Metall. **64**, 213 (1987).
279. B. A. Gurskiy and Z. A. Gurskiy, Ukr. Fiz. Zhur. **21**, 1603 (1976).
280. L. Dageus, J. Phys. F **6**, 1801 (1976).
281. A. A. Yurev, N. A. Vatolin *et al.*, Proceedings of Scientific Reports of the Fifth All-Union Conf. on the Structure and Properties of Metallic and Shale Melts, Sverdlovsk, 1983, Chap. 1, p. 123.
282. A. Zunger and M. L. Cohen, Phys. Rev. **18**, 5449 (1978).
283. E. V. Chulkov, I. Yu. Sklyadneva, and V. E. Panin, Fiz. Met. Metall. **56**, 3 (1983).
284. W. Kohn and L. Sham, Phys. Rev. A **140**, 1133 (1965).
285. L. Hedin and B. I. Lundquist, J. Phys. C **4**, 2062 (1971).
286. W. Harrison, *Pseudopotentials in the Theory of Metals* (Mir Press, Moscow, 1968) (Russian trans.); *Solid State Theory* (Mir Press, Moscow, 1972) (Russian trans.).
287. N. W. Ashcroft, Phys. Lett. **23**, 48 (1966).
288. V. Heine and J. V. Abarenkov, Philos. Mag. **9**, 451 (1964).
289. J. V. Abarenkov and V. Heine, Philos. Mag. **12**, 529 (1965).
290. G. L. Krasko and Z. A. Gurskiy, Pis'ma ZhETF **9**, 596 (1969) [JETP Lett. **9** (1969)].
291. M. Rasolt and R. Taylor, Phys. Rev. B **11**, 2717 (1975).
292. R. W. Shaw, Phys. Rev. **174**, 769 (1968).
293. P. Rennert, Phys. Status Solidi B **50**, 37 (1972).
294. V. S. Stepanyuk, A. A. Katsnelson, and L. I. Yastrebov, Metallofizika **8**, 5, 108 (year).
295. A. O. E. Animalu, Phys. Rev. B **8**, 3542 (1973).
296. A. A. Katsnelson and V. S. Stepanyuk, *Interparticle Interactions and Metallic Properties* (Moscow State University Press, Moscow, 1987).
297. Sh. M. Khannanov, Fiz. Met. Metall. **43**, 431 (1977).
298. G. Paasch and A. Heinrich, Phys. Status Solidi B **102**, 323 (1980).
299. G. Paasch and P. Rennert, Phys. Status Solidi B **83**, 501 (1977).
300. N. Almbladh, Phys. Rev. **14**, 2250 (1976).
301. G. L. Z. Kreske, Z. Naturforsch. Teil A **36**, 1129 (1981).
302. V. A. Krymov and D. L. Fuks, Metallofizika **9**, 99 (1987).
303. J. Hafner, J. Phys. Status Solidi B **57**, 101 (1973).
304. E. V. Chulkov, Abstracts of candidate's dissertations, TGU, Tomsk, 1980.
305. V. A. Krashanin, Abstracts of candidate's dissertations, UGU, Sverdlovsk, 1988.
306. V. A. Krymov, Abstracts of candidate's dissertations, MGU, Moscow, 1988.
307. Ya. I. Dutchak, P. N. Yakbchuk, and M. I. Zhovtanetskiy, Ukr. Fiz. Zhur. **2**, 34 (1976).
308. A. N. Pilyankevich and D. A. Zakaryan, Ukr. Fiz. Zhur. **30**, 18 (1985).
309. I. A. Anishchenko, O. T. Malyuchkov, and L. I. Yastrebov, Izv. Vuzov Ser. Chernaya Metall. (1988).
310. V. E. Egorushkin and Yu. A. Khon, *Electronic Theory of Transition Metal Alloys* (Nauka Press, Novosibirsk, 1985).
311. V. M. Kuznetsov, P. L. Kaminskiy, and V. F. Perevalov, Fiz. Met. Metall. **63**, 213 (1987).

312. T. Suzuki, A. V. Granato, and F. Thomas, Phys. Rev. **175**, 766 (1968).
313. A. A. Smirnov, *Molecular-kinetic Theory of Metals* (Nauka Press, Moscow, 1966).
314. *Alloy Phase Diagrams*, edited by L. Bennett *et al.* (Mir Press, Moscow, 1986) (Russian trans.).
315. V. N. Aleksandrov and N. V. Dalakova, Fiz. Met. Metall. **59**, 83 (1985).
316. V. N. Aleksandrov and N. V. Dalakova, Fiz. Met. Metall. **65**, 88 (1988).
317. A. A. Katsnelson, O. V. Krisko, V. M. Silonov, and T. V. Skorobogatova, Metallofizika **10**, 103 (1988).
318. M. A. Krivoglaz, *Theory of X-ray and Thermal Neutron Scattering by Real Crystals* (Nauka Press, Moscow, 1968).
319. E. Mrosan and G. Lehmann, Phys. Status Solidi B **78**, 159 (1976).
320. A. A. Katsnelson, V. S. Stepanyuk, I. S. Shpotin, and L. I. Yastrebov, Metallofizika **2**, 28 (1980).
321. A. A. Katsnelson, *Introduction to Solid State Physics* (Moscow State University Press, Moscow, 1984).
322. V. M. Silonov, O. V. Krisko, and A. A. Katsnelson, Metallofizika **10**, 12 (1988).
323. K. Hirata, Y. Waseda, A. Juin, and R. Srivastava, J. Phys. F **7**, 419 (1977).
324. F. A. Khavadzha and V. M. Silonov, Fiz. Met. Metall. **50**, 710 (1980).
325. A. A. Yurev and N. A. Vatolin, Izv. AN SSSR Ser. Metall. **5**, 44 (1984).
326. J. A. Moriarty, Phys. Rev. B **1**, 1363 (1970); **6**, 1239 (1973).
327. E. I. Kharkov, V. I. Lyusov, and V. E. Fedorov, *Liquid Metal Physics* (Vishcha shkola, Kiev, 1979).
328. L. V. Meisel and P. Cote, Phys. Rev. B **16**, 2978 (1977).
329. *Metallic Glasses*, edited by H. F. Guntherodt and G. Bena (Mir Press, Moscow, 1983), Vol. 1 (Russian trans.).
330. V. V. Nemoshkalenko, *X-ray Emission Spectroscopy of Metals and Alloys* (Naukova Dumka Press, Kiev, 1972).
331. E. P. Domashevskaya, Dissertation for doctorate degree in physics and mathematics, Voronezh, 1979.
332. A. Szasz, J. Kojnok, L. Kertesz, Z. Paal, and Z. Hegedus, Thin Solid Films **116**, 279 (1984).
333. E. Belin, A. Traverse, H. Szasz, and F. Machizand, J. Phys. F **17**, 1913 (1987).
334. V. V. Nemoshkalenko and V. G. Aleshin, *Electron Spectroscopy of Crystals* (Naukova Dumka Press, Kiev, 1976).
335. V. V. Nemoshkalenko and V. G. Aleshin, *Theoretical Principles of X-ray Emission Spectroscopy* (Naukova Dumka Press, Kiev, 1974).
336. *Methods of Calculating the Energy Structure and Physical Properties of Crystals* (Naukova Dumka Press, Kiev, 1982), p. 53.
337. O. B. Sonder and Y. A. Babanoi, Phys. Status Solidi **51**, 46 (1972).
338. B. L. Györffy and G. U. Stocks, in: *Band Structure Spectroscopy of Metals and Alloys*, edited by D. F. Fabian and L. M. Watson (Academic Press, New York, 1972), p. 641.
339. F. Bassani, R. Pastori, and F. Parravichini, *Electronic States and Optical Transitions in Solids* (Nauka Press, Moscow, 1982) (Russian trans.).
340. A. N. Basilev and V. V. Mikhaylin, *Introduction to Steady-state Spectroscopy* (Moscow State University Press, Moscow, 1987).
341. E. O. Kane, Phys. Rev. **159**, 624 (1967).
342. G. D. Mahan, Phys. Rev. B **2**, 4334 (1970).
343. J. F. Janak, A. R. Williams, and V. L. Moruzzi, Phys. Rev. B **11**, 1522 (1975).
344. I. Petroff and O. K. Viswanathan, Phys. Rev. B **4**, 799 (1971).
345. J. Rowe and N. V. Smith, Phys. Rev. B **10**, 320 (1974).
346. M. Sugarton and N. Skerchek, Phys. Rev. B **17**, 3859 (1978).
347. N. V. Sherchek and N. Liebowits, Phys. Rev. B **18**, 1618 (1978).
348. N. V. Smith, Phys. Rev. B **19**, 5019 (1979).
349. V. V. Nemoshkalenko, E. E. Krasovskiy, V. N. Antonov, Yu. V. Kudryavtse, and I. N. Mishchenko, DAN U.S.S.R. 7, 50 (1986).
350. E. V. Stepanova, V. S. Stepanyuk *et al.*, Fiz. Tverd. Tela **31**, 5 (1989) [Sov. Phys. Solid State].
351. E. V. Stepanova, V. S. Stepanyuk *et al.*, Izv. Vuzov Fiz. **7**, 82 (1988).
352. V. S. Stepanyuk, A. A. Grigorenko *et al.*, Fiz. Tverd. Tela **31**, 5 (1989) [Sov. Phys. Solid State].
353. A. A. Grigorenko, E. V. Stepanova, V. S. Stepanyuk *et al.*, Vestnik MGU **30**, 61 (1989).

354. L. Hedin, Phys. Rev. A **139**, 769 (1965).
355. L. Hedin and S. Lundquist, Solid State Phys. **23**, 181 (1969).
356. M. S. Hybersten and S. G. Louie, Phys. Rev. Lett. **55**, 1418 (1985).
357. M. S. Hybersten and S. G. Louie, Phys. Rev. B **32**, 7005 (1985).
358. L. Sham, Phys. Rev. B **32**, 3876 (1985).
359. L. Sham and N. Shlüter, Phys. Rev. B **32**, 3883 (1985).
360. M. Lanu, M. Schlüter, and L. Sham, Phys. Rev. B **32**, 3890 (1985).
361. J. P. Perdew, Int. J. Quantum Chem. **19**, 497 (1986).
362. J. P. Perdew and M. Levy, Phys. Rev. Lett. **51**, 1884 (1983).
363. M. Lanu, M. Schlüter, and L. Sham, Phys. Rev. Lett. **56**, 2415 (1986).
364. P. W. Godby, M. Schlüter, and L. Sham, Phys. Rev. B **35**, 4170 (1987).
365. M. S. Hybersten and S. G. Louie, Phys. Rev. B **34**, 5390 (1986).
366. N. I. Kulikov, M. Alouani, M. A. Khan, and M. Magnitskaya, Phys. Rev. B **36**, 929 (1987).
367. I. I. Mazin *et al.*, Fiz. Tverd. Tela **29**, 2629 (1987) [Sov. Phys. Solid State].
368. S. N. Rashkeev, Yu. A. Uspenskiy, and I. I. Mazin, ZhETF **88**, 1687 (1985).
369. V. Ya. Asanovich, V. A. Gorbunov, and I. T. Sryvalin, Fiz. Met. Metall. **46**, 415 (1978).
370. V. Ya. Asanovich, V. A. Gorbunov, and I. T. Sryvalin, Fiz. Met. Metall. **4**, 1096 (1978).
371. V. A. Gorbunov, Electronic structure and properties of results of aluminides of the iron subgroup, Ph.D. dissertation, Volgograd State Univ., 1987.
372. *Problems of High-temperature Superconductivity*, edited by V. L. Ginzburg and D. A. Kirzhnits (Nauka Press, Moscow, 1977).
373. W. L. McMillan, Phys. Rev. **167**, 33 (1968).
374. G. M. Eliashberg, ZhETF **38**, 966 (1960).
375. A. E. Karakozov, E. G. Maksimov, and S. A. Mashkov, ZhETF **68**, 1937 (1975).
376. N. N. Bogolyubov, V. V. Tolmachev, and D. V. Shirkov, *New Method in Theory of Superconductivity* (Acad. Sci. U.S.S.R., Moscow, 1958).
377. G. D. Guspari and B. L. Gyorffy, Phys. Rev. Lett. **28**, 801 (1972).
378. S. V. Vlasov, Electronic structure calculation of rare earths and their monohalogenides in local density functional approach, Ph.D. dissertation, Vorowerk State Univ., 1984.

Printed in the United States
By Bookmasters